INTERNATIONAL CONFLICT

To Frank Zagare

Who taught me about international conflict

INTERNATIONAL CONFLICT

LOGIC AND EVIDENCE

Stephen L. Quackenbush
University of Missouri

Los Angeles | London | New Delhi
Singapore | Washington DC

Los Angeles | London | New Delhi
Singapore | Washington DC

FOR INFORMATION:

CQ Press
An Imprint of SAGE Publications, Inc.
2455 Teller Road
Thousand Oaks, California 91320
E-mail: order@sagepub.com

SAGE Publications Ltd.
1 Oliver's Yard
55 City Road
London, EC1Y 1SP
United Kingdom

SAGE Publications India Pvt. Ltd.
B 1/I 1 Mohan Cooperative
Industrial Area
Mathura Road, New Delhi 110 044
India

SAGE Publications Asia-Pacific Pte. Ltd.
3 Church Street
#10-04 Samsung Hub
Singapore 049483

Acquisitions Editor:	Sarah Calabi
Developmental Editor:	Nancy Matuszak
Editorial Assistant:	Davia Grant
Project Editor:	Bennie Clark Allen
Production Editor:	Stephanie Palermini
Copy Editor:	Diane DiMura
Typesetter:	Hurix Systems Pvt. Ltd.
Proofreader:	Sally Jaskold
Indexer:	Jennifer Pairan
Cover Designer:	Anupama Krishnan
Marketing Manager:	Amy Whitaker

Printed in the United States of America

Library of Congress Cataloging-in-Publication Data

Quackenbush, Stephen L.

International conflict : logic and evidence / Stephen L. Quackenbush, University of Missouri-Columbia.

pages cm

Includes bibliographical references and index.

ISBN 978-1-4522-4098-5 (pbk. : alk. paper) 1. War—Causes. 2. Conflict management. 3. International relations. I. Title.

JZ6385.Q34 2015

355.02'7—dc23 2014008335

This book is printed on acid-free paper.

SFI label applies to text stock

14 15 16 17 18 10 9 8 7 6 5 4 3 2 1

Brief Contents

Detailed Contents

Preface

War and other forms of international conflict shape the world as we know it. War has played a crucial role in the formation, expansion, and disappearance of countries around the world. It has had a broad range of political, social, and economic effects, from the creation of international organizations and alliances and the empowerment of women to rapid technological innovations, to name just a few from the past century. War is also an extremely costly endeavor, in terms of lives lost, property destroyed, and money spent.

While many have wished for world peace, a necessary prerequisite for achieving it is to understand how international conflict works—why it happens, how it is terminated, and so on. *International Conflict: Logic and Evidence* is intended to provide an introduction to the study of the topic. Written for advanced undergraduate students, it is also appropriate for graduate courses on international conflict and as a reference for international conflict researchers.

Unlike other books on the market, which are organized around case studies or present a collection of readings, *International Conflict* is scientific research based. There is often a large disconnect between the way that political science is studied and the way it is taught. This does a great disservice to both students and faculty. This book seeks to remedy this problem by introducing students to key theories and empirical findings stemming from scientific research on international conflict, while covering them at an appropriate level for advanced undergraduate students. The key theme is that a proper understanding of international conflict only comes from logic and evidence, not opinion and anecdote. Our primary focus is on conflict between states, rather than within them. This is ideal for places where interstate and civil conflict are covered in separate courses. For those wishing to cover both topics at once, the book can be supplemented with additional readings to cover civil wars in more detail.

Organization of the Book

As a whole, this book covers the entire process of interstate war, not just the causes. The process of war, as developed by Bremer (1995), is a cycle that begins with international context, which can lead to interstate conflicts of interest, which can then lead to militarized interstate disputes and war, and finally feeds back to the context. I find that organizing the course around different substantive topics—such as contiguity, power, the democratic peace, for example—works better for students' learning. Accordingly, this book is organized topically, with each chapter focusing on a different substantive topic.

The first part of the book, Foundations, covers in chapters 1–3 essential information about the scientific study of international conflict and introduces the process of war theoretical model, which provides the framework for the remainder of the book. Part II, Causes of War and Militarized Disputes, focuses in chapters 4–9 on different causes of international conflict. Contiguity, power, alliances, democratic peace, deterrence, and the escalation of disputes to war are addressed as I attempt to examine a broad variety of causes of war identified by conflict scholars.

Chapters 10–12 in Part III, The Conduct and Aftermath of War, cover the evolution of war itself—military strategy and the duration, outcomes, and termination of war, as well as war's consequences. This is an increasingly important part of the literature on international conflict, yet is often ignored by textbooks on the subject. Chapter 13 covers recurrent conflict and international rivalry—thereby completing the process of war.

In Part IV, Conclusion, chapter 14 wraps up the book by exploring the cumulation of knowledge and the importance of theory. With so many studies in the field of international conflict, it is important that a synthesis of research helps us to determine which empirical patterns hold over a broad range of research and which ones do not. Although a brief chapter, it presents an important perspective on how to approach and process the myriad and often conflicting studies and data.

Specific examples are used throughout the chapters to illustrate various concepts and theories. More in-depth examples, such as in chapter 10's discussion of Military Doctrine and Strategy about the role of Israel's high combat effectiveness in the Six-Day War, are used to expand on key points in the text in ***Case in Point*** boxes. ***Concept in Focus*** boxes appear in some chapters to delve further into abstract terms that provide important background for understanding international relations, such as the differences between a state and a nation and tips for understanding quantitative research. Additionally, a few general boxes introduce students to other important topics such as pioneers in the scientific study of war and the Correlates of War Project. ***Key concepts*** appear throughout in bold and are listed at the end of each chapter as a convenient study aid, while ***chapter conclusions*** help students synthesize all that they have learned in the chapter and place it into the wider context of understanding international conflict.

The ***Appendix*** presents a list of wars and major crises over the past two hundred years, along with brief summaries of each and a few recommended sources for further information. I have found such a reference helpful when I teach my classes, as it provides a useful starting point for student assignments to research different conflicts, allowing them to apply concepts and theories covered in the class to explain the outbreak or outcome of a particular conflict. Finally, ***References*** at the back of the book include full citations for all of the research examined and noted in the chapters.

Versions 4 of the militarized interstate dispute and alliance data sets were released by the Correlates of War Project late in the book's production process. The book was revised to account for these updates to the fullest extent possible considering the timing of their arrival and limitations imposed by other data sets (such as contiguity and power) that have not yet been further updated.

Acknowledgments

As with any project, there are a number of people who deserve thanks. The entire team at CQ Press has been a pleasure to work with. Acquisitions editor Elise Frasier was excited about the project as soon as I shared the idea with her, and she guided me through the early stages of writing. Publisher Charisse Kiino and executive director Brenda Carter supported the book from inception. Developmental editor Nancy Matuszak painstakingly edited the entire manuscript and greatly improved the final product. Production editors Stephanie Palermini and Bennie Allen shepherded the book through its final stages toward publication. Copy editor Diane DiMura had a great eye for detail, making many improvements in clarity and style. Sarah Calabi arrived at CQ Press during the production phase and has been an excellent addition to the team.

Chuck Boehmer, Johanna Cox, Cooper Drury, Paul Hensel, Pat James, Sara Mitchell, Amanda Murdie, and Mike Rudy all provided encouragement about the project as I was working on it. *International Conflict: Logic and Evidence* has been class tested, and feedback from students and reviewers led to a variety of improvements. My thanks to all, particularly to those who provided excellent feedback in reviews of the proposal and draft manuscript:

Philip Arena, University at Buffalo, SUNY
Andrea Bartoli, Seton Hall University
Mark Crescenzi, University of North Carolina, Chapel Hill
Theodora-Ismene Gizelis, University of Essex
Jeannie Grussendorf, Georgia State University
Patrick James, University of Southern California
Nate Jensen, Washington University in St. Louis
Alynna Lyon, University of New Hampshire
Brian J. Phillips, Center for Research and Teaching in Economics (CIDE)
Nil S. Satana, Bilkent University, Turkey
Brent J. Steele, University of Utah
Barry H. Steiner, California State University–Long Beach
Brandon Valeriano, University of Glasgow

Nils B. Weidmann, University of Konstanz, Germany

Steve A. Yetiv, Old Dominion University

Finally, I have dedicated the book to Frank Zagare. Frank was my advisor in my PhD program at the University at Buffalo, and has been a trusted mentor and friend for the past decade and a half. I hope that this book will enable students to learn about international conflict as well as I learned from him.

About the Author

Stephen L. Quackenbush is Associate Professor of Political Science at the University of Missouri. His research and teaching focuses on international conflict. Specific areas of interest include understanding the dynamics of deterrence, the effect of settlements on recurrent conflict, and the effect of strategy and other factors on war outcomes. He is the author of *Understanding General Deterrence: Theory and Application* (Palgrave Macmillan), and his research has been published in journals such as the *Journal of Politics*, *Journal of Conflict Resolution*, *Political Research Quarterly*, *Journal of Peace Research*, *International Interactions*, *Conflict Management and Peace Science*, and *Review of International Studies*. He served as an Army officer, including a year in Iraq in 2004, receiving the Bronze Star Medal.

Part I

Foundations

1 The Scientific Study of War

American military commanders study a battle map in Italy in 1944. Like them, we are interested in studying war, although our methods are different.

Source: © Hulton-Deutsch Collection/CORBIS

Since the Cold War, the United States has fought two wars against Iraq (in 1991 and 2003), invaded Afghanistan, and fought an aerial war against Serbia. It has also fought counterinsurgency campaigns in both Afghanistan and Iraq and had lesser military interventions in Libya, Haiti, Somalia, and Bosnia-Herzegovina. There have, of course, been many more conflicts around the world that the United States has not been involved in. Why do these conflicts happen? Why do crises sometimes escalate to war, as the July Crisis of 1914 that escalated to World War I, while other times they do not, as the Cuban Missile Crisis in 1962 that did not lead to World War III? Why are wars sometimes quite short (like the Six-Day War in 1967), while other times they last for years (such as World War II, from 1939 to 1945)?

These are just some of the many questions that we could ask about international conflict. And these are important questions to answer because war and other forms of international conflict shape the world as we know it. War has played a crucial role in the formation, expansion, and disappearance of countries around the world. It has had a broad range of political, social, and economic effects: The creation of international organizations and alliances, the empowerment of women, and rapid technological innovations are just a few of the effects that wars in the past century have had. War, however, is also an extremely costly endeavor, in terms of lives lost, property destroyed, and money spent.

Because of these high costs, many people in history have wished for world peace. But a necessary prerequisite for achieving peace, or even simply for reducing the frequency and severity of conflict, is understanding how international conflict works—why it happens, how it is terminated, and so on. Accordingly, this book is about war, and international conflict more broadly. But it is more than that; it is about the scientific study of war. For some readers, this may seem to be quite surprising. After all, many people have joked over the years that political science is an oxymoron, a contradictory term. Obviously this book takes a different view.

We will first examine the meaning of science, followed by a comparison between the scientific and classical approaches to studying international relations. We then examine the importance of levels of analysis in international relations before turning attention to two specific studies that provide a framework for the examination of international conflict throughout the book: dangerous dyads and the process of war.

What Is Science?

Before moving on to exploring the scientific study of international conflict, it is necessary to discuss briefly what we mean by science. *Science* is a term that is often used, but not as often understood. Sometimes *science* is treated as a synonym for *technology*, which it certainly is not. *Science* is often treated as an umbrella term to refer to physics, chemistry, biology, and related subjects—the natural sciences. While those are scientific fields of study, they are not the same as science.

Science is a way of trying to understand things. Science is primarily an empirical method of inquiry that can be used to study how the world works. By empirical, we mean that science is based on observations. As Hoover and Donovan (2004, 34) state, the "scientific method seeks to test thoughts against observable evidence in a disciplined manner, with each step in the process made explicit." We believe that gravity works because we can observe its effects. If objects did not fall toward the earth when free to do so, then theories of gravity would be contradicted. Similarly, if something is thought to be a cause of war, then we need to be able to observe evidence that it does.

Science seeks to form generalizations. In other words, we are more interested in being able to explain war in general than any specific war in particular. We would certainly like to know what caused the Vietnam War, the Iran-Iraq War, the Russo-Japanese War, or any other particular war in history. If we do not understand what causes war in general, however, then how can we have confidence that we know the causes of any specific war?

Many people have argued that those who do not learn from history are destined to repeat it.[1] The idea that one can learn lessons from history depends on generalizations. If there are no general trends, then it is not possible to learn from history because every situation is entirely unique. But if general trends do exist in international politics, then the scientific method is appropriate for uncovering them.

The scientific method cannot tell us what is good or what values should be pursued. For example, science can help us understand what causes war, what factors explain how long wars last, what the consequences of war are, and so on. With a better understanding of war's causes, we could work toward changing those causes so as to reduce or eliminate war. Science, however, cannot tell us whether or not eliminating war would be a good thing. In other words, science is a tool for answering questions about "what is" or "what could be," but it does not apply to questions about "what ought to be." This is not to say that such questions are not worth asking or answering, just that our answers must come from somewhere besides scientific analysis.

Ultimately, science insists that understanding is based on logic and evidence. A key element of science is that it requires a particular logic of confirmation. It means demanding that our expectations about the causes of something (such as war) be tested against reality. This involves following specific procedures to test a hypothesis and assess its adequacy in light of the evidence.

---

[1] This idea has been attributed to the British philosopher Edmund Burke in the 1700s, the Spanish philosopher George Santayana in the early 1900s, and the British prime minister Winston Churchill.

Elements of Scientific Theories

The centerpiece of scientific explanation is theory. Because our focus in this book is on being able to explain international conflict, we will examine a variety of different theories. We begin by examining important components of theories, including variables and hypotheses. We will consider each in turn.

Variables are key elements of scientific studies. A variable is something that varies, as opposed to a constant. Although the term *variable* often conjures up images of numbers, variables do not have to be numbers. For example, gender is a variable that can either be female or male. Similarly, war is a variable because sometimes there is a war and other times there is not.

A **dependent variable** is something that we want to explain. An **independent variable** is some factor that we expect to provide all or part of the explanation of the dependent variable. Variables can be either independent or dependent; they are not inherently one or the other. For example, many studies seek to explain war, in which case war is the dependent variable. Other studies, however, look at the effect of war on other factors, such as international trade, alliances, or birth rates. In these cases, war is an independent variable.

In order for our variables to be useful for examining the real world, we must have some way of measuring them. Measurement is the process by which observations of reality are counted or categorized based on specific rules and procedures. Fundamental to measurement are two different types of definitions: conceptual definitions and operational definitions. A **conceptual definition** is a definition in terms of ideas. This is similar to what we typically think of for definitions for words like those that we find in a dictionary. For example, *Webster's Dictionary* provides several definitions of war, including "a state of usually open and declared armed hostile conflict between states or nations" and "a state of hostility, conflict, or antagonism." These are conceptual definitions because they focus on the underlying ideas of what we mean by war. They do not, however, help very much in determining whether a particular event is a war or not.

Measurement requires us to be more specific about what we mean by a concept, leading us to use an operational definition. An **operational definition** specifies observable characteristics of a concept to allow the concept to actually be measured. In other words, we must explicitly specify criteria that we can observe in order to identify categories or levels of a concept. For example, one of the variables Stam (1996) uses in his analysis of war outcomes is military quality. Higher-quality militaries have better trained troops and higher-technology equipment, but how can we measure that? Stam (1996, 95) operationalizes "the quality of military forces for a country as military expenditures (in constant US dollars) divided by number of military personnel." Stam focuses on spending per soldier because training and technology are expensive; while having more troops leads to higher levels of absolute spending, higher levels of technology and greater training lead to higher spending per soldier. This is not a perfect measure of military quality, but it makes a lot of sense. Furthermore, data on military expenditures and military personnel are available annually for nearly the past two centuries. For basically the same reasons, a similar method is often used to measure quality of education in school districts (spending per student).

Two key considerations for measuring variables are reliability and validity. Reliability refers to whether repeat measurements will produce the same (or very similar) readings. Of course, just because a measurement tool is reliable does not mean that it provides correct measurements. Validity refers to whether the measurement actually measures the underlying concept that it is intended to measure. For example, how does one measure success in war? This is a question that posed great difficulty for the United States in the Vietnam War. Because the United States was fighting against an insurgency rather than a conventional military, standard measures such as territory gained and lost were not appropriate. Accordingly, the US Army relied mostly on body count to measure progress in its counterinsurgency efforts. This measure, however, faced a couple of important limitations. First, it was not a valid measure of progress in counterinsurgency because it did not get at the political and psychological elements that were of crucial importance in the war. Second, it was not a reliable measure either, because it was difficult to distinguish guerrilla casualties from civilian casualties and there were strong incentives for soldiers in the field to overestimate the number of casualties inflicted.[2]

Hypotheses are the predictions of theories to be tested. They specify expected relationships between independent and dependent variables. For instance, "arms races cause war" could be a hypothesis. This hypothesis indicates that "arms races" is the independent variable, "war" is the dependent variable, and the relationship between them is positive. A positive relationship between variables indicates that as one variable occurs or increases in level, the other variable increases as well; thus, the variables move in the same direction. In contrast, a negative relationship indicates that as one variable increases, the other decreases; the variables move in opposite directions.

A theory represents what we think happens in the world. Theories organize, explain, and predict knowledge. A theory is a set of interrelated hypotheses and also provides the logic supporting those hypotheses. Theories are not true or false, in the sense that "true" theories exist out in the world just waiting to be discovered. On the contrary, theories are useful (or not) for helping us to understand how the world works. Theories are intellectual tools that are products of human imagination and hard work.

Deduction Versus Induction

How are these theories created? Two important logical tools used in theory development are deduction and induction, and people have often debated about which is better. Deduction means going from the general to the specific, whereas induction involves going from the specific to the general.

---

[2] For an in-depth examination of efforts to measure progress in the Vietnam War, see Daddis (2011).

With induction, we start with observations of particular cases of interest, uncover regularities from those observations, and then a theory is formed based on those observations.

The primary weakness is that correlation can never prove a causal connection. What it can do is give support to an explanation that you can satisfy on logical grounds.

Such logical grounds are the focus of deduction. With deduction, we start with assumptions about how the world works and use these to logically derive hypotheses about the relationship between variables. This logical structure and set of hypotheses is then our theory. The problem is that we cannot establish whether there is any empirical support for our theory through deduction alone.

Some argue that the **scientific approach** is "based either upon logical or mathematical proof, or upon strict, empirical procedures of verification" (Bull 1966, 362). This statement implies that we only need deduction or induction, not both. In contrast, Bueno de Mesquita (1981, 9–10) argues the following:

> Because a theory's usefulness can only be judged empirically, while an empirical generalization's truthfulness can only be judged logically, neither inductive nor deductive reasoning by itself is adequate. Inductively derived generalizations that cannot be derived logically must be spurious, just as deductively derived generalizations that do not explain events of interest to the researcher must be trivial.

What we need, then, is to demonstrate (deductively) that our theory is logically correct, and then to test it (inductively) to demonstrate that it is empirically supported. Ultimately, science requires both logic and evidence. Without logic, we cannot know that the empirical patterns that we observe are real. Without empirical evidence, our theories do not help us understand the real world, no matter how logically consistent they are.

Criteria for a Good Theory

Many different theories have been developed to explain various aspects of international conflict. Some theories are very broad, seeking to explain virtually all of international politics, while others are quite narrow, focusing only on some particular aspect of international conflict (e.g., the role of alliances). Theories are very important, but not all theories are equal. There are several criteria that we look for when evaluating a theory.

First of all, logical consistency is very important. A good place to judge a theory is to determine whether some parts of the theory contradict others. If they do, then there will be considerable confusion as to what the theory predicts, or what its explanation is. Presence of inconsistencies means that part of the theory is false on logical grounds. There is no need to look at reality to judge a theory's usefulness if it is illogical. Such contradictions are sometimes overlooked or sidestepped when scholars construct theories about international politics. It is important, however, to guard against inconsistencies by insisting that the theory's assumptions and logic are stated clearly.

For an example of logical consistency, consider the influential writings of Morgenthau. As part of his realist theory of international relations, Morgenthau (1967, 202) argues that "the desire to attain a maximum of power is universal." In other words, all states, without exception, want to acquire as much power as possible. But two pages later, Morgenthau (204) states that "the status quo nations, which by definition are dedicated to peaceful pursuits and want only to hold what they have, will hardly be able to keep pace with the dynamic and rapid increase in power characteristic of a nation bent upon imperialistic expansion." Thus, there are two types of states in the world. Status quo powers are content with what they have, so do not pursue increases in power. In contrast, imperialist powers are dissatisfied with the amount of power they have and try to gain more.

Thus, any action by states contradicts one aspect of Morgenthau's theory while being consistent with another. Therefore, the theory cannot be proven to be wrong, even in principle. Logical inconsistencies in a theory rule out the possibility of the theory making reliable predictions. Thus, Morgenthau's theory fails to provide any guidance about one of the very phenomena that it was intended to address.

The second key thing that we look for in theory is accuracy. By accuracy, we mean whether or not the theory makes accurate predictions about behavior in the empirical world. Predictions in the natural sciences are often deterministic and certain. Prediction in the social sciences, however, tends to be more difficult than in the natural sciences because predicting social phenomena such as international conflict involves elements of human choice that predicting physical phenomena such as condensation or chemical reactions does not have to deal with. Thus, while predictions of international relations theories are sometimes deterministic, more often they are probabilistic. Probabilistic predictions are predictions using probability rather than a single definitive prediction. Although such predictions are less certain, they are nonetheless quite useful. For example, weather forecasts are probabilistic. If the forecast calls for a 75 percent chance of snow rather than a 10 percent chance, that is important information to know.

Many of the events that we would like to predict in the study of international politics have already occurred. Thus, our "prediction" could often be termed *postdiction*. When predicting past events, it is important to make sure that we use information from before the event taking place. Hindsight may always be 20/20, but we cannot use the outcome to evaluate a decision; decisions must be evaluated within the context that they took place. Essentially what we are looking for is explanation, which is basically showing why something could not be otherwise.

The remaining criteria are less central to theory evaluation, but they can still be important. The hypotheses of scientific theories must be falsifiable. Falsifiability refers to whether or not a hypothesis could ever be proven wrong. Some hypotheses are not falsifiable. One consequence of logical inconsistency is that it undermines the falsifiability of the theory because whatever we observe will be consistent with some aspect of the theory, as in the previous example of Morgenthau's work. Valid theories cannot be falsified, but they must be falsifiable.

We also need our theories to be nonspurious. This is important because accurate predictions can be achieved even without meaningful explanation. A relation between two variables is called a *correlation*, but an important adage to keep in mind is that correlation does not equal causation. This is because the relationship between two variables might occur because of chance or because of some other variable; in other words, the relationship is spurious. For example, cities with larger police forces experience more crime. This is a spurious relationship, however, since cities with larger populations have both larger police forces and more crime. More particular to international relations, early empirical evidence showed that alliance partners were more likely to fight each other; subsequent study showed this relationship to be spurious, as neighboring states are more likely to fight and are also more likely to be allies (Bremer 1992). We need empirical evidence to establish that there is a relationship between the independent and dependent variables, and we need logic to establish that the relationship makes sense.

There are a variety of other considerations that people look for when evaluating theories. Theories should have organizing power, or the ability to bring together information and show how it fits together. Theories should have heuristic power. This refers to the fruitfulness of a theory, the theory's ability to raise new questions. Does the theory lead anywhere? Finally, parsimony is valued in theory. *Parsimony* means simplicity; when possible, we want theories to be simple. We might, however, need to use more complicated theories in order to generate useful predictions.

Ultimately, theory evaluation is a competitive endeavor. The one that explains more is better. This is essentially the idea of the twentieth-century philosopher of science Imre Lakatos. Bueno de Mesquita (2010, 400) calls this the first principle of wing walking: "If you are out on a wing of an airplane in flight . . . , don't let go of what you are holding on to unless you have something better to hold on to." This is highly applicable to theory evaluation because there is no reason to reject (let go of) a theory even when you have determined that it has various weaknesses unless there is a better theory to grab hold of.

The Scientific and Classical Approaches

People began to study war scientifically in the aftermath of World War I. The Great War, as it was known at the time, had shattered the long period of great power peace in Europe that had lasted for much of the century since Napoleon's final defeat in 1815. World War I shaped the world in a number of ways. One of those ways was that scholars began to explore international conflict in a systematic manner in order to understand what causes war. If the causes of war are understood, then one can take steps to prevent future wars. Two scholars in particular—Wright and Richardson—pioneered the scientific

Box 1.1

Pioneers of the Scientific Study of War

The scientific study of international relations has advanced through the efforts of many scholars over the past century. Although many scholars today—in the United States and across the world—are continually making advances in our understanding of international relations, three pioneers of the scientific study of international conflict stand out.

The first is Quincy Wright (1890–1970), a professor at the University of Chicago. In 1942 Wright published his seminal work, *A Study of War,* in which he examined war historically, legally, and culturally. Quincy Wright did more than pile up information about war; he developed a basic theory of war, arguing that the key factors for understanding it are technology, law, social organization, and opinions and attitudes concerning basic values. In addition to *A Study of War,* Wright published a further twenty books and nearly four hundred journal articles during his career.

The second pioneer of scientific study is Lewis Fry Richardson (1881–1953). Richardson was an English mathematician who made contributions to several fields of study, particularly meteorology, solving systems of linear equations, and fractals. Importantly, Richardson also devoted his attention to understanding international conflict. He developed a revolutionary mathematical model of arms races. His best known works on international conflict were *Arms and Insecurity* (1960) and *Statistics of Deadly Quarrels* (1960), each published posthumously.

The final pioneer highlighted here is J. David Singer (1925–2009). Singer was a professor of political science at the University of Michigan. Building on the work of Wright and Richardson, Singer founded the Correlates of War Project (discussed in Box 2.1) in 1963. He recognized that advancing the scientific study of international conflict required collection of data, which he embarked on. Singer authored or edited nineteen books and authored well over one hundred articles and book chapters.

Wright, Richardson, and Singer each made tremendous contributions to the scientific study of international relations in general, and international conflict in particular.

study of war in the interwar period. Although he did not begin his career as a political scientist until the 1950s, Singer was another important pioneer.

The scientific approach that we have been discussing is not the only approach that has been used to study international relations. That approach is contrasted with the classical approach. As its name implies, the **classical approach to theory** has a long history among scholars of international relations, hearkening back to writers such as Thucydides and Machiavelli from centuries ago.

One of the staunchest advocates of the classical approach to theory was Bull, a British scholar. In Bull's (1966) words, the classical approach is

> the approach to theorizing that derives from philosophy, history, and law, and is characterized above all by explicit reliance upon the exercise of judgment and by the assumptions that if we confine ourselves

> to strict standards of verification and proof, there is very little of significance that can be said about international relations, that general propositions about this subject must therefore derive from a scientifically imperfect process of perception and intuition, and that these general propositions cannot be accorded anything more than the tentative and inconclusive status appropriate to their doubtful origin. (361)

Bull not only advocated a classical approach to theory but also fiercely criticized the scientific approach. He lists seven "propositions" critiquing the scientific approach, arguing that it has "done a great disservice to" the study of international relations (Bull 1966, 370). His first argument is that, "by confining themselves to what can be logically or mathematically proved or verified according to strict procedures, the practitioners of the scientific approach are denying themselves the only instruments that are at present available for coming to grips with the substance of the subject" (366). Thus, while advocates of science argue that logic and evidence are the keys to knowledge, Bull argues that they cannot be relied upon. The analyst's judgment is the important thing, not logic or evidence.

Bull follows this up by arguing that "where practitioners of the scientific approach have succeeded in casting light upon the substance of the subject it has been by stepping beyond the bounds of that approach and employing the classical method" (Bull 1966, 368). Employing science, however, does not mean that you cannot use judgment to come up with ideas. As Singer (1969, 68) explains, "Classical concepts and historical insights are very much *within* (and not beyond) the bounds of the scientific spirit." It does not matter where the idea behind a given theory comes from; science simply says that we must use logic and evidence to assess ideas.

Finally, Bull argues that "practitioners of the scientific approach, by cutting themselves off from history and philosophy, have deprived themselves of the means of self-criticism, and in consequence have a view of their subject and its possibilities that is callow and brash" (Bull 1966, 375). Certainly, history is very important. In many ways, history is our laboratory in the study of international relations, and therefore it is important to know a lot about the history and geography of international politics.

The scientific and classical approaches have very different perspectives on the study of international relations, but ultimately the difference between them can be seen by their answers to this one simple question: *Why should we believe someone's argument about how international politics work?* Is it because they present sound logic and solid evidence in support of it? Or because we trust their judgment and experience?

There have been other arguments against a scientific approach than those addressed by Bull. Feminist theorists have argued that scientific principles were created by men and remain male dominated. Thus, according to this view, science is greatly limited as a path to knowledge because it is gender specific (Tickner 1997). Nonetheless, gender and feminist worldviews can be incorporated into scientific and quantitative analyses (Caprioli 2004). For example, Caprioli and Boyer (2001) examine the impact of gender on international conflict, finding that countries with higher levels of gender equality are less likely to escalate international crises to higher levels of conflict.

The debate over science in international relations is often taken to be a debate over the relative merits of qualitative and quantitative research. **Quantitative research** uses numbers and statistics to test hypotheses, seeking to uncover general trends. On the other hand, **qualitative research** is an umbrella term describing a variety of approaches that seek to describe or understand events without the use of statistical comparisons. Focusing on science, however, does not require one to use a quantitative analysis. Rather, the same concerns about logic and evidence that are central to the scientific method apply to qualitative research as well (King, Keohane, and Verba 1994).

The debate over science is a much broader one than we can fully cover here, but our goal is to seek understanding of international conflict. With this basic understanding of the issues concerning a scientific approach to international conflict, we now turn our attention to another important preliminary issue, levels of analysis.

Levels of Analysis

Like other topics of interest, international relations can be studied on many different levels. We will examine four of the most common **levels of analysis** used in studies of international relations. In his book *Man, the State, and War,* Waltz (1959) conducted one of the earliest examinations of levels of analysis in international politics. Waltz considers three: individuals, the state, and the international system.

At the smallest scale, one could focus on individual people. Waltz calls this the first image. The individual level of analysis explicitly recognizes that states and other collective actors do not do anything—people do. Thus, when we say that one country invaded another, what we actually mean is that the leader or leaders of the country decided to launch an invasion of another country. For instance, in his book *Essence of Decision,* Allison (1971) examined explanations of the Cuban Missile Crisis at different levels of analysis and argued that the individual level of analysis works best. In particular, for the different key individuals, "where you stand depends on where you sit" (176), meaning that individual decision makers' preferences regarding a situation are driven by their role within the government.

At a larger scale, one could also focus on the state, which Waltz considers to be the second image. Essentially, state level analyses seek to identify characteristics of states that make them more or less likely to engage in international conflict. At the state level of analysis, for example, scholars have looked at national attributes such as population, economic development and business cycles, national culture, and domestic conflict; regime characteristics such as government centralization, bureaucratic attributes, democracy, and election cycles; capabilities factors such as power status and militarization; as well as other factors such as status quo orientation (Geller and Singer 1998).

The third image in Waltz's classification is the international system, the broadest level of analysis. The international system entails all of the states and institutions that make up the world at a given point in time. Realist theory often focuses on the international system as a whole, in particular arguing that the distribution of power in the system is an important determinant of world politics. For instance, Waltz (1964) argues that a bipolar system (where there are two major powers much stronger than anyone else), such as the Cold War system in which the United States and the Soviet Union dominated international politics between 1945 and 1990, is more stable than a multipolar system (where there are multiple major powers), such as in the first half of the twentieth century.

There is a fourth level of analysis that is important to consider, even though traditional discussions such as Waltz's do not mention it: the dyadic level of analysis. A **dyad** is a pair of states, and the dyad is a popular level at which to focus analyses because it takes two states to fight, two states to form an alliance, and so on. Most and Starr (1989, 76–8) argue that because war (and most other events in international relations) is "the outcome of the interactions of at least two parties, the attributes of all those parties—not just one of them—must be considered in one's attempts to understand and explain when wars will and will not occur." Given this, the dyadic level of analysis is usually the most appropriate level of analysis for studies of international politics.

A variation of the dyad unit of analysis is the directed dyad. Within a directed dyad, the direction of interaction between the states is important. Thus, for example, United States→Japan is one directed dyad and Japan→United States is another. Nondirected dyads, where we observe each pair of states only once, are appropriate when we just want to analyze the occurrence of conflict. Directed dyads are needed for analyzing choices to initiate conflict, rather than just its occurrence. In other words, if we just want to know whether a pair of countries fights in a given year, then we should use nondirected dyads; if we want to know not just whether they fight, but also which state initiates the conflict, then we need to use directed dyads.

The modern study of international conflict focuses a lot on the dyadic level of analysis, while the state and system levels of analysis have become increasingly less common (Geller and Singer 1998). In this book we will share this focus, although we will certainly also look at other levels of analysis as appropriate.

Dangerous Dyads

At this point, it is useful to examine a specific scientific study of war. Bremer introduced the concept of dangerous dyads in a seminal article

Box 1.2

Concept in Focus: States Versus Nations

In the United States, we often talk about different countries, such as Germany, India, and Brazil, and different states, such as Missouri, New York, and Florida. We usually think of "nation" as being synonymous with "country," so we can talk of the nation of Canada, for example. It is important, however, to be familiar with the technical meanings of these terms when studying international politics.

A *state* is a government entity that controls a given territory. States are sovereign, meaning that they have legal and political supremacy within their borders. On the other hand, a *nation* is a group of people that share common characteristics, usually a common language, culture, religion, or values. States and nations are not necessarily the same thing. Some states are multinational, such as the former Soviet Union, which was one state composed of many different nations (Russians, Ukrainians, Belarusians, etc.). Also, some nations have no state of their own, such as the Kurds, who are separated among Iran, Iraq, Syria, and Turkey. When a nation and a state coincide, we have a *nation-state*—a political unit within which the people share an identity.

published in 1992. The idea of **dangerous dyads** is that some pairs of states (dyads) at some points in time have a higher likelihood of getting involved in a conflict (i.e., are more dangerous) than other dyads or times. Furthermore, certain contextual factors have a very important impact on the likelihood of conflict between states.

Bremer focused on six key determinants of war and sought to rank order them empirically based on their importance as causes of war. They are listed in rank order in Table 1.1. He found that the strongest single factor that increases the likelihood of war is the presence of contiguity. There is a substantial amount of research supporting this finding. States sharing a border form more dangerous dyads for two primary reasons. First, borders provide opportunities to fight. Second, contiguous states are more likely to disagree about territory, which is an important cause of war. We will examine the importance of contiguity and territory in chapter 4.

Table 1.1 Most Important Factors Creating Dangerous Dyads

FACTOR	RANK
Presence of contiguity	1
Absence of alliance	2
Absence of more advanced economy	3
Absence of democratic polity	4
Absence of overwhelming preponderance	5
Presence of major power	6

The absence of an alliance is the second most important factor forming dangerous dyads according to Bremer's (1992) results. States that share an alliance are less likely to fight each other. This is because an alliance is generally a useful indicator of shared interests, and when states have common interests, they have less to fight about. We will examine the origins and effects of alliances more closely in chapter 6, looking not only at the impact of alliances on conflict between allies, but also the impact of alliances on conflict with other states.

The factor that Bremer (1992) finds to be the third most important is the absence of a more advanced economy in the dyad. War is always a costly endeavor for participants, but richer countries have more to lose. Therefore, they have more reasons to seek to avoid fighting. Bremer finds economic development matters only when both states in a dyad are advanced; if even one of them is less advanced, then the risk of conflict is much greater. Others have found that economic interdependence—how tightly integrated countries are through trade and other economic ties—is the key economic factor affecting conflict.

Although the democratic peace has been one of the most studied areas of international relations in the past few decades, Bremer (1992) finds that the absence of democratic polity is only the fourth most important factor forming dangerous dyads. Although Bremer's analysis supports the monadic democratic peace—the idea that democracies are more peaceful than nondemocracies in general—more extensive analyses reveal no empirical support for the monadic democratic peace. Nonetheless, the dyadic democratic peace—that democratic dyads are more peaceful than other dyads despite the fact that democracies are not generally more peaceful—has received strong empirical support through many studies. We will return to look at the democratic peace, as well as economic interdependence, in chapter 7.

Bremer (1992) finds that the fifth most important factor creating dangerous dyads is the absence of overwhelming preponderance. In other words, when a pair of states is relatively equal in power, they are much more likely to fight than when there is an imbalance of power between them. This finding supports the basic argument of power transition theory and contradicts the balance of power argument of realism.

The final, and least important, factor that Bremer (1992) examines is the presence of major power in the dyad. Because of their status as the strongest states in the international system, major powers have interests across a broader range of issues and places than do weaker states. Furthermore, major powers are better able to project power over long distances than are minor powers. Both preponderance and major power status are examples of the importance of power as a cause of conflict, which is the focus of chapter 5.

This concept of dangerous dyads can be applied to specific pairs of states. For example, consider which dyad is more dangerous: the United States and Canada, or Israel and Syria? Most readers probably have an instinct that Israel and Syria is the more dangerous dyad without being able to specify exactly why. But Bremer's analysis helps us do exactly that. The United States and Canada share a border and the United States is a major power, each of which increases the probability of conflict within the dyad. Both countries, however, are democracies, they are allies, they both have advanced economies, and the United States has overwhelming preponderance over Canada; each of these factors makes the dyad less conflict prone.

Turning our attention to Israel and Syria, we can see that they are contiguous, are not allied, do not both have advanced economies, are not jointly democratic, and there is no preponderance of power between the two. On the other hand, neither state is a major power, making the dyad somewhat less dangerous, although this is the least important of the six factors. Each of the five most important factors producing dangerous dyads makes the Israel-Syria dyad more conflict prone. Therefore, the evidence indicates that the likelihood of war is much greater between Israel and Syria than it is between the United States and Canada.

Not only does Bremer's study illustrate the usefulness of the scientific study of international conflict, it also provides a partial roadmap for this book to follow. The factors that Bremer identifies in making dyads more dangerous are the primary ones that we will focus on in exploring the causes of conflict. Yet this book is not only about the causes of conflict; it is about the entire process of war, a concept to which we now turn.

A Process Model of War

A useful conceptualization of international conflict is the process of war, also introduced by Bremer (1995). The **process of war** characterizes dyadic relationships, focusing on the basic steps through which war between states can emerge. This process begins with the contextual factors characterizing the dyadic relationship, as shown in the upper left of Figure 1.1. This

Figure 1.1 **The Process of War**

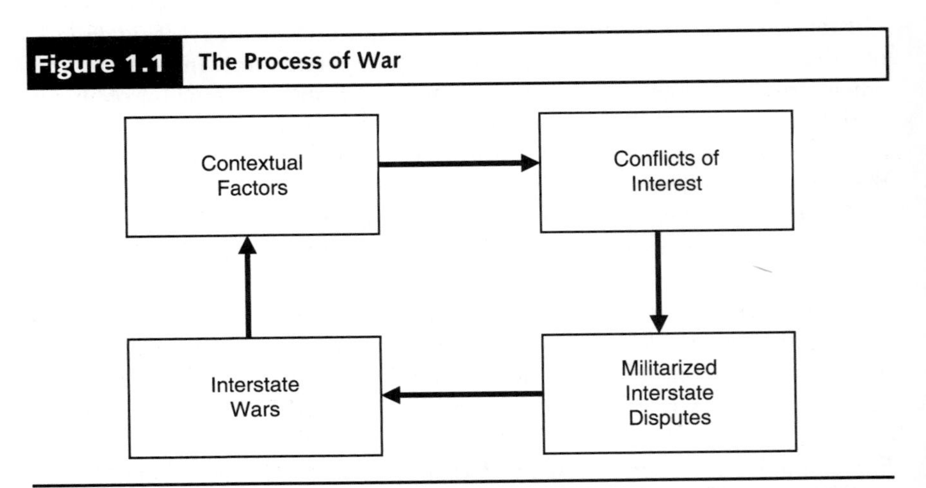

Source: Adapted from Stuart A. Bremer, "Advancing the Scientific Study of War," in Stuart A. Bremer and Thomas R. Cusack, eds., *The Process of War: Advancing the Scientific Study of War* (Amsterdam: Gordon and Breach, 1995), 4.

context includes political, economic, military, social, and geographic conditions in and between the states that can shape dyadic relationships in important ways.

Sometimes conflicts of interest between states arise out of this dyadic context, as shown by the top arrow in the figure. These conflicts of interest may involve disagreements about territory, ideological differences, policy disagreements, and a host of other possibilities. Most of the time, such conflicts between states either remain peaceful or are resolved through some nonmilitarized process of negotiation and compromise. At some times, however, states decide to pursue a militarized resolution to the conflict of interest, leading to international conflict.

Scholars have generally agreed upon two levels of militarized international conflict: dispute and war. Militarized interstate disputes are situations between states involving the threat, display, or use of military force. The next stage of the process of war occurs when a conflict of interest between the states in a dyad breaks out into a militarized dispute through one of these three actions, as shown in the lower right corner of Figure 1.1. The vast majority of disputes end in some manner without further escalation to war. However, about 5 percent of militarized interstate disputes do escalate to war.

Wars are the most severe form of international conflict. We explore the differences between militarized disputes and wars in the next chapter, as well as discuss how explicitly to define and measure them. Wars are very costly events in terms of lives lost, financial expenditure, and destruction of infrastructure. The process of war, however, does not end with the onset of war.

The evolution of war once it starts is part of the process. In addition, every war must end. Wars (and militarized disputes) have important consequences for international politics. A variety of changes can occur at both domestic (economic, demographic, and sociopolitical changes) and international (changes in geopolitical orientations, relative capabilities, alignment patterns, and economic conditions) levels. These consequences of conflict provide feedback to the contextual factors underlying relations between states, as indicated by the upward arrow on the left side of Figure 1.1. From that point, the process of war continues, with the potential for new conflicts of interests to arise, or old ones to rise again, into further militarized disputes and wars.

Most research on international conflict can be organized within this process of war (Bremer and Cusack 1995). Certainly the greatest amount of attention within the international conflict field has been focused on the causes of conflict. Some studies have examined specific factors, such as territory, alliances, or democracy, which affect the likelihood of conflict. Other studies seek to provide comprehensive theories of conflict outbreak. The common element in these studies is the underlying idea that contextual factors and conflicts of interest can lead to interstate disputes and wars as illustrated by the depiction of the process of war in Figure 1.1. We consider these explanations of the outbreak of international conflict in chapters 4 through 8, and to some extent in chapter 3.

Usually, particular studies focus either on the outbreak of militarized disputes or the outbreak of wars. Other studies focus on the escalation from disputes to war. This is the critical step moving from right to left along the bottom of Figure 1.1. We focus on explaining the escalation of disputes to war in chapter 9.

While the majority of studies on international conflict focus on the causes of disputes and wars, the process of war shows that the outbreak of conflict is only part of the story. An increasingly prominent literature seeks to explain the evolution of war. The three primary characteristics that studies focus on are the outcomes, duration, and costs of war. These occur within the interstate war box at the bottom left of Figure 1.1. We examine the evolution of war in chapters 10 and 11.

In order to move up the left side of Figure 1.1 from war back to contextual factors, the war must end. Accordingly, several scholars have focused on explaining war termination. These are closely related to studies of war outcomes and duration—the primary difference is that war termination studies explicitly seek to explain the dynamics through which states agree to cease their fighting.

Finally, following a war (or dispute) the process cycles back to (re) shape the underlying context between states, as illustrated by the upward arrow on the left side of Figure 1.1. These consequences of war have been less studied than other aspects of international conflict, but they can have dramatic, far-reaching impacts on international relations. Some of the primary consequences of war include changes in power and alliances, technological innovation, and a variety of changes in the economic and social realms. We look at war termination and consequences in chapter 12, and then more closely at the consequences of recurrent conflict and rivalry in chapter 13.

The process of war is a mental model of international conflict that provides a useful theoretical framework for organizing seemingly very disparate ideas about international conflict. Furthermore, it helps us not only with understanding the genesis of disputes and wars, but also with their escalation, evolution, termination, and consequences. Finally, the process of war reminds us that there are multiple paths to war. Some analysts have argued that there is one key thing, such as the balance of power or human nature, that causes war, but that simplistic viewpoint simply cannot be maintained.

Conclusion

The purpose of this book is to introduce readers to the scientific study of international conflict. Science is an empirical method of inquiry that insists that understanding comes through logic and evidence. Accordingly, both deduction and induction are vital to scientific theories. We need our theories to establish the logical relationships between independent and dependent variables, and we need empirical testing of our theories in order to determine whether or not there is evidence to support them.

While history and judgment—the focus of classical approaches to studying international relations—are important, they do not rule out the need for scientific rigor. To test our theoretical expectations, we can use either quantitative methods—statistical analyses to examine large numbers of cases—or qualitative methods—case studies of a small number of cases. In either case, we need to ensure that our understanding of international conflict is based on logic and evidence, not opinion and anecdote.

Bremer's (1992) idea of dangerous dyads marked a tremendously important contribution to the study of international conflict, and it provides a useful framework for our approach to the causes of conflict in this book. In particular, it was one of the early works establishing the dyad, or pair of states, as the primary level of analysis for studies of international politics. Characteristics of dyadic relationships—such as contiguity, alliances, power, and democracy—can have important effects on the likelihood of conflict, as we will examine in the coming chapters.

Our focus on international conflict in this book does not imply an exclusive focus on war. Rather, international conflict occurs in cycles, going from the underlying context, through various stages of conflict, and back again (Bremer 1995). This process of war provides a theoretical framework that guides our examination of international conflict, looking not only at the outbreak and escalation of international conflict, but the evolution and consequences of war as well.

Key Concepts

Classical approach to theory

Conceptual definition

Dangerous dyads

Dependent variable

Dyad

Independent variable

Levels of analysis

Operational definition

Process of war

Qualitative research

Quantitative research

Science

Scientific approach

2 Identifying Wars and Militarized Disputes

Afghan fighters battling alongside the United States rest at a former al-Qaida base near Tora Bora in 2001 behind a string of ammunition found after the retreat of al-Qaida members from the area. While the interstate portion of the Afghanistan War lasted less than three months, the insurgency that followed (an extrastate war) has endured for more than a decade.

Source: Associated Press

We are interested in understanding international conflict—the entire process of war, from its outbreak to the evolution of war itself, as well as the outcomes and consequences of war. Because this book focuses on scientific understanding, it is necessarily focused on various theories of conflict rather than on history.

Nonetheless, history is of crucial importance. Whatever we are interested in—whether it is the impact of territorial issues or democracy on the likelihood of conflict, the effect of military strategy on war outcomes, the effect of different types of settlement on the duration of peace following a conflict, or any number of other questions—we must turn to the history of international conflict to uncover patterns in the real world. Additionally, if we are to study international conflict scientifically, we must be able to define war and other types of conflict in a clear and consistent manner. This enables us to identify when conflicts have actually occurred so that we can conduct empirical analyses as discussed in the previous chapter.

These issues are the focal points of this chapter. In the next section, we examine the history of international conflict. Then we examine the definition of war and the various types of war that exist. The chapter ends with an examination of militarized interstate disputes and other measures of international conflict.

International Conflict in the Past Two Centuries

Humans are estimated to have been on Earth for about two million years. Scholars disagree about how long warfare has existed. Some argue it has been around since the earliest humans, while others estimate that warfare has been around for approximately twelve thousand years (Keeley 1996; Otterbein 2004). Accordingly, in considering the history of warfare, we could look back to that time. Or more realistically, we could go back to ancient Greece, where Thucydides wrote about the Peloponnesian Wars in 431–404 BCE.

Most studies of international conflict, however, look at the past two centuries—the time since the Napoleonic Wars. Following the 1789 French Revolution, France fought a series of wars called the French Revolutionary

Wars, from 1792 to 1802 against various European states. Napoleon seized power in France in 1799 and declared himself emperor in 1803 (prior to that, he had been a general in the French Army). France's subsequent wars across Europe—in Spain, Germany, Austria, Russia, and elsewhere—from 1803 until Napoleon's final defeat at the Battle of Waterloo in 1815 are called the Napoleonic Wars.

The Concert of Europe was established in 1815 by Austria, Prussia, Russia, and the United Kingdom to manage great power relations following the Napoleonic Wars. Accordingly, the year 1816 ushered in the modern era of international relations and is the starting point of many data collection efforts on international relations. Our focus in this book is therefore on events since that time.

Major Nineteenth-Century Wars

The first notable war following the Napoleonic Wars was the Mexican-American War from 1846 to 1848, which was an important step on the road to the United States becoming a major power. Mexico was upset that the United States annexed Texas in 1845, and the two sides disagreed about the proper location of the border. War broke out in April 1846, and after initial battles along the Rio Grande, the United States invaded Mexico. After the fall of Mexico City in September 1847, Mexico sued for peace. In the aftermath, the United States annexed much of what is now the southwestern United States—California, Arizona, and New Mexico.

The Crimean War, fought from 1853 to 1856, is notable for being the first to be fought between major powers since the Napoleonic Wars. Russia sought to push the Turks out of Europe and expand its access to the Mediterranean. To do so, it attacked the Ottoman Empire (modern Turkey) on 23 October 1853. The United Kingdom and France entered the war to defend the Ottoman Empire in 1854, with Sardinia/Piedmont (modern Italy) joining the following year. After a long campaign best known for the Charge of the Light Brigade and nursing innovations by Florence Nightingale, Russia sued for peace on 1 March 1856.

Other wars in the post-Napoleonic era are important for their results: unification. Both Italy and Germany were not unified into the nation-states that we think of today until the second half of the nineteenth century. Each of their unification processes occurred through a series of wars. Italy was divided into six different states: Sardinia/Piedmont, Papal States, Two Sicilies, Modena, Parma, and Tuscany. In addition, France and Austria both held territory in what is now Italy. The leading proponent of Italian unification was the Kingdom of Sardinia/Piedmont, while Austria was the major opponent.

The process began with two unsuccessful wars. The Austro-Sardinian War began in March 1848 when Sardinia/Piedmont attacked Austria in an attempt to eliminate Austrian rule of Lombardy and Venetia. Tuscany and Modena fought alongside Sardinia, but the war ended in March 1849 with an Austrian victory. The next war occurred when Italian liberals declared a

Roman Republic in February 1849. France sent an army to restore the Pope to power. In the resulting War of the Roman Republic, France was joined by the Kingdom of Two Sicilies and Austria, and defeated the burgeoning republic from April to July 1849.

Success finally came a decade later. In the War of Italian Unification, France fought alongside Sardinia/Piedmont and they defeated Austria from April to July 1859. Austria ceded Lombardy to Sardinia (through France), and in 1860, Modena, Parma, and Tuscany were incorporated into Sardinia/Italy. While foreign influence had been successfully reduced, independent states in southern Italy stood in the way of unification. In the Italian-Roman War of September 1860, Sardinia/Piedmont defeated the Papal States, and then in the Neapolitan War of October 1860 through February 1861, Sardinia/Piedmont defeated the Kingdom of the Two Sicilies. Italy (apart from Venetia, still under Austrian rule) was finally unified from north to south.

Like Italy, German unification was accomplished through a series of wars. Through the mid-nineteenth century, a number of German states existed—Prussia, Austria, Bavaria, Hannover, Baden, Württemberg, and others. Germany's wars for unification began with two wars against Denmark. In the First Schleswig-Holstein War (1848–1849), Prussia fought against Denmark after the German-speaking majorities in Schleswig and Holstein revolted against Danish rule. Prussian troops drove the Danish from the two duchies, but ultimately the issue was unresolved. In the Second Schleswig-Holstein War (1864), Austria fought alongside Prussia against Denmark. This time, they succeeded in taking the two duchies, with Prussia gaining control of Schleswig and Austria gaining control of Holstein.

Austria and Prussia were the two leading German states at this time. Prussian prime minister Otto von Bismarck wanted to create a unified Germany under Prussian leadership. Because Austria had a multinational empire, however, including not only Germans, but also Hungarians, Czechs, Poles, Ukrainians, and others, Austria would have to be excluded from a unified Germany. Accordingly, Bismarck sought war with Austria to establish Prussia as the leading German state. The ensuing Seven Weeks War (or Austro-Prussian War) of 1866 pitted Prussia and Italy against Austria and minor German states such as Bavaria and Hannover. Prussia invaded Austria and, taking advantage of superior rifles and tactics, decisively defeated the Austrian forces. Following the culminating Battle of Königgrätz, Austria sued for peace.

France now remained the sole obstacle to German unification. In the Franco-Prussian War of 1870–1871, Prussia and minor German states Bavaria, Baden, and Württemberg invaded France following the French declaration of war on 15 July 1870. They quickly defeated the French Army at the Battle of Sedan on 1 September 1870 and then laid siege to Paris beginning on 19 September. Germany was unified on 18 January 1871 before the war finally ended on 26 February 1871.

More than twenty-five years later, the Spanish-American War of 1898 would mark the emergence of the United States as a global power. Tensions between the United States and Spain had been rising over Spain's repression in its colony of Cuba and then boiled over after the US battleship *Maine* was sunk in Havana harbor on 15 February 1898. Rallying around the cry of "Remember the *Maine*," the United States went to war on 22 April 1898. Following a series of naval battles, the United States invaded the Philippines, Puerto Rico, and Cuba, all territories in the Spanish colonial empire. An armistice was signed on 12 August, and the United States annexed Puerto Rico and the Philippines and gained a naval base at Guantanamo Bay, Cuba, while Cuba was granted independence.

The Russo-Japanese War of 1904–1905 saw the emergence of Japan as a major power. Following wars with China in the previous decade, Russia and Japan had increased their influence in Manchuria and Korea, respectively. Each sought to expand their influence in the area, but Russia also sought to contain Japan's influence in Korea. In response, Japan launched a surprise attack on the Russian fleet at Port Arthur in modern-day China on 8 February 1904, followed up by a siege of the port, land battles in Korea and Manchuria, and a series of naval engagements. The war ended on 15 September 1905 and resulted in a decisive Japanese victory, after which Korea was occupied by Japan.

The Emergence of Modern Warfare

All of these wars paled in comparison to the world wars of the twentieth century. The early part of the twentieth century was marked by a series of crises. Although Britain and France nearly went to war in the Fashoda Crisis in 1898, they were otherwise strong allies, along with Russia, in the Triple Entente. They were faced by the Triple Alliance of Germany, Austria-Hungary, and Italy as Europe was engulfed in opposing alliance blocs.

Germany was quickly rising in power, embarking on a naval arms race with Britain and attempting to increase its colonial influence. This led to a series of crises, particularly the First Moroccan Crisis in 1905 between France and Germany and the Second Moroccan (Agadir) Crisis in 1911, this time with Britain alongside France against Germany. These crises further served to fuel rising tensions between the opposing sides.

War came with the July Crisis of 1914. The crisis was sparked by the assassination of Archduke Franz Ferdinand, heir to the Austrian throne, by a Serbian nationalist. The ensuing war, called the Great War at the time but later known as World War I, pitted the Central Powers of Germany, Austria-Hungary, and the Ottoman Empire against the Allies of France, Russia, the United Kingdom, Serbia, Italy, Belgium, Greece, the United States, and others. The war lasted from August 1914 until November 1918, and was dominated—particularly on the decisive western front—by trench warfare and stalemate. Attempts to restore mobility to the battlefield in the face of massive firepower, particularly from artillery and machine guns, led to the emergence of modern warfare.

There was hope that the Great War would have been the war to end all wars, but it was not to be. The League of Nations, an international organization aimed at keeping peace, was founded with this goal in the wake of World War I but was ultimately ineffective. After a relatively peaceful decade of the 1920s, the world descended down the path toward war once more in the 1930s. Three countries were particularly aggressive.

Japan was a member of the Allies in World War I. In the Second Sino-Japanese War (also known as the Manchurian War), Japan sought to expand its influence in Manchuria while China was distracted in civil war. The ensuing conflict lasted from 1931 to 1933, in which Japan defeated China and established a puppet state, called Manchukuo, in Manchuria. The League of Nations complained about Japanese aggression, but Japan merely responded by withdrawing from the League, illustrating the organization's weakness.

In July 1937, Japan invaded the rest of China, launching the Third Sino-Japanese War. It captured Beijing and much of China's urban areas and coastal provinces. The brutal campaign was highlighted by a variety of abuses of the Chinese population, especially the Rape of Nanking, where over 600,000 civilians were killed in 1937 alone. Nonetheless, China continued fighting, and after Japan's attack on Pearl Harbor in 1941, which brought the United States into combat, this became the Chinese front of World War II.

Meanwhile, Italy embarked on its own adventures under fascist leader Benito Mussolini. In the Italo-Ethiopian War (also known as the Abyssinian War), Italy invaded Ethiopia from its colony in Italian Somaliland. In a relatively short campaign from October 1935 to May 1936, Italy conquered Ethiopia and made it a colony. The League of Nations was once again powerless to stop aggression. Italy then conquered Albania in a short campaign in April 1939.

The most serious threat, however, was posed by Germany. Adolf Hitler gained leadership on 30 January 1933 and began to reestablish German power. In the March 1936 Rhineland Crisis, Germany moved its army west of the Rhine River, territory that had been demilitarized in the Versailles Treaty that ended World War I. France, Britain, and Belgium did no more than complain. In March 1938, Germany absorbed Austria in what is known as the Anschluss.

The Munich Crisis came next in September 1938 as Hitler demanded that the Sudetenland, territory of Czechoslovakia mostly composed of ethnic Germans, be united with Germany. In a meeting at Munich, France and the United Kingdom agreed to cede the Sudetenland to Germany, sparking British prime minister Neville Chamberlain's infamous claim that "I believe it is peace for our time." When Germany occupied the remainder of Czechoslovakia in March 1939, it finally became clear that Hitler would not be appeased.

World War II began on 1 September 1939 when Germany invaded Poland. Britain and France had had enough with Hitler's aggressiveness and declared war two days later. After quickly defeating Poland, Germany turned its attention west. In 1940, Germany conquered Denmark, Norway, the Netherlands, Belgium, Luxembourg, and France. Italy joined the war as Germany's ally in June 1940. The United Kingdom held on as the only state opposing Germany.

Hitler next turned his attention eastward. After rapidly conquering Yugoslavia and Greece, Germany invaded the Soviet Union in June 1941. The geographic scale of the war expanded greatly when Japan attacked the US naval base at Pearl Harbor, Hawaii, on 7 December 1941. Japan swept through the South Pacific and Southeast Asia, taking the Philippines, Malaya, Singapore, the Dutch East Indies, Thailand, Burma, and other locations in a massively successful series of campaigns over the next several months. The United States handed Japan its first defeat at the Battle of Midway in June 1942, sinking four Japanese aircraft carriers. The United States began its campaign to push back Japan by invading the island of Guadalcanal in August 1942.

The Soviets defeated German forces at the Battle of Stalingrad, which lasted from September 1942 to January 1943. Stalingrad is generally considered to be the turning point of World War II. The western Allies invaded Sicily in July 1943 and then mainland Italy in September 1943. Italy surrendered, but the Allied advance in Italy was slow and arduous against continued German defense.

The United States and Britain opened up a western front by invading France in June 1944. At the same time, the Soviets continued pushing in from the east. Germany launched one final offensive in December 1944 before finally surrendering on 8 May 1945. In the Pacific, the United States conducted an island hopping campaign, attacking only crucial locations. Battles in the Philippines, Iwo Jima, and Okinawa showed that defeating Japan, while perhaps inevitable, would be incredibly costly. In an attempt to avoid having to invade Japan itself, the United States dropped atomic bombs on Hiroshima and Nagasaki in August 1945, which led directly to the Japanese surrender. (A Case in Point box in chapter 12 provides more detail about World War II.) World War II was the largest war in history, with the most significant and farthest reaching consequences of any war. One of these consequences was the emergence of the United States and Soviet Union as the dominant states in the world.

Cold War

International politics in the decades after World War II were dominated by the Cold War. The Cold War was not actually a war; rather, it is the name given to the enduring rivalry between the United States and Soviet Union from 1945 to 1989. The intensity of the rivalry led many to believe that an actual "hot" war could break out any moment. Fortunately, it remained "cold," in terms of not involving actual war between the United States and the Soviet Union. (A Case in Point box in chapter 8 provides more detail about the Cold War.)

Although there had already been a series of disagreements between the two sides, Cold War tensions greatly escalated with the Berlin Blockade between the Soviet Union and the United States, United Kingdom, and France. Berlin, like the rest of Germany, was divided into occupation zones among the four powers after World War II. Berlin, however, was in the midst of the Soviet zone of occupation. The crisis began on 24 June 1948 when the Soviet Union blockaded all Western transportation into and out of Berlin. The Western powers responded by airlifting supplies to the city. The blockade ended on 12 May 1949, and the division of Germany into two states was further solidified.[1]

The Korean War served as another early setting for the Cold War. It began in June 1950 when North Korea invaded South Korea. The United States, along with other countries fighting under the authority of the United Nations, intervened on South Korea's behalf. After US forces began moving north across the 38th parallel in October, China also intervened, on the side of North Korea. The front line stabilized near the original border by June 1951 and remained largely stationary until an armistice was finally reached in July 1953, ending the war.

The Cold War nearly turned hot in October 1962 after the United States discovered on a routine reconnaissance flight that the Soviets were installing nuclear missiles in Cuba. The ensuing Cuban Missile Crisis was probably the closest the world has come to nuclear war. President John F. Kennedy estimated the chances of nuclear war as "between one out of three and even." However, war was successfully avoided as the Soviets backed down in the face of an American naval quarantine (a polite term for a blockade). As US Secretary of State Dean Rusk stated, "We're eyeball to eyeball and the other fellow just blinked."

The Vietnam War was the next battleground in the Cold War and became the most controversial war in US history. Following its defeat at Dien Bien Phu in 1954, France withdrew from French Indochina and the former colony was split into North Vietnam, South Vietnam, Laos, and Cambodia. Similarly to Korea, Vietnam was split into communist North Vietnam and noncommunist South Vietnam. The South Vietnamese government was very unpopular and faced an insurgency supported by the north. After years of providing aid and military advisors, the United States became directly involved in 1965, beginning the Vietnam War. Fighting ground battles in the south against Viet Cong insurgents and the North Vietnamese Army and an air campaign in the north, the United States struggled to make progress. The war became immensely unpopular on the American home front, and the United States withdrew in January 1973. In January 1975, North Vietnam invaded South Vietnam, and by the end of April had unified the country under communism.

Although the Cold War was dominated by tensions between the United States and Soviet Union, friction with China also played a role. China had been wracked by civil war between the Nationalists and Communists from 1930 to 1936 and again from 1946 to 1950 (the break in the middle was taken up by war against Japan). The Communists eventually won, forcing the Nationalists

[1]The American, British, and French occupation zones became West Germany (and West Berlin), while the Soviet occupation zone became East Germany (and East Berlin). Germany was later reunified on 3 October 1990.

to flee to the island of Formosa (Taiwan) in December 1949. Tensions between the two sides remained, and China and Taiwan fought two wars over the next decade. In the Off-shore Islands War from September 1954 to April 1955, China attacked Taiwan's hold on islands off the mainland coast. Taiwan evacuated all of the islands except for Quemoy and Matsu. The unresolved status of these islands led to the Taiwan Straits War in 1958, which ended in stalemate after the US Navy assisted in supplying Taiwan. Tensions between China and Taiwan have remained over the ensuing decades. Most notably, in the Taiwan Missile Crisis from July 1995 through March 1996, China test-launched a series of missiles over Taiwan and conducted military maneuvers off of its coast in an attempt to influence Taiwan's first democratic presidential election.

Other Conflicts in the Cold War Era

The post–World War II era also witnessed two important series of wars that were largely unrelated to the Cold War. The first of these occurred between Israel and its neighbors, called the Arab-Israeli Wars.[2] Israel was created from the British protectorate of Palestine. The presence of a Jewish state in the midst of the otherwise Arab Middle East was an immediate source of conflict. Israel became an independent state on 14 May 1948, and the Arab states attacked the next day. The Palestine War lasted from May 1948 to January 1949, as Israel was able to prevail against a coalition of Egypt, Syria, Jordan, Lebanon, and Iraq, gaining control of almost 80 percent of the Palestine mandate.

The next war was the only Arab-Israeli conflict to involve countries from outside the Middle East. Egypt nationalized the Suez Canal, which had been operated by Britain and France, in July 1956. Britain and France decided they needed to punish Egypt for this action, and so developed a plan alongside Israel to invade. The ensuing Sinai War began on 29 October 1956 when Israel invaded the Sinai Peninsula, quickly routing the Egyptian troops. The United Kingdom and France began air strikes on 31 October and landed troops to seize the canal on 5 November. A cease-fire was put in place on 6 November, ending the war, and a United Nations peacekeeping force was deployed to the Sinai.

War came to the region again a decade later in the Six-Day War. Fearing an attack by its Arab neighbors, Israel preemptively attacked Egypt, Jordan, and Syria. In a lightning campaign during 5–10 June 1967, Israel decisively defeated the Arab states. Following its victory, Israel occupied the Gaza Strip, Sinai Peninsula, West Bank, and Golan Heights. (A Case in Point box in chapter 10 provides more detail about the war.) Two years later, the War of Attrition commenced between Israel and Egypt. Lasting from March 1969 to August 1970, it was characterized by artillery duels, air strikes, periodic land raids, and general stalemate.

Three years later, the Yom Kippur War, also known as the October War, began as the Arab states, dissatisfied with Israel's control of the territories occupied since the Six-Day War, decided to retake the territory by force. On

---

[2] Although the Arab-Israeli Wars were not part of the Cold War, Israel was an American ally and the Arab states were allied with the Soviet Union.

6 October 1973, Egypt and Syria launched a coordinated surprise attack on Israel. Jordan, Iraq, and Saudi Arabia joined them within days. Initially, the Arab forces succeeded in pushing into the Sinai Peninsula and Golan Heights before Israel was able to gain the upper hand through a series of counterattacks. Israel won another decisive victory, crossing to the west side of the Suez Canal before agreeing to a cease-fire on 24 October.

The Yom Kippur War marked the last war between Israel and Egypt to date, as the two sides agreed to the Camp David Accords in 1977. Jordan later signed a peace treaty with Israel in 1994. Conflict between Israel and Syria, however, as well as other Arab states, persists.

The final interstate Arab-Israeli war to date was the Israel-Syrian War in Lebanon from April through September 1982. The Palestine Liberation Organization (PLO) had moved into Lebanon after being expelled from Jordan, and in 1982, Israel invaded Lebanon to remove the PLO. Syria, who occupied part of Lebanon, fought against Israel. Although Israel dominated a series of aerial battles with Syria, in the end the war was a stalemate.

The second series of wars waged at this time was the Indo-Pakistani Wars.[3] India had been the crown jewel of the British Empire. Upon decolonization, it was split into two major states: The areas with a Muslim majority became Pakistan, while the rest (with a Hindu majority) became India.[4] The ruler of Kashmir, on the border between India and Pakistan, joined India even though the majority of Kashmir's population was Muslim. This decision was hotly contested by Pakistan, leading to war. The First Kashmir War between India and Pakistan lasted from October 1947 to January 1949 and ended in a stalemate, with Pakistan in control of about one-third of Kashmir.

A line of control was established between the two sectors, although India and Pakistan continued to claim all of Kashmir as their own. Nonetheless, the area remained relatively peaceful for nearly two decades. In August 1965, however, Pakistan invaded Indian-controlled Kashmir, launching the Second Kashmir War. Fighting continued through September, with Pakistan gaining a small victory against India.

At the time of decolonization, Pakistan was divided into two parts, East and West Pakistan, on either side of India. Although they were both Muslim, significant cultural and language differences—in addition to geography—divided East and West Pakistan. East Pakistan began rebelling against the government of West Pakistan in the late 1960s, and in December 1971, India went to war to support East Pakistan's independence. In the resulting Bangladesh War, India defeated Pakistan in only two weeks, and East Pakistan gained independence as the new state of Bangladesh.

The two sides went to war again over Kashmir in 1999. In the Kargil War, Pakistan infiltrated troops past the line of control. In the two-month war from May through July 1999, India defeated Pakistan and reestablished its positions along the line of control. (A Case in Point

[3] Pakistan was an American ally, and India's major arms supplier was the Soviet Union. These wars were even less connected to the Cold War than the Arab-Israeli Wars.

[4] A third state, Sri Lanka, was created on the island of Ceylon off the southeast coast of India.

box in chapter 13 provides more detail about the rivalry between India and Pakistan.)

The last decade of the Cold War experienced a variety of conflicts, largely unconnected to broader US-Soviet conflict. The Iran-Iraq War was initiated by Iraq in September 1980. Although Iraqi forces were initially successful, Iran counterattacked in January 1981, and by May 1982 had driven Iraqi forces out of Iranian territory. The war bogged down in a long stalemate, finally ending in August 1988.

The Falklands War occurred from March through June 1982 when Argentina seized the British-controlled Falkland Islands. The United Kingdom fought back, and after a difficult naval and air campaign, successfully retook the islands. (A Case in Point box in chapter 4 provides more detail about the war.)

The United States had a series of relatively minor conflicts in the 1980s. In October 1983, a military coup overthrew the government of Grenada, an island state in the Caribbean. In response, the United States invaded Grenada and restored the constitutional government to power. The United States also had conflict with Libya. Although it was traditionally viewed as international waters, Libya under Muammar Qaddafi claimed the entire Gulf of Syrte as its territorial waters. Under the Reagan administration, the US Navy challenged these claims by holding naval exercises in the gulf, leading to crises between the United States and Libya in August 1981 and March through April 1986. Two Libyan aircraft were shot down by American aircraft in 1981; in 1986, the United States responded to Libyan attempts to shoot down American aircraft by conducting air raids on Tripoli and Benghazi, the two largest Libyan cities.

In another conflict closer to home, the United States invaded Panama in December 1989 to remove Manuel Noriega from power and bring him to trial on charges of direct involvement in drug trafficking from South America to the United States. In a quick campaign from 20 December 1989 through 3 January 1990, called Operation Just Cause, the United States succeeded in its objectives.

The United States became involved in another war months later when Iraq invaded Kuwait on 2 August 1990, beginning the Gulf War of 1990–1991. The United States led a coalition of the United Kingdom, France, Saudi Arabia, and other Arab states to liberate Kuwait. The campaign, called Operation Desert Storm, began on 16 January 1991 with a six-week air campaign preceding a short, decisive ground offensive beginning on 24 February.

The Post–Cold War Era

Later that year, the collapse of the Soviet Union left the United States as undoubtedly the world's strongest power. While the threat of major power war had seemingly receded, the United States still faced challenges from a variety of different directions. This led to interventions in widespread locales, such as Somalia, Bosnia-Herzegovina, Haiti, and Libya, and three interstate wars. The first of these interstate wars was the Kosovo War, during which the United States and its NATO allies waged an air campaign against Yugoslavia (essentially just Serbia) from March through June 1999 to end the Serbian campaign of ethnic cleansing in Kosovo. (A Case in Point box in chapter 9 provides more detail about the war.)

Two other wars were initiated by the George W. Bush administration in connection with the "Global War on Terrorism" following the 11 September 2001 terrorist attacks on the World Trade Center and Pentagon that killed nearly three thousand people. Afghanistan had provided a haven for the extremist group al-Qaida who was responsible for the attacks, and so was an obvious target of retaliation. The Afghanistan War was launched on 7 October 2001 by the United States, United Kingdom, and other allies. The interstate portion of the war was a decisive success, as the Taliban government was removed from power, and the war ended on 22 December 2001. It was followed, however, by an insurgency against the United States and the new Afghan government that remains ongoing.

In a much more controversial move, the United States, United Kingdom, and Australia invaded Iraq. The ensuing Iraq War lasted from 19 March through 2 May 2003, removing Saddam Hussein's government from power in a quick and decisive campaign. As in Afghanistan, however, the United States, the new Iraqi government, and their allies faced a strong insurgency campaign in the years following the war. The United States withdrew its remaining forces in December 2011. (A Case in Point box in chapter 11 provides more detail about the war.)

Defining War

To this point, we have talked a lot about war but have been relatively silent on the question of what war is. This is an important question that must be explicitly answered. How does one know when there is a war? Was World War II a war? Was the US invasion of Grenada in 1983 a war? What about the fighting between India and Pakistan around Kargil in the spring of 1999? Was the NATO air campaign against Libya in 2011 a war? Some answers are obvious, but some are unclear.

What makes a war a war? In other words, how do we define war? Bull (1977, 184) defines *war* as "organized violence carried on by political units against each other." This definition helps conceptualize what we mean by war, eliminating many of the ways that the word *war* is commonly used, such as the war on poverty or war on drugs. Unfortunately, it does not help to identify specific wars in history. We need a definition of war that points to specific, measureable criteria so that we can look at various events in history and determine whether or not they are wars. There are a variety of different factors that we might use as criteria for identifying wars. Whatever criteria we use, we want to be certain that we know what criteria we are using so that we can agree on what a war is.

A traditional way that people have separated wars from other types of international conflict is to focus on a formal declaration of war. Accordingly, we would label a conflict as a war only if the states involved make a public declaration of war. For example, when the United States was fighting in Korea in the early 1950s, President Truman claimed that the United States was not in a war because there had been no declaration of war. Rather, the media at the time often referred to it as the Korean Police Action.

We could also focus on a certain amount of time as a criterion for defining war. Since we think of wars as extended clashes between militaries, perhaps we only want to think of a conflict as a war if it lasts longer than some amount of time. It is not clear, however, what one should consider the minimum length of time to qualify for a war.

Since part of war is extensive violence, we might focus on a certain number of deaths as a requirement to consider a conflict a war. In other words, we might consider a clash between armies as a war (rather than a skirmish or dispute) only if casualties reach a certain level. This also accounts in part for the length of time, since the longer a conflict lasts the more opportunity it has to exceed a given casualty threshold.

Each of these criteria may seem reasonable, but we can clearly identify the ramifications of using them. The Vietnam War and many other conflicts would not be considered wars if one focused on a declaration of war. For instance, the United States has not declared war since World War II; indeed, most conflicts since World War II have not been declared wars. The Six-Day War between Israel and its neighbors in 1967 would probably not be considered a war if we focused on the length of time criterion. The invasion of Grenada by the United States in 1983 would likely not be considered a war under the casualty criterion.

Another important consideration is the belligerents involved in a conflict. Are the participants in the conflict sovereign states or are they colonies, insurgent groups, or some other kind of actor? While this criterion is not necessarily useful in identifying whether a particular conflict is a war or not, it is of crucial importance in determining the type of war.

The **Correlates of War Project (COW)** defines **interstate war** as fighting between the regular military forces of two or more countries, directed and approved of by central authorities, where at least one thousand battle deaths occur. Sarkees and Wayman (2010) discuss the latest update of the COW war data, while Singer and Small (1972) and Small and Singer (1982) provide discussions of earlier versions. This definition is the most common definition of war in use by political scientists today. Note that it focuses on the casualty criterion in order to distinguish wars from other types of conflict. The type of belligerents is also important in differentiating interstate wars from other types of war.

Box 2.1

The Correlates of War Project

The Correlates of War Project (or COW, for short) was founded by J. David Singer at the University of Michigan in 1963. The goal of the project was, and continues to be, the systematic accumulation of scientific knowledge about war. The project began by assembling a more accurate data set on war in the post-Napoleonic period.

But collecting data on wars was only the beginning. Since COW's focus was on identifying causes of war, data on potential causes was also needed. Accordingly, data on factors such as national capability, alliances, and geography were also collected. More generally, the COW project promoted cumulative science in the field of international relations. By helping to establish a clear temporal and spatial domain for research, promoting the use of clearly defined concepts and common variable operationalizations, and allowing replication of research, the project has been a mainstay of rigorous international relations scholarship.

In 2001, COW moved to Pennsylvania State University under the leadership of Stuart Bremer. Following Bremer's death in 2002, the project was led by interim director Scott Bennett before moving to the University of Illinois in 2005, where Paul Diehl became director. The current director of the project is Zeev Maoz, at the University of California–Davis, where the project moved in 2013. The project's data sets, as well as other information, are available from their webpage at http://correlatesofwar.org.

One might question whether 1,000 battle deaths is an arbitrary threshold to define a war. Fortunately, most conflicts do not hover around that mark. Rather, conflicts tend to either have many thousands of deaths, or have very few (less than 200) deaths. Table 2.1 shows the distribution of interstate wars by casualty level. Wars most frequently have between 5,000 and 20,000 battle deaths, with between 1,100 and 5,000 following closely

Table 2.1 Battle Deaths of Interstate Wars

BATTLE DEATHS	FREQUENCY	PERCENTAGE
1,000–1,100	10	10.53
1,101–5,000	29	30.53
5,001–20,000	30	31.58
20,001–100,000	12	12.63
100,001–1,000,000	10	10.53
1,000,001 +	4	4.21

Source: Compiled by author.

behind. Four wars (World War I, World War II, Vietnam War, and Iran-Iraq War) have more than one million battle deaths.

Ten wars have less than 1,100 battle deaths, and therefore barely cross the threshold to qualify as a war. For example, in the Falklands War, COW codes Argentina as having 746 battle deaths, with 255 for the United Kingdom, for a total of 1,001. Nine other wars are coded as having exactly 1,000 battle deaths.[5] By this measure, these conflicts are somewhat questionable regarding whether they fully meet the defining criteria for war, illustrating the impact of the 1,000 battle death threshold. Ultimately, the 1,000 battle death threshold is arbitrary; however, it is useful because it allows us to develop an agreed upon list of wars and ensure comparability between different studies.

The COW war data cover more than just interstate war. The typology of war used by the project is shown in Table 2.2. COW records 95 interstate wars from 1816 through 2007, with a total of 337 participants, or an average of 3.55 states participating in each war. More than half of the wars—57—are bilateral, involving only two states, while the other 38 are multilateral wars involving at least three states. Although there are only 95 wars, there are 442 war dyads, or pairs of states involved in a war on opposite sides.

Table 2.2 Correlates of War Typology of War

WAR CATEGORY			TYPE NUMBER
I. Interstate wars			1
II. Extrastate wars			
	A. Colonial—conflict with colony		2
	B. Imperial—state vs. nonstate		3
III. Intrastate wars			
	A. Civil wars		
		1. For central control	4
		2. Over local issues	5
	B. Regional internal		6
	C. Intercommunal		7
IV. Nonstate wars			
	A. In nonstate territory		8
	B. Across state borders		9

Source: Adapted from Meredith Reid Sarkees and Frank Whelon Wayman, *Resort to War: A Data Guide to Inter-state, Extra-state, Intra-state, and Non-state Wars, 1816–2007* (Washington, DC: CQ Press, 2010), 46.

[5] The wars with 1,000 battle deaths according to COW are Franco-Spanish War, Italian-Roman War, Neapolitan War, Ecuadorian-Colombian War, Naval War, Second Central American War, Third Central American War, Fourth Central American War, and Lithuanian-Polish War.

Extrastate war occurs when a state fights a war outside of its borders against a nonstate actor. There are two distinct types of this kind of conflict. In colonial wars, a state is fighting against a colony of that state. This is war type 2. Examples include the First Boer War (1880–1881) and Second Boer War (1899–1902), in which the United Kingdom fought against Boers (descendents of Dutch settlers) in the British South African colony. In imperial wars, a state is fighting against a nonstate actor that is not its colony. This is war type 3, and often includes attempts to establish colonies. An example is the Soviet war in Afghanistan from 1980 to 1989. COW records 163 extrastate wars, of which 61 are type 2 and 102 are type 3.

Intrastate war occurs between state and nonstate actors within the territory of a state, and this category has several different types. The most common type is civil wars, which are intrastate wars in which the national government of the state is one of the participants. Two different categories of civil war exist, depending on the purpose of the group(s) fighting against the government. In a civil war for central control (war type 4), the group is fighting to overthrow the national government and replace it with one it prefers. An example is the Spanish Civil War from 1936 to 1939, in which Francisco Franco eventually succeeded in taking control of the Spanish government. In a civil war over local issues (war type 5), the group is fighting to modify the government's treatment of a particular region or secede from the state altogether. The American Civil War from 1861 to 1865, fought over the right of the southern states to secede from the United States, is an example of this type of intrastate war.

The other two types of intrastate wars also occur within the borders of a state but do not involve the national government as one of the participants. A regional internal war (war type 6) occurs when a local or regional government—rather than the national government—is fighting against nonstate forces over local issues. An example is the first phase of the Cultural Revolution (1967) in China, as Mao Zedong used Red Guard forces to attack regional military forces and bring them under control of the central government. In an intercommunal war (war type 7), the government is not involved at all; rather, different factions within the state are fighting against one another. The Second Lebanese War of 1975–1976 serves as an example here, as Christians and Muslims fought against each other although the Lebanese government was uninvolved in the fighting. COW records 335 intrastate wars, of which 307 are civil wars. Of these, 175 are type 4 and 132 are type 5, while only 11 are type 6 and 17 are type 7.

Finally, there are **nonstate wars**, which involve nonstate actors fighting against each other outside of a particular state's borders—no state is involved. The first subtype is a nonstate war in nonstate territory (war type 8). An example is the Palestine War of 1947–1948, in which Arabs fought Zionists following the United Nations resolution calling for the division of Palestine into Jewish and Arab states. The second subtype is a nonstate war across state borders (war type 9). COW records 62 nonstate wars from 1816 to 2007, of which 61 are type 8 and only one is type 9.

Figure 2.1 shows the trend of wars over time. In this figure, wars of all types are displayed together. The decade with the most wars was from 1857 to 1866, with 50 war onsets; the next three decades in terms of war onsets are 1967–1976, 1987–1996, and 1997–2007. Of course the number of war onsets does not tell us what type of wars are being started or how serious they are; the decade with the fewest war onsets was 1937–1946, the decade dominated by World War II.

But not all of those wars are interstate conflicts. Figure 2.2 shows the trend of wars by major type over time. We can see from this figure that the number of nonstate and extrastate wars has decreased over time. This is a natural product of decolonization, as there are fewer nonstate actors around to become involved in wars. In contrast, there has been a sharp increase in the number of intrastate wars over time, particularly in the past five decades. Interstate wars have remained a minority of wars throughout, although there is a slight upward trend in their number over time.

Figure 2.1 War Onsets by Decade, 1816–2007

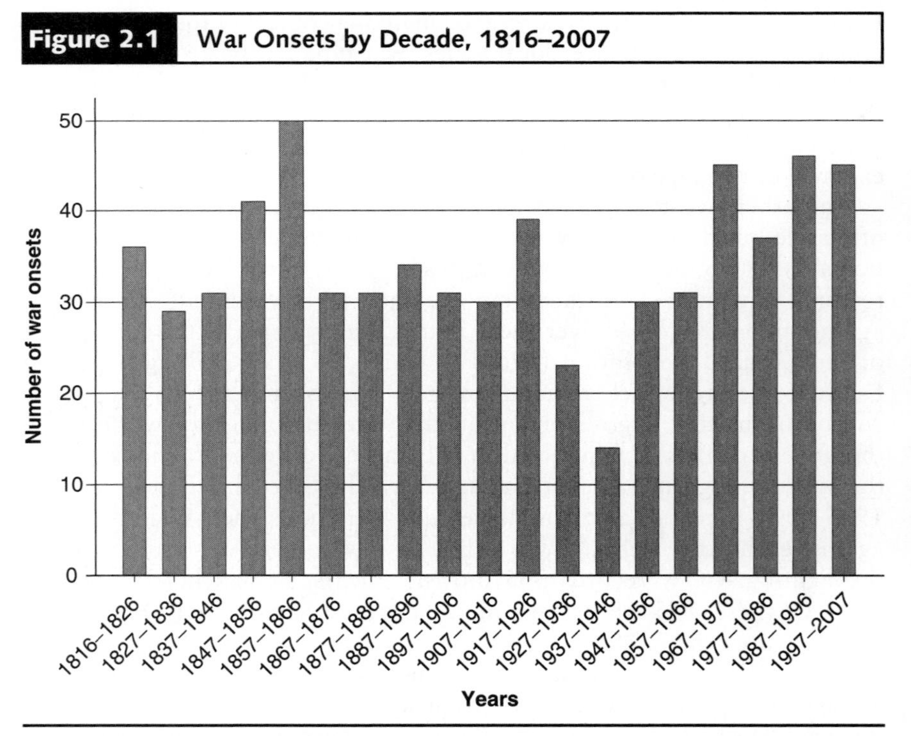

Source: Adapted from Meredith Reid Sarkees and Frank Whelon Wayman, *Resort to War: A Data Guide to Inter-state, Extra-state, Intra-state, and Non-state Wars, 1816–2007* (Washington, DC: CQ Press, 2010), 564.

Figure 2.2 War Onsets by Type per Decade, 1816–2007

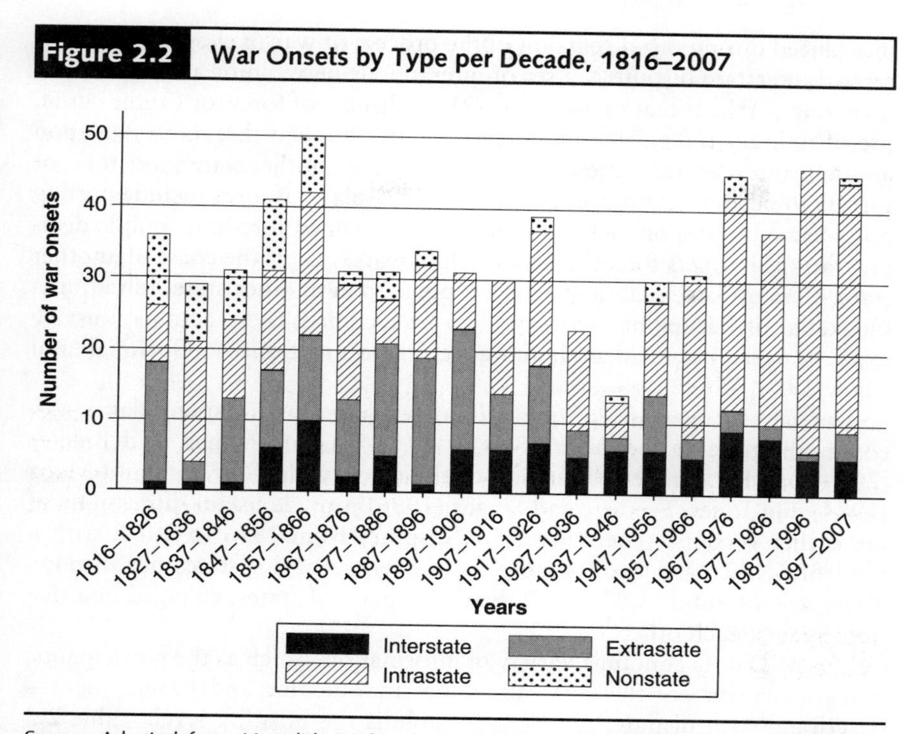

Source: Adapted from Meredith Reid Sarkees and Frank Whelon Wayman, *Resort to War: A Data Guide to Inter-state, Extra-state, Intra-state, and Non-state Wars, 1816–2007* (Washington, DC: CQ Press, 2010), 564.

We can see that interstate wars are only a portion of all wars. Nonetheless, our focus in this book is on interstate conflict. That does not mean that wars of other types are unimportant. Rather, it is recognition that they are so important that they—especially intrastate wars—deserve our full attention elsewhere. Although there are similarities among these different forms of conflict, there are important differences as well. Some factors that we will examine that have important effects on interstate conflict, such as contiguity, do not make much sense in the context of civil wars. Other theories that we will examine, such as the bargaining model of war that will be introduced in chapter 3, can and have been applied to explaining intrastate conflict. We will examine examples from extrastate and intrastate conflict as appropriate.

Militarized Interstate Disputes

Many international conflicts never reach the level of war. A commonly used category of conflict is a **militarized interstate dispute** (MID), which was introduced during our discussion of the process of war in chapter 1. A militarized interstate dispute is a set of interactions between or among nations involving (1) the threat to use force, (2) the display of force, or (3) the actual use of military force. Threats of force include explicit threats to fire upon another state's armed forces, threats to blockade another state's territory, or threats to occupy another state's territory. Displays of force include putting one's armed forces on alert, mobilizing one's armed forces, or public demonstrations of one's forces, such as sailing warships off the coast of another state. Uses of force include firing weapons upon another state, initiating a blockade, or occupying territory. "To be included, these actions must be explicit, overt, nonincidental, and government sanctioned" (Gochman and Maoz 1984, 586).

The current version of the militarized interstate dispute data, 4.0, covers the time period from 1816 to 2010. Ghosn, Palmer, and Bremer (2004) provide detailed discussion of the data, while Gochman and Maoz (1984) and Jones, Bremer, and Singer (1996) provide useful discussions of the earlier versions. There are 2,586 disputes from 1816 to 2010, with a total of 6,132 participant states, or an average of 2.4 states per dispute. There are a total of 3,841 MID dyads, or pairs of states, engaged in a dispute against each other.

The MID data contain a variety of information—such as the participants, dates of involvement, the level of casualties, the outcome, and the settlement—describing each dispute. One key variable is the hostility level. Table 2.3 shows the different categories of hostility level. The hostility level is recorded for each state in a dispute. Only the highest level of hostility for the state is recorded; for example, if a state first threatened to use force and then later used force in the same MID, only the use of force is recorded. Note that war is the highest level of hostility; thus wars are a subset of militarized disputes, not a wholly separate category of conflict.

Examples of specific episodes from history help illustrate the different types of actions that make up militarized disputes. On 23 March 2000, North Korea unilaterally declared new navigation zones and waterways in the Yellow Sea in disputed waters near five South Korean–held islands. North Korea threatened military action against intruders without warn-

Table 2.3 Militarized Interstate Dispute Hostility Levels

HOSTILITY LEVEL	DESCRIPTION
1	No militarized action
2	Threat to use force
3	Display of force
4	Use of force
5	War

ing. Because this is an explicit threat of force, this is coded as a dispute, with North Korea as the initiator and South Korea and the United States as targets.

The Taiwan Missile Crisis discussed earlier in the chapter provides an example of a dispute involving the display of force. From July 1995 through March 1996, China test-launched a series of missiles over Taiwan and conducted military maneuvers off of the Taiwan coast. This MID is coded with China as the initiator and Taiwan and the United States as targets; each state had a display of force as its highest level of hostility.

An example of the use of force is the United States's bombing of targets in Sudan and Afghanistan on 20 August 1998 in retaliation for the bombing of American embassies in Kenya and Tanzania. This is coded as two separate MIDs, each involving the use of force by the United States; however, Sudan and Afghanistan each have a hostility level of 1 because they did not have any observable response to the attacks.

Table 2.4 shows the states that are most frequently involved in MIDs. This list is essentially a top-ten list of the most conflict-prone states, although it actually includes thirteen states because it shows every state that has been involved in at least one hundred disputes. The United States and Russia/Soviet Union (USSR) have been involved in the most disputes. Eight of the top ten—the United States, Russia, Britain, China, Germany, France, Japan, and Italy—are states that have spent at least part of the 1816–2010 period as major powers, while India is emerging as a major power. The other four—Turkey, Iran, Israel, and Iraq—are located in the Middle East.

Table 2.4 States Involved in the Most MIDs, 1816–2010

STATE	DISPUTES
United States	372
Russia/USSR	361
United Kingdom	266
China	245
Germany	181
France	178
Japan	177
Turkey	174
Iran	164
Italy	132
Israel	128
Iraq	124
India	104

Source: Compiled by author.

Table 2.5 shows the states that have initiated the most disputes. This list shows the thirteen states that have initiated at least fifty MIDs. The top two remain the same, although their order has switched: Russia/Soviet Union has initiated more disputes than any other state. The rest of the countries are largely the same as in dispute involvement, although their order is somewhat different. Japan, however, is no longer on the list, indicating that it is much more frequently a target than an initiator. Syria, although it has not been involved in at least one hundred disputes, appears on the list of most frequent initiators.

Militarized interstate disputes provide us with a much broader measure of international conflict than wars do. They also eliminate difficulties arising from the one thousand battle deaths threshold. Accordingly, data on disputes are often used in studies of the causes of international conflict.

Other Measures of International Conflict

While the COW data on wars and militarized interstate disputes are the most broadly used measures in political science, other measures of international conflict have been developed. They differ in their time frames as well as their specific coding rules. We will examine two of the most prominent data sets here. The first covers crises and the second covers armed conflicts.

Crises are a common focus in international conflict. Our earlier discussion of history covered events such as the Agadir, Munich, and Cuban

Table 2.5 States Initiating the Most MIDs, 1816–2010

STATE	DISPUTE INITIATIONS
Russia/USSR	213
United States	187
United Kingdom	128
Iran	108
China	107
Germany	105
France	91
Turkey	73
Iraq	66
Italy	65
Syria	57
Israel	57
India	52

Source: Compiled by author.

Missile crises. There are many additional crises that we could have discussed. But what do we mean by crisis? How do we identify one?

Michael Brecher and Jonathan Wilkenfeld founded the International Crisis Behavior (ICB) project in 1975 to answer these questions and systematically collect data on international crises. The ICB definition of an **international crisis** is a situation that (1) threatens one or more basic values, (2) allows only a finite time for response to the value threat, and (3) has a heightened probability of involvement in military hostilities (Brecher and Wilkenfeld 2000, 3). The ICB data have been continually updated, and now cover the time period from 1918 to 2007. There are 455 international crises in the data. The data are freely available on the web at www.cidcm.umd.edu/icb/, and in the ICB Data Viewer at www.cidcm.umd.edu/icb/dataviewer/.

In the Cuban Missile Crisis, for example, the Soviet emplacement of nuclear missiles—within striking range of Washington, DC and much of the US east coast—was clearly perceived as threatening to basic values of security in the United States. Furthermore, the United States faced a finite time to respond because the missiles, while not yet operational, would soon be. Once they became operational, the situation would be clearly changed. Finally, there was a heightened probability of military hostilities because each of the primary responses considered by the Kennedy administration, whether air strikes on the missile emplacements, an invasion of Cuba, or even a naval blockade (or quarantine), either used military force or made its use more likely. Thus, each of the criteria to be a crisis according to the ICB project was met in the Cuban Missile Crisis.

Another important project collecting data on international conflict is the **Uppsala Conflict Data Program (UCDP)** based at Uppsala University in Sweden. The project is led by Peter Wallensteen and Lotta Themnér. Gleditsch et al. (2002) describe the data set and coding procedures. The data are updated annually and currently cover the post–World War II period from 1946 to 2012. The data are available online at www.pcr.uu.se/research/UCDP/.

An *armed conflict* is defined by the UCDP as "a contested incompatibility that concerns government or territory or both where the use of armed force between two parties results in at least 25 battle-related deaths. Of these two parties, at least one is the government of a state" (Gleditsch et al. 2002, 618–9). This provides a measure of conflict that is more encompassing than the COW war data, since the casualty threshold is twenty-five deaths rather than one thousand. It is more restrictive, however, than the MID data, since disputes do not have any casualty threshold for inclusion or even require the use of force.

The UCDP data identify a total of 252 armed conflicts. The data include interstate, extrasystemic, intrastate, and internationalized intrastate conflicts. Thus, unlike the militarized interstate dispute and international crisis behavior data sets, the armed conflict data are useful for identifying extrastate and intrastate conflicts below the threshold of war.

Conclusion

We are interested in examining the logic and evidence of explanations of international conflict. The history of international conflict is important because it provides the record of events that is crucial for testing theories about the outbreak, outcomes, and consequences of war and other forms of conflict. This is our primary source of evidence for evaluating theories. We focus on history since the Napoleonic Wars because this is the time period that most data collections cover.

Although examining detailed definitions of war and other forms of international conflict may seem tedious, it is important because explicit definitions and coding rules are necessary for the collection of data. Our focus here is on interstate war (and conflict more broadly), but it is useful to know that other types of war—extrastate, intrastate, and nonstate—exist and to understand the distinctions between them. In addition, the militarized interstate dispute, interstate crisis, and armed conflict data sets provide us with a broader conception of international conflict than wars and represent an important element of the process of war.

Standardized data sets provide an important contribution to our understanding of international conflict. These data sets are freely available for everyone to examine and analyze, which save people from considerable duplication of effort. Furthermore—and more importantly—when different studies all examine the same data, this enhances their comparability since we know that differences between them are not driven by defining concepts such as war differently. As we move on in the book, we will examine a variety of studies that make use of these data in their analyses.

Key Concepts

Correlates of War Project (COW)

Extrastate war

International crisis

Interstate war

Intrastate war

Militarized interstate dispute

Nonstate war

Uppsala Conflict Data Program (UCDP)

3 Rational Choice Theory

Soviet General Secretary Mikhail Gorbachev and US President Ronald Reagan sign the Intermediate-range Nuclear Forces (INF) Treaty on 8 December 1987, in which the two countries agreed to eliminate their entire arsenals of the weapons. Rational choice theory can help us understand decisions such as these.

Source: © Reuters/CORBIS

Why do things happen in international relations? Wars start and end, alliances are formed, states build weapons, and so on. We are interested in understanding why events like these happen. Ultimately, things happen in international relations—as in the political, economic, and social worlds more generally—as a result of decisions made by people. These people may be leaders of states, members of the legislature or military, members of nongovernmental organizations, or just simply citizens of a country.

Many people argue that to explain international conflict we need to focus on explaining decisions. But how do we explain them? To do so, many rely upon the assumption of rationality. Assuming rationality opens the door to various methodologies to explain decision making, particularly expected utility theory and game theory. This leads to an approach to studying international conflict (and other areas of social science) known as rational choice theory, and perhaps more accurately as the strategic perspective (Bueno de Mesquita 2010). Rational choice has led to a number of advances in theorizing about international conflict.

We begin by taking a closer look at what is meant by rationality, as well as discussing nonrational approaches to studying foreign policy decision making. We then explore expected utility theory and game theory, tools that become available through the assumption of rationality. Finally, we explore the bargaining model of war, an important theory that helps us understand international conflict throughout multiple stages of the process of war.

Rationality

What is rationality? Should one assume rationality? Answers to these questions have been subject to a great deal of debate. Decisions about whether an assumption of rationality is useful depend on what the assumption entails. Many different assumptions of rationality have been used. We will examine three of the most common: procedural rationality, bounded rationality, and instrumental rationality.

Procedural rationality is similar to the common, everyday conception of rationality in which omniscient actors are said to make a "cool and clear-headed ends-means calculation" (Verba 1961, 95) in the course of considering all available options and choosing the best one.[1] This is called maximizing behavior because the procedurally rational actor chooses the alternative with the maximum gain. In addition, the best option is the alternative that

[1]The label *procedural rationality* comes from Simon (1976).

is objectively in the best interest of the decision maker. Therefore, different rational actors that are in the same situation will always make the same choice.

Nonlogical influences—influences which the actor is unaware of, and if he were aware of it, he would discount it—are not allowed within procedural rationality. For instance, George and George (1956) argue that Woodrow Wilson's decision making at several crucial points, including during his efforts to get the League of Nations treaty ratified by the United States Senate following World War I, was affected in large part by Wilson's childhood relationship with his father. Other potential nonlogical influences include misperceptions, emotions, and psychological or cognitive limitations. To the extent that real actors are influenced by factors like these when making decisions, they are not procedurally rational.

Herbert Simon (1957) developed a different conception of rationality, which is called **bounded rationality**. The idea of bounded rationality is that "human behavior is intendedly rational, but only boundedly so" (Simon 1997, 88). Bounded rationality involves what is called satisficing behavior, rather than maximizing behavior. Satisficing is a sequential process. Instead of examining all possible alternatives and choosing the very best

Box 3.1

Concept in Focus: The Unitary Actor Assumption

We often talk about states doing things: Germany invaded Poland in 1939, Iraq attacked Kuwait in 1990, and so on. This is known as the **unitary actor assumption**, where states are assumed to be single actors. The unitary actor assumption has a long history in studies of international relations, particularly within realism, a perspective focused on power and the structure of the international system.

Snyder, Bruck, and Sapin (1954, 1962) created what is called the decision making perspective and launched an important challenge on the unitary actor assumption. They argued that foreign policy consists of decisions, made by identifiable decision makers; the making of decisions, therefore, is the behavioral activity that requires explanation. They also highlighted the importance of preferences and perceptions in explaining those decisions. Thus, in many ways, the ideas of Snyder, Bruck, and Sapin are fundamental to rational choice theory in international relations.

The decision making approach was revolutionary because it challenged the tendency that realists have to reify the state (assuming that states have goals, fears, aspirations, etc., as if they were people). But states do not have character, and so on; states are legal entities. The individuals within each state are what matters. The state itself is not some autonomous actor apart from its people.

Nonetheless, we might talk about one state invading another, but what we really mean is that the leader(s) of one state decided to invade another. Although Allison (1971) argues that the unitary actor assumption is a fundamental part of rational choice theory, the two are separate issues. Rational choice theory is grounded in the decision making perspective and is rooted in the idea that people, not states, make decisions.

one, options are looked at sequentially. The decision maker then evaluates each alternative to determine if it is acceptable. If so, that option is chosen; otherwise, the decision maker will keep looking until an alternative is found that is "good enough."

This is a very different notion of rationality that seems to be a much more realistic description of how people actually make decisions. Leaders of states are often affected by misperceptions when making decisions (Jervis 1976). Decision makers usually do not have enough time available to consider all possible courses of action (Allison 1971). Furthermore, people sometimes lack the cognitive ability to understand the ramifications of their decisions (George and Smoke 1974). Because of such constraints on decision making, Braybrooke and Lindblom (1963) argue that most decision making is "disjointed incrementalism," where small decisions are made with little understanding.

Psychological and Cognitive Approaches

Given these limitations of procedural rationality, a number of scholars have sought to explain foreign policy decision making without assuming rationality. These **psychological approaches** instead employ social or cognitive psychology in their attempts to explain international politics. Most common are studies focusing on cognitive psychology to examine how individuals' beliefs, biases, and ability to process information lead to deviations from procedurally rational behavior. There are numerous examples of this approach. Lampton (1973) argues that the image that American leaders had of Chinese leadership played an important role in determining the outcome of crises between the United States and China. Jervis (1976) argues that perceptions and misperceptions have important impacts on international politics. Hermann and Kegley (1995) argue that leaders' images and beliefs can explain peace between democracies.

Other scholars have used elements of both psychology and rationality to explain foreign policy decision making. George (1969) introduces the idea of the **operational code**, which refers to a political leader's beliefs about the nature of politics and political conflict, his views regarding the extent to which historical developments can be shaped, and his notions of correct strategy and tactics. Walker, Schafer, and Young (1998) develop systematic procedures for identifying leaders' operational codes. Additionally, operational codes can be used to identify leaders' preferences and then game theory used to explain the resulting decisions.

Prospect theory is a similar framework to expected utility theory (which we will examine in more detail shortly) for analyzing decision making with risk, but there are important differences. Prospect theory highlights the importance of the decision-maker's frame of reference when confronting a situation. People tend to be averse to risk when they anticipate gains but tend to be risk acceptant when they anticipate losses (Levy 1997). Although it is sometimes characterized as being contrary to rational choice, Carlson and

Dacey (2006) show that prospect theory can be incorporated in a game-theoretic model.

Finally, **poliheuristic theory** is an approach that specifically integrates cognitive and rational choice together. Poliheuristic theory views decision making as a two-stage process. At the first stage, decision makers use non-rational means to reduce the set of possible options to a manageable level, and at the second stage, they make a rational choice among the remaining options (Mintz 2004).

Instrumental Rationality

Discussions of bounded rationality as well as psychological and cognitive approaches seem to demonstrate that the assumption of rationality in rational choice theory is very problematic. These scholars have critiqued procedural rationality, however, but that is not the rationality of rational choice theory. What then do we mean by rationality?

Rational choice theory is rooted in the assumption of **instrumental rationality.**[2] An instrumentally rational actor is one who, when confronted with "two alternatives which give rise to outcomes . . . will choose the one which yields the more preferred outcome" (Luce and Raiffa 1957, 50). Note that the definition does not say which alternative is better; the decision maker decides. This definition avoids many problems by not trying to dictate what decision makers should prefer. Instrumentally rational players, then, are those who *always* make choices they believe are consistent with their interests and objectives *as they define them*. In other words, instrumentally rational players are purposeful players.

Additionally, a rational actor must have complete and transitive preferences. For preferences to be complete, for any two alternatives x and y, either xPy (x is preferred to y), yPx (y is preferred to x), or xIy (the person is indifferent between x and y). Transitivity involves preferences among at least three different outcomes. Considering the most basic case with three alternatives x, y, and z, transitivity means that if xPy and yPz, then xPz. Thus, if I prefer chocolate ice cream to vanilla, and vanilla ice cream to strawberry, then it only makes sense that I prefer chocolate ice cream to strawberry.

Instrumental rationality makes no normative judgments about preferences. Whether one's preferences are "good" or "evil," "instrumental" or "expressive," or anything else has no impact on one's instrumental rationality. Thus, a person such as Adolf Hitler who prefers to launch an attempt to conquer entire continents and wipe out entire races, and acts accordingly, is just as rational as someone like Woodrow Wilson who seeks to promote democracy and acts accordingly. Similarly, a job seeker who prefers to maximize leisure, with no regard for income, and acts accordingly, is just

[2]For detailed discussions of instrumental rationality, see Zagare (1990), Quackenbush (2004), and Hindmoor (2006).

as rational as another who acts according to her preference for maximizing income. Saying that someone is instrumentally rational is not paying them a compliment; it is simply saying that they act according to their preferences, whatever they may be. Even so, many have argued that a variety of psychological, informational, or structural factors interfere with actors'—particularly states'—ability to act rationally. These factors, however, only interfere with procedural rationality, not instrumental rationality. Instrumental rationality is compatible with a wide variety of supposedly limiting factors (Zagare 1990).

If decisions are central to international relations, it makes sense that decision makers make those decisions with a purpose. Such purposive decision making is what instrumental rationality is about. In addition, assuming instrumental rationality allows us to use expected utility theory and game theory, useful tools for explaining decision making. We explore each in turn.

Expected Utility Theory

Expected utility theory is a model of rational decision making. We will examine the basics of expected utility theory and then turn our attention to a specific application, the expected utility theory of international conflict developed by Bueno de Mesquita (1981).[3] Modeling decisions using expected utility requires that we know several pieces of information about the situation we want to examine.

First, we must know the possible choices that could be made. These are the various courses of action that an individual can choose from: starting a war or not, putting more money into the defense budget or into the education budget, and so on. Situations can have many alternatives. Part of setting up an expected utility model is deciding what options we should include, which means judging what the most important or salient alternatives were to the leaders when they made their decision.

Second, we must know the possible outcomes of the situation. These are the various things that could occur in the world: various states of being or ways in which things could turn out. For example, if a state decides to start a war, it may win or it may lose. There can of course be many more outcomes than two. When we develop models we choose how many outcomes and how much detail about them we want to deal with.

Third, we need to know the players' utility or preferences for the different outcomes. Utility is the value that people attach to different outcomes. People want, or prefer, outcomes with higher utility. Essentially, utility is net benefits, or the benefits of a particular outcome minus the costs of achieving that outcome. There are different ways that utility can be measured.

---

[3] For a more detailed introduction to expected utility theory, see Morrow (1994) or Resnik (1987).

The most basic way to measure utility is at the ordinal level. Sometimes we just know that one outcome is preferred to another. For example, I like chocolate ice cream more than vanilla; I prefer winning a war to losing a war. At the ordinal level, we can simply determine the order of the actors' preferences.

A more detailed way to measure utility is at the cardinal level. Sometimes we can judge exactly how much outcomes are valued on a continuous scale. This will allow us to tell how much more one outcome is worth than another. For example, we may know that winning a war is worth 1,000, while losing a war is worth -500 and remaining at peace is worth 250. By measuring utility on a cardinal scale, decision makers can make rather complicated comparisons and judgments within their own cardinal scale.

Finally, we need to know the probability, or likelihood, that the different outcomes will occur in some situation. Often when we make a choice, we do not know for sure what the outcome will be, but we usually have some idea, that is, a rough estimate of the probability. When a very strong country starts a war, for instance, there may be 90 percent chance that it will win. Other times, if countries are closely matched, there may be 50 percent chance of winning. We sometimes discuss probability as a percentage, such as 75 percent, or as a decimal, such as 0.75. These mean the same thing. Probabilities of different outcomes that might follow some action must add up to 1 (or 100 percent). Often, we see the options of "win" and "lose." With only two outcomes, we only need to know one probability, since we can calculate the other: p(lose) = 1 - p(win).

Using expected utility suggests that people actually think not only about what they want—their preferences or the utility they get—but also the odds that they will get it. Expected utility is calculated simply be multiplying the utility of outcomes by the probability that they will actually happen and summing the results over the set of possible outcomes. So, if there are two possible outcomes to a choice, then we could write

$$EU_{choice} = p_1 U_1 + p_2 U_2$$

where EU_{choice} is the expected utility of a given choice, p_1 is the probability of outcome 1, U_1 is the utility of outcome 1, p_2 is the probability of outcome 2, and U_2 is the utility of outcome 2.

We have a utility of an outcome, but an expected utility of a choice. We assume that any individual outcome, such as winning a war, has a fixed utility without any probability involved. Once we move to the possibility that multiple outcomes might follow from a particular choice, and hence to a situation where probability comes into play, then it becomes expected utility.

A simple example of expected utility can be from a state considering a military attack on another state. In this example, $U_{peace} = 100$, $U_{win} = 500$, $U_{lose} = -1000$, and $p_{win} = 0.5$. Therefore, we can calculate the expected utility of attacking as

$$EU_{attack} = p_{win}U_{win} + p_{lose}U_{lose}$$

$$EU_{attack} = 0.5(500) + 0.5(-1000)$$

$$EU_{attack} = 250 - 500$$

$$EU_{attack} = -250$$

The potential attacker has two options: to attack or not to attack. A rational state will choose the preferred alternative. In this example, $EU_{attack} = -250$, whereas $EU_{not\ attack} = U_{peace} = 100$. Thus, not attacking is much better for the state than attacking, so we would predict that the state would not attack. If the state's likelihood of winning was greater, say $p_{win} = 0.75$, however, and everything else remained the same, then $EU_{attack} = 125$. This new expected utility for attacking is now larger than the expected utility of not attacking, so we would predict that the state would attack.

This example is useful for illustrating the logic of expected utility. A substantial limitation, however, is the question of where the numbers come from. Saying that peace is worth 100 while winning is worth 500 is nice, but what do these numbers mean? It is possible to develop ways to measure the utilities and probabilities in a meaningful way.

In an early development of rational choice theory, Bueno de Mesquita (1981) did exactly that. In his book *The War Trap,* Bueno de Mesquita developed an expected utility theory of international conflict. Rationality implies that leaders would only start wars that they expect to gain from in some way; if they expect to suffer a net loss, then they will avoid war. Accordingly, Bueno de Mesquita developed a theory to determine whether states would expect a gain or loss against certain opponents at different points in time.

If a state attacks a given opponent, they are likely to wind up fighting a war involving only two states: the initial attacker and the initial target. This is a bilateral war and is essentially the situation assumed in our example above. Some wars, however, do expand to include additional countries, so Bueno de Mesquita accounts for this possibility as well. In calculating the expected utility of a multilateral war, Bueno de Mesquita seeks to answer the following questions: Who is likely to join the war? On which side will they join? And what effect will their joining have on the outcome of the war?

The resulting expected utility for a conflict in Bueno de Mesquita's (1981) model depends on these individual expected utilities for a bilateral war and a multilateral war:

$$EU(\text{conflict}) = EU(\text{bilateral war}) + EU(\text{multilateral war}).$$

To this point, we have focused on the logic of Bueno de Mesquita's theory, but it is also important to look at the evidence regarding it. Bueno de Mesquita does this by examining regional dyads (pairs of states within the same region of the world) between 1816 and 1974. For each conflict observed, he measures each state's expected utility for attacking the opponent. In the previous chapter, we saw how conflict can be observed. The more difficult obstacle is measuring expected utility. Bueno de Mesquita does so by developing techniques to measure the probability of victory and utility.

To measure probability of victory, Bueno de Mesquita focused on the relative power between states. There is a common measure of power used in studies of international politics called the Composite Indicator of National Capabilities (or CINC score), which we will discuss more fully in chapter 5. Bueno de Mesquita assumes that a state's probability of victory is essentially equivalent to its relative power in comparison with its potential opponent. Thus, if a state has 60 percent of the power within a dyad, then its probability of victory is 0.60, whereas if the state has only 35 percent of the power in the dyad, then its probability of winning is 0.35.

To measure utility, Bueno de Mesquita used states' alliance portfolios. A state's alliance portfolio refers to its entire configuration of alliance commitments to other states. We can look at not only whether a state has an alliance with another country, but also what kind of alliance it is.[4] Then we compare the alliances of state A (the potential initiator) and state B (the potential target). It is assumed that the more similar two states' alliance commitments are, the greater is their foreign policy similarity. For example, if state A has alliances with ten other countries, and state B also has alliances with ten other countries, and they are the same ten countries, then state A and B seem to have a lot in common. On the other hand, if B is allied to ten different countries than A is allied to (with no overlap), this would indicate that A and B have much less in common. Of course the possibilities are endless, but the basic idea remains the same.

We can visualize the results by looking at a series of tables that compare conflict initiation with expected utility. Results for interstate wars are shown in Table 3.1. Bueno de Mesquita's theory indicates that states will only initiate conflict if their expected utility for conflict is greater than zero. There were 76 cases of interstate war initiation; of these, the initiator's expected utility was greater than or equal to zero 65 times (85.5 percent) and negative only 11 times. For the target, the trend is exactly opposite: Targets had a nonnegative expected utility only 11 times and a negative one 65 times. This is a statistically significant relationship, indicating support for Bueno de Mesquita's theory.

Table 3.1 Interstate War Initiation and Expected Utility

EXPECTED UTILITY SCORE	INITIATOR	OPPONENT
Greater than or equal to zero	65	11
Less than zero	11	65

Source: Bruce Bueno de Mesquita, *The War Trap* (New Haven, CT: Yale University Press, 1981), 129.

[4] We will examine alliances in more detail in chapter 6.

Bueno de Mesquita (1981) also examines lesser levels of conflict. Results of comparing expected utility with uses of force are shown in Table 3.2. Out of 102 disputes involving the use of force, the initiator had a nonnegative expected utility 78 times (76.5 percent) and a negative expected utility 24 times. On the other hand, targets had an expected utility greater than or equal to zero only 14 times, and a negative one 88 times. Once again, this relationship is unlikely to arise from chance.

Bueno de Mesquita examines one final level of conflict, interstate threats, results for which are shown in Table 3.3. Out of 73 interstate threats, the initiator's expected utility was greater than or equal to zero 50 times (68.5 percent), and less than zero 23 times. Targets had a nonnegative expected utility 13 times and a negative one 60 times. As with the previous analyses, this relationship is again unlikely to arise from chance. One might also note that as we went down in the level of hostility from war, to use of force, and finally to interstate threat, expected utility had a slightly weaker effect on decisions to initiate. Bueno de Mesquita argues that this is to be expected because states are more willing to bluff (threatening even when they prefer not to fight) at lower hostility levels.

Bueno de Mesquita also conducts these analyses looking at different periods of time as well as different regions of the world and the results are essentially the same. Furthermore, he examines the relationship between expected utility, relative power, and conflict. While the relationship between power and conflict is quite similar to the relationship between expected utility and conflict, his expected utility theory greatly outperforms a power explanation in cases where the predictions between the two differ.

Table 3.2 Uses of Militarized Force and Expected Utility

EXPECTED UTILITY SCORE	INITIATOR	OPPONENT
Greater than or equal to zero	78	14
Less than zero	24	88

Source: Bruce Bueno de Mesquita, *The War Trap* (New Haven, CT: Yale University Press, 1981), 130.

Table 3.3 Interstate Threats and Expected Utility

EXPECTED UTILITY SCORE	INITIATOR	OPPONENT
Greater than or equal to zero	50	13
Less than zero	23	60

Source: Bruce Bueno de Mesquita, *The War Trap* (New Haven, CT: Yale University Press, 1981), 130.

In summary, Bueno de Mesquita's expected utility theory of international conflict is well supported by the historical record, even if its predictions are not perfect. This indicates empirical support for the basic logic of expected utility theory and rational choice. It also indicates that measuring the utilities and probabilities necessary for calculating expected utility is possible, even if the measures developed are not perfect.

Expected utility theory, however, has one crucial limitation: It does not account for strategic interaction between decision makers. Rather, expected utility theory focuses on each actor's choice in a vacuum. Of course in the real world, leaders of states must take into account a variety of other potential decision makers: other states, members of their own government, international organizations, the domestic population, and many others. While most theories do not account for all of these different decision makers, it is important to be able to understand this strategic interaction between multiple actors. To do so, we turn our attention to game theory.

Game Theory

Game theory is the analysis of how decision makers interact in decision making to take into account reactions and choices of the other decision makers. Although its title uses the word *theory*, game theory not a unified empirical theory. Rather, it is a methodology for examining strategic behavior among interacting and interdependent units.[5]

We call it game theory because we use it to examine games of strategy, as opposed to games of skill (such as a 100-meter dash) or games of chance (such as Roulette). A strategic game consists of a set of players, a set of actions available to each player, and the preferences for each player over the outcomes that may arise. We call them games because we think of actors making decisions as anticipating how the other will respond, much as we do in a game of chess or checkers.

In game theory, the interacting units are called *players*. Who or what constitutes a player, however, is a determination left to the individual analyst. Normally, the identification of players depends on the analyst's purpose and the specific research questions addressed. One could, for example, conceive of the 2011 conflict in Libya as a two-person game between Libya and the National Transitional Council, as a two-person game between the NATO alliance and Libya, or as a three-person game among Libya, NATO, and the National Transitional Council. An even more detailed analysis that recognized important policy differences that separated the United States from France and Germany, however, might specify five (or more) players. In other words, both the number and identification of players are extra-game-theoretic decisions that, in principle, should be driven by theoretical considerations, empirical factors, or both rather than by methodological considerations.

[5]Numerous textbooks provide an overview of game theory. Game theory texts that focus on applications in political science include Gates and Humes (1997), McCarty and Meirowitz (2007), and Morrow (1994).

A realist who was also a game theorist would likely limit the specification of players to states. But players need not be so restricted. For example, one could employ game theory from within Allison's (1971) organizational process model to focus on, say, the decision-making process within the United States government. In this game, the players—such as the White House, the Departments of State and Defense, and other important bureaucratic departments and agencies—would be different organizations within the government, rather than entire governments. But the analysis, nonetheless, could still be game theoretic.

Regardless of how they are specified, the players are assumed to make *choices* that, along with the choices of the other players, lead to specific *outcomes*. The specification of outcomes, like that of the players, is another important judgment call that an analyst must make. Thus, a general analysis of conflict might simply specify "negotiation" as a possible outcome. By contrast, a more fine-grained study might distinguish between a negotiated outcome that favored one side and one that favored the other.

In game theory, the players are assumed to evaluate (subjectively) the *utility* (or worth) of each possible outcome, and to make choices that are instrumentally rational. This is not to say, though, that the players evaluate the outcomes in the same way. In fact, many conflicts are traceable to disputes about the value of specific outcomes. For example, prior to the 1999 Kosovo War, the Kosovo Liberation Army (KLA) believed that by pressing its case against Serbia, an independent Kosovo would result. Serbian leader Slobodan Milosevic disagreed. Milosevic thought he could use force to stamp out the separatist movement in Kosovo and, in the process, consolidate his internal political standing.

To minimally specify a game, the players, the choices available to them, the consequences of their choices (i.e., the outcomes), and the players' utility for each outcome must be specified. Morrow (1994, 57) correctly observes that "the design of the game . . . [is] . . . the single most important decision in modeling." Note, however, that these are decisions that must be made *before* the analytic tools of game theory can be used. This means that the explanatory power of any game-theoretic analysis depends less on the methodology itself than on the theoretical sensitivity of the analyst. Nonetheless, game theory provides a potentially useful methodological framework for analyzing international conflict.

Games can be shown in two basic forms. First, there is the strategic form, also known as the normal form. If a game is reduced from the actual display of what paths actors may follow to a matrix of choices of outcomes, it is said to be in strategic form. Second is the extensive form, where the game is represented as a tree of choices. Generally, the strategic form is useful for illustrating simultaneous games, whereas the extensive form is useful for illustrating sequential games. Any game, however, can be shown in either form. We will examine the logic of game theory, including the basics of how to solve games, before examining applications of game theory to explaining international conflict.

Strategic Form Games

We begin our examination of strategic form games by looking at Prisoner's Dilemma, the most famous of all games in game theory. The basic story of this game is that the police arrest two people suspected of committing a serious crime. The district attorney, however, does not have enough evidence to convict them without a confession. So the police place the suspects into separate cells and make the following offer to each individually: "If you confess and your partner does not, then you'll be the state's witness and get off scot-free and we'll throw the book at your partner. But if your partner squeals and you don't, then I'll throw the book at you. If you both refuse to confess, then I'll still be able to convict you of more minor charges, so you'll still go to jail. If you both confess, then you'll both be convicted of the more serious charges, but I'll make sure that you get a lighter sentence than if you keep your mouth shut."

Each of the suspects, A and B, has two options. They can either cooperate with the other player by keeping quiet and telling the police nothing, or they can defect from cooperation with their partner by confessing to the police. Since there are two players each with two strategies, four outcomes are possible. Each player's best outcome is to defect while the other cooperates—the temptation outcome. Each player's worst outcome is to cooperate while the other defects—the sucker outcome. For the two outcomes in between, mutual cooperation would be best for both suspects since they would only be convicted of a more minor crime. The resulting game is shown in Figure 3.1. Within each cell are two numbers: The number before the comma identifies the payoff for the row player (A in this case), while the number after the comma identifies the payoff for the column player (B).

What will the players do? We determine that by solving the game, or more formally, by finding the equilibrium or equilibria of the game. There are different types of equilibria (or equilibrium concepts) that we can use, each of which is appropriate under different circumstances. The most fundamental equilibrium concept is the Nash equilibrium, named after John Nash, an important pioneer of game theory and subject of the 2001 movie *A Beautiful Mind*. A **Nash equilibrium** is a set of strategies where each player plays her best response to the other's strategy; neither player has an incentive to unilaterally switch strategies.

Figure 3.1 Prisoners' Dilemma

		B	
		Cooperate (keep quiet)	Defect (confess)
A	Cooperate (keep quiet)	(3, 3)	(1, 4)
	Defect (confess)	(4, 1)	(2, 2)*

We can identify this in a straightforward manner. Starting with suspect A, we can see that if B chooses to cooperate, A will get 3 by also cooperating but 4 by defecting. Thus, choosing to defect is his best response to cooperation. If B chooses to defect, A will get 1 by cooperating but 2 by defecting. Once again, choosing to defect is A's best response. Thus, A has what is known as a dominant strategy. A **dominant strategy** is a strategy that is always a player's best response to whatever strategy the opponent might do. In contrast, a dominated strategy is a strategy that is never a best response to the opponent's strategy.

Since Prisoner's Dilemma is a symmetric game, it is easy to see that B also has a dominant strategy to defect. Therefore, there is only one Nash equilibrium: mutual defection, which we could also write as (defect, defect). Note that we should always write an equilibrium in terms of strategies, not in terms of utilities, that is, not as (2, 2), because there can be more than one outcome with the same payoffs to the players.

So why is this game called a dilemma? It is a very straightforward game to solve, where each player's rational choice is obvious because each has a dominant strategy. If you look more closely at the game, you can see that if the players would both cooperate, then they would each get their second-best outcome, 3, but by both defecting, then they both get their second-worst outcome, 2. They would both be better off than they are by defecting; being rational leads them to a worse outcome.

Prisoner's Dilemma has been applied to a variety of topics in international relations, including arms races, deterrence, and international trade. One of the common questions is how can we get cooperation between the players in a Prisoner's Dilemma situation? The short answer is that, if the players only play the game once, then there is no way to get them to cooperate short of changing their payoffs and thus the game. If, however, the game is repeated again and again, then cooperation emerges as a rational possibility (Axelrod 1984). Exploring cooperation between states in this way has enabled scholars to better understand the impact of international institutions, such as the United Nations, on international cooperation and conflict (Keohane 1984; Oye 1986).

We will examine one more strategic form game before moving on. Chicken is another classic game that helps to further illustrate game theory. The basic story of Chicken is two people (teenage boys, most likely) drive cars straight at each other until one of them swerves. If one swerves while the other drives straight, then he is the chicken (loses), while the one who drives straight wins. Thus, the best outcome is to drive straight (defect) while the other swerves (cooperates). The second best outcome for each player is if they both swerve; although neither wins, they aren't the chicken either. Being the chicken—swerving while the other drives straight—is the second-worst outcome. The worst outcome for both players is if neither swerves; although neither is a chicken, they both likely die in a head-on collision. Apparently, it is better to be a live chicken than a dead duck. The resulting game is shown in Figure 3.2.

Figure 3.2 Chicken

		B	
		Cooperate (swerve)	Defect (drive straight)
A	Cooperate (swerve)	(3, 3)	(2, 4)*
	Defect (drive straight)	(4, 2)*	(1, 1)

Considering A's choices first, we can see that if B cooperates then A's best response is to defect (getting 4 rather than 3). On the other hand, if B defects, A gets only 1 by also defecting but gets 2 by cooperating; A's best response is to cooperate. Thus, we can see that A's best response is to choose the opposite of what B does. Similarly, B wants to choose the opposite of what A chooses. Thus, there are two pure-strategy Nash equilibria in Chicken: (defect, cooperate) and (cooperate, defect). Chicken has been applied to models of deterrence, crisis bargaining, and other topics in international conflict.

This covers the basics of strategic form games and Nash equilibria. However, there is much more to these games than we have examined here. First, not all games are 2 × 2 games like Prisoner's Dilemma and Chicken because the players might have more than two strategies or there might be more than two players. The logic of finding Nash equilibria, however, remains the same. Second, we have only been looking at pure strategies, when a player chooses a particular strategy (such as cooperate or defect) with certainty. Yet it is also possible for players to employ mixed strategies, where they randomly choose between their pure strategies according to specified probabilities (e.g., cooperate two-thirds of the time and defect one-third of the time).

Extensive Form Games

The second primary way to illustrate a game is in the extensive form. Extensive form games are also called dynamic games because they explicitly model the sequence of decisions—an aspect of time—within the game. Extensive form games also make it easier to examine situations in which there are three or more players or there are multiple options available to the decision makers. For example, instead of simply having options to cooperate or defect, states may also have options to escalate or to select not only whether to make a demand, but how much of one to make. Although the strategic form is the classical way to represent games, the extensive form has become more common in recent years.

To illustrate the extensive form, we will revisit Prisoner's Dilemma. Recall that in the story above, the suspects were separated into separate rooms and presented with their options separately. Thus, each player had to decide whether to cooperate or defect without knowing the choice of the other. In effect, their moves were simultaneous, which nicely fits the strategic form of Figure 3.1. But what if their moves were sequential, that is, what if suspect A had to decide first, and then suspect B could decide after observing what A had done?

This new situation is illustrated in the extensive form game of Figure 3.3. Extensive form games are divided into decision nodes, where the player making the decision at that point must choose one of the alternatives, which are represented as branches extending from the node. At node 1, suspect A has to choose between C (cooperate) along the left branch or D (defect) along the right branch. If A cooperates, then it is B's choice at node 2. B can also choose C along the left branch or D along the right branch. If A defects, then B chooses at node 3, where B can again choose between C and D. Regardless of B's choice, a terminal node is reached where the game ends and payoffs are assigned. Payoffs (or utilities) are separated by a comma; the number before the comma is the payoff for the first mover (suspect A in this game), and the number after the comma is the payoff for the second mover (suspect B in this game). The payoffs are the same for the different outcomes as in Figure 3.1; mutual cooperation is (3, 3), mutual defection is (2, 2), and one player cooperating while the other defects is either (4, 1) or (1, 4) with the defecting player getting 4.

We can solve extensive form games using a technique called backward induction. Backward induction involves determining the rational choice that each player will make at each node of the game, beginning at the end of game and working backward to the beginning. Beginning at node 3, we can see that B will choose D to get 2 rather than 1. At node 2, B once again will choose D, getting 4 rather than 3. Now we can move to A's choice at node 1. If A chooses D, she knows that B will also choose D, so A will get 2; if A chooses C, she knows that B will again choose D, so A will get only 1. Thus, A's rational choice is to choose D.

Figure 3.3 Sequential Prisoners' Dilemma

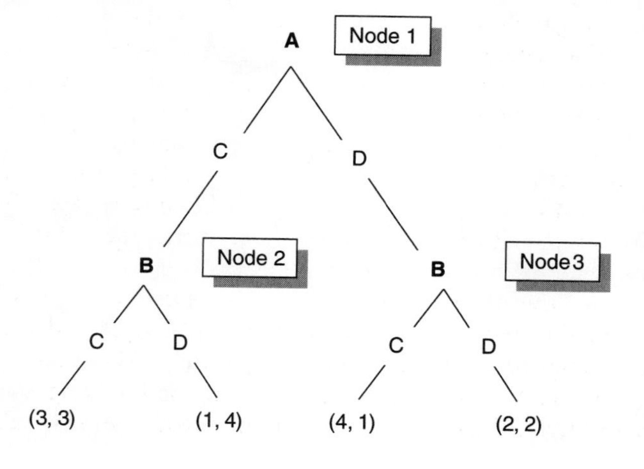

When we use backward induction to solve a game, we find the **subgame perfect equilibrium.** In the sequential Prisoner's Dilemma game, the subgame perfect equilibrium is (defect; defect, defect); A's strategy is defect, and B's strategy is defect, defect (in game theory, a strategy is a complete plan of action for play of a game, so a player's strategy must indicate her decision at every possible decision node in the game). So even though we have made the game sequential, the equilibrium outcome remains mutual defection.

The subgame perfect equilibrium concept was developed by Reinhard Selten (1975) and is a refinement of Nash equilibrium that is more consistent with the assumption of instrumental rationality. Nash equilibrium is the basis of all solution concepts for noncooperative games because only outcomes that are associated with Nash equilibria are consistent with instrumentally rational choices by all players in a game. Some Nash equilibria, however, are based on instrumentally irrational choices.

Consider the example game developed by John Harsanyi (1977) and shown in Figure 3.4. There are two players, A and B, each of whom makes one choice. Backward induction allows the course of rational play to easily be seen. At node 2, player B will choose D, securing $u = 2$ rather than the 0 received by choosing C. Knowing this, player A must make a choice at node 1 between C ($u = 1$) or D ($u = 2$, since B will choose D). Thus, the rational choice is for each player to choose D, as indicated by the arrows in Figure 3.4. This is the subgame perfect equilibrium.

Although the outcome DD is the only subgame perfect equilibrium, there are two Nash equilibria. (I indicate outcomes by using two letters, indicating the strategies used by A and B to reach the outcome. Thus, outcome CD indicates that A cooperates and B defects, while outcome DD indicates that both defect.) The Nash equilibria, which are more easily identified in strategic form (Figure 3.5), are CC and DD. The equilibrium CC, however,

Figure 3.4 Extensive-Form Representation of Harsanyi's Game

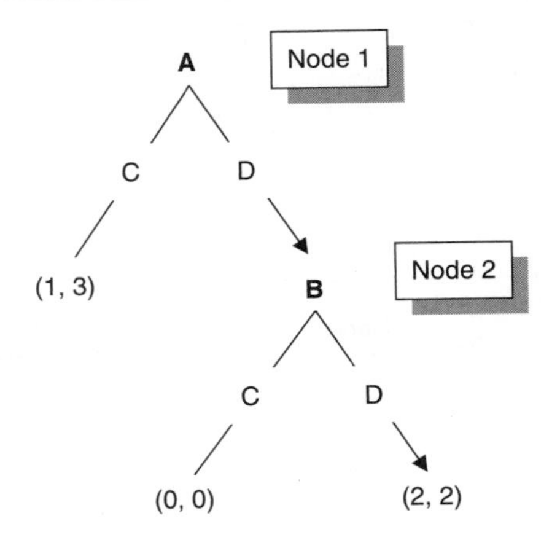

Figure 3.5 **Strategic-Form Representation of Harsanyi's Game**

		B: C	B: D
A	C	(1, 3)*	(1, 3)
	D	(0, 0)	(2, 2)*

relies on player B's irrational threat to choose C if A chooses D. This occurs because Nash equilibria do not exclude irrational threats off the equilibrium path. The equilibrium path is the path from the start of the game through each decision node until the equilibrium outcome is reached. For the DD equilibrium, the equilibrium path goes from the start of the game (node 1) through node 2 to the end. The equilibrium path for the CC equilibrium, however, only includes node 1 because A's choice of C at node 1 ends the game; thus, B's choice at node 2 is off the equilibrium path.

Subgame perfect equilibria are those Nash equilibria that are based on instrumentally rational choices at every decision point. If there is a difference between the Nash and subgame perfect equilibria of a particular game, we should always rely on the subgame perfect equilibrium as the correct solution.

One simplification that we have assumed in our examination of game theory is that each player knows the opponent's preferences. This is known as having complete information. We can also examine games of incomplete information, or asymmetric information, where at least one side has some uncertainty about the opponent, usually their preferences. With games of incomplete information, we need to use a different solution concept. A common one is the Perfect Bayesian equilibrium, which ensures that actors make rational choices at each decision node and update their beliefs about opponents using Bayes's rule when possible.

Using Game Theory

Game theory can be used in several different ways. First, game theory could be used as a strictly normative tool to evaluate the efficacy of competing policy prescriptions. For example, Zagare (1983) uses a game-theoretic framework to analyze the 1973 crisis in the Middle East, finding that the decision of the Nixon administration to place US strategic forces on alert was not only justifiable but was also more efficacious than the less provocative approach of reassuring the Soviets of a US willingness to compromise.

Game-theoretic models could also be employed descriptively to explain single cases that are intrinsically interesting or otherwise important. Bates et al. (1998) call this approach "analytic narratives," where the theory and a case study are tightly integrated. Game theory is frequently utilized in this way. For example, Zagare's (1979) analysis of the 1954 Geneva Conference helps to reconcile the well-known but unexplained discrepancy between the public and private policy pronouncements of US decision makers both before and during the negotiations that ended the Franco-Vietminh War. Zagare (1982) uses the same game-theoretic framework to eliminate competing, seemingly plausible, explanations of the Geneva Conference advanced by Thakur (1982).

A game-theoretic analysis of a single historical case could also be thought as an inductive step taken to facilitate the development of a general theory. If they are to be useful, formal models, game-theoretic or otherwise, cannot be fashioned out of whole cloth. Game theory provides a useful framework for developing comparable case studies, which, in turn, could serve as a guide in the construction of more refined models or more powerful theories.

Game theory is most commonly used within the study of international relations to develop general theories. Such theories strive to model general political processes such as those associated with crisis bargaining, alliance formation, and war. There have been a large number of such applications in the study of international conflict. We now turn our attention to one important example.

The International Interaction Game

A prominent example of a game-theoretic model of international conflict is the International Interaction Game (IIG) developed by Bueno de Mesquita and Lalman in their 1992 book *War and Reason*. In this model (see Figure 3.6), two actors (State A and State B) choose whether to make

Figure 3.6 International Interaction Game

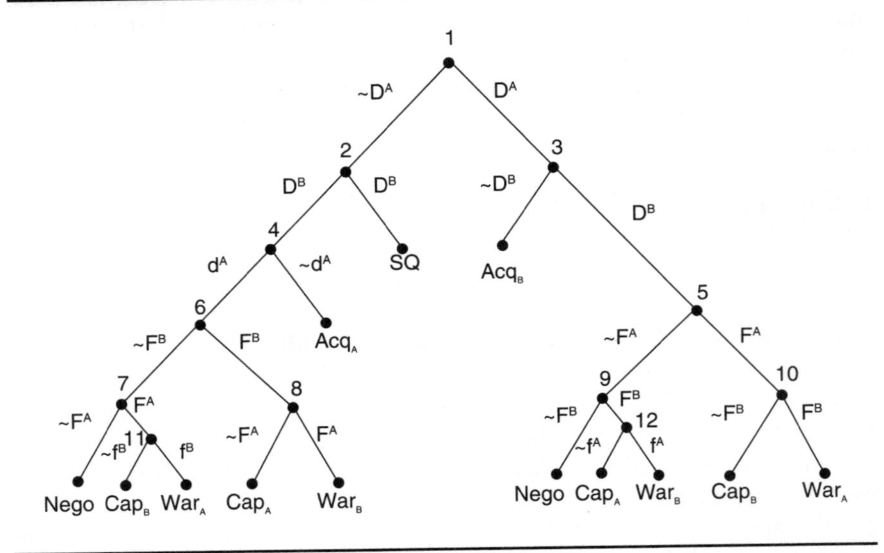

Source: Bruce Bueno de Mesquita and David Lalman, *War and Reason: Domestic and International Imperatives* (New Haven, CT: Yale University Press, 1992), 30.

a demand on the other, and given that demands have been made, whether or not to use force to pursue their aims.[6] If neither state makes a demand, the status quo results (SQ). If both states make demands and choose to use force, then war commences (War_A or War_B, depending on who used force first). If a state backs down to a demand, it is said to acquiesce (Acq_A or Acq_B), but if it backs down to a use of force, it is said to capitulate (Cap_A or Cap_B). Finally, if both states make demands but neither resorts to the use of force, negotiation results (Nego). In the domestic constraints variant of the game, which receives the most empirical support, the decision to make a demand is the result of unspecified domestic processes.

There are seven basic assumptions that form the heart of Bueno de Mesquita and Lalman's theory. They assume that decision makers are instrumentally rational; that negotiations and war involve a lottery in which each player has a chance to win or lose with some probability; that achieving one's demands represents a gain from the status quo; and that each outcome has a set of costs, benefits, or both associated with it. (For a complete discussion of the assumptions, see Bueno de Mesquita and Lalman 1992, 40–6; or Bueno de Mesquita 2010, 151.) These assumptions limit the possible preference orderings as shown in Table 3.4. There are eight different outcomes that can occur from the game. The most preferred outcome takes a value of 8, while the least preferred outcome takes a value of 1.

Table 3.4 Preference Restrictions for the International Interaction Game

OUTCOME	RESTRICTION ON ORDERING	POSSIBLE PREFERENCE RANK
SQ	> Acq_i, Cap_i	7 to 3
Acq_j	> all other outcomes	8
Acq_i	> Cap_i	5 to 2
Nego	> Acq_i, Cap_i, War_i, War_j	7 to 5
Cap_j	> War_i, War_j	7 to 3
War_i	> War_j	5 to 2
Cap_i	—	4 to 1
War_j	—	4 to 1

Note: All restrictions are in terms of Nation *i*'s preferences.

Source: Bruce Bueno de Mesquita and David Lalman, *War and Reason: Domestic and International Imperatives* (New Haven, CT: Yale University Press, 1992), 47.

[6] A demand placed by State i is labeled D^i, whereas the choice to not make a demand is $\sim D^i$. Similarly, the use of force is F^i, and the decision to not use force is $\sim F^i$, where i denotes the state making the decision. Lowercase letters are used when a state has a second opportunity to make a demand or use force (for example, State A's choice at node 4).

These preference restrictions follow from Bueno de Mesquita and Lalman's assumptions. Forcing the opponent to acquiesce is a low-cost way to achieve one's demand, and is therefore preferred over all other outcomes; its value is always 8. Negotiation is assumed to be preferred to four other outcomes, so it can fall anywhere between the second-best outcome (preference rank 7) and the fourth-best outcome (preference rank 5). Given these preference restrictions, there are 2,704 different pairs of preferences for the two players. Thus, the International Interaction Game provides a general theory that accounts for 2,704 different strategic situations.

Bueno de Mesquita and Lalman define several different player types that identify important characteristics of states in the International Interaction Game. A hawk is a state that would like to force his opponent to capitulate rather than negotiate (Cap_j > Nego), while a dove is a state that prefers to negotiate with its opponent (Nego > Cap_j). A retaliator prefers to respond to an attack with a war rather than capitulate (War_j > Cap_i), while a surrenderer prefers to give in (capitulate) rather than fight (Cap_i > War_j). Finally, a pacific dove is a state who is both a dove and a surrenderer.

The conditions leading to war in the International Interaction Game are shown by the basic war theorem. The basic war theorem comes directly from examining the subgame perfect equilibria of the International Interaction Game. The theorem tells us that war is the sole equilibrium (in complete and perfect information) if

Player A prefers to initiate a war than acquiesce (War_A > Acq_A),

Player A prefers to capitulate than fight a war started by B (Cap_A > War_B),

Player B prefers to fight a war started by A than capitulate (War_A > Cap_B), and

Player B prefers to force A to capitulate than negotiate (Cap_A > Nego).

So war can only occur with a hawk-retaliator against a surrenderer. The surrenderer's weakness, rather than making war less likely, actually makes it more likely.

The IIG can help us understand the relationship between uncertainty and war. War can occur as an outcome of the IIG through several different paths; there are many different preference orderings that can lead states into war. War can happen even without uncertainty, as the basic war theorem demonstrates. Many scholars have argued that uncertainty makes war more likely while certainty makes war less likely. By contrast, the IIG identifies conditions under which increases in uncertainty make war more likely and conditions under which increases in uncertainty make war less likely. If at least one of the four conditions of the basic war theorem are not met, but decision makers perceive that all four are satisfied because each has incomplete information about the preferences of the other, then uncertainty makes war more likely. If, however, the four conditions are met, then uncertainty about the preferences of an opponent can lead to choices that do not result in war even though war would result if there was complete information. Thus, uncertainty can either make war more likely or less likely, depending upon the circumstances.

Bueno de Mesquita and Lalman also use the IIG to examine other relationships. Controversially, they argue that dissatisfaction with the status quo has no effect on the likelihood of conflict. They do so because war can occur in the IIG even when both sides value the status quo very highly and does not necessarily occur even when the status quo is poorly valued by the two sides. They argue that the IIG also sheds light on the impacts that democracy and power have on international conflict.

As with any other theory, it is important to test Bueno de Mesquita and Lalman's theory with empirical evidence. Bueno de Mesquita and Lalman (1992) develop techniques similar to the techniques Bueno de Mesquita (1981) used to measure expected utility. This enables them to conduct a series of quantitative tests of their theory using data from Europe between 1816 and 1970. Bennett and Stam (2000a, 2000b, 2004) conduct a broader series of quantitative tests of the International Interaction Game across all regions of the world from 1816 to 1992. Through a series of empirical tests, the predictions made by the International Interaction Game are generally supported by the evidence.

Bargaining Model of War

The **bargaining model of war** refers to a variety of individual models that, despite having certain differences, share important common features, particularly the inclusion of the size of demands as part of the game and the inclusion of war as part of the bargaining process (Reiter 2003). It is a rational choice framework that is a focus of much recent research on international conflict that has been effective in explaining the entire process of war. The bargaining model focuses squarely on an important puzzle regarding international conflict. War is an extremely costly way for states to settle their disputes. And although countries have conflicts of interest all the time, only some disputes are resolved by force. So why go to war? Given the human and material costs of military conflict, why do states sometimes wage war rather than resolve their disputes through negotiations?

Fearon (1995) used a bargaining model to examine rationalist explanations of the causes of war. As long as war is costly, some bargain should exist that both sides would prefer to actually fighting. If so, war can only arise as a result of some bargaining failure. To see this, consider the bargaining model shown in Figure 3.7. There are two states, A and B, engaged in bargaining over some issue, such as territory. The good is divided between the two states, with x indicating the share for State A (therefore, $1 - x$ is the share for State B). Although x is shown to the left of point p in the figure, x can take any value between 0 and 1. State A's **ideal point**—an actor's most

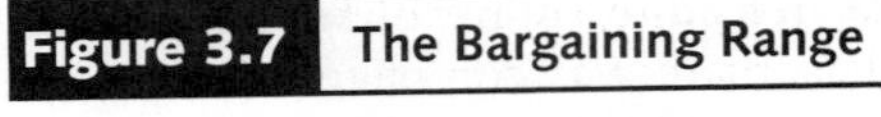

Figure 3.7 The Bargaining Range

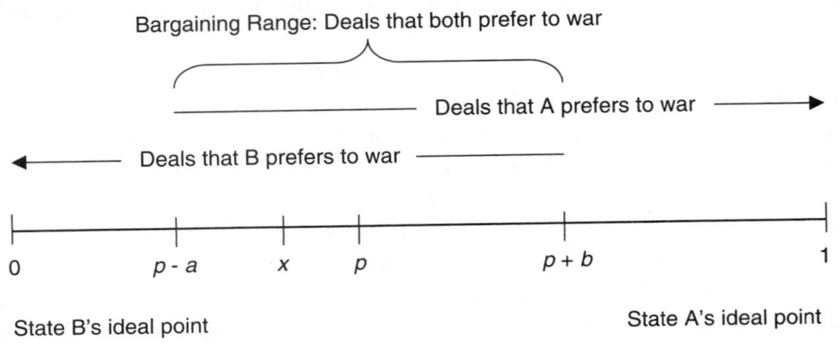

preferred division of the issue—is at the right edge of the figure, where $x = 1$, while state B's ideal point is to the left where $x = 0$. The further we go away from someone's ideal point, the less satisfied they are. We assume that State A would win a war between the two sides with probability p. Since war is costly, there are also cost terms: a is the expected cost that A would have to pay in war, while b is the expected costs for State B.

For example, India and Pakistan disagree about Kashmir; although each controls only a portion of the territory, they claim the entirety as their own. India's ideal point is for India to control all of Kashmir while Pakistan controls none of it ($x = 1$), whereas Pakistan's ideal point is exactly the opposite ($x = 0$). Although each side wants to have all of the territory, neither side actually does. Why not?

Examining this question more generally gets to the heart of the bargaining model of war. We will assume that the winner of a war would be able to take all of the good in question. Since the probability that A wins is p, the expected value following a war is p. Since war is costly, however, each state's expected value for war is diminished. Accordingly, the expected value of war for State A is $p - a$, while the expected value of war for State B is $p + b$ (notice that the cost terms move each state's expected value of war away from their ideal point).

Since State A's expected value of war is $p - a$, A prefers any value of x that is greater than $p - a$ to war. In other words, A prefers any deal that is to the right of $p - a$ over war. Similarly, B prefers any value of x that is less than (i.e., to the left of) $p + b$ to war. Therefore, both sides prefer any division of the good that is between $p - a$ and $p + b$ over war: This is the bargaining range. As long as war is costly, a bargaining range will exist where both sides prefer settlement to war. Thus, war can only result from some breakdown of bargaining. Bargaining failures can arise from three sources: incomplete information, commitment problems, and indivisible issues. We will explore each in turn.

Fearon (1995) argues that the most important source of bargaining failure is incomplete information with incentives to misrepresent. Incomplete information can come from several sources. States can be uncertain about each other's willingness to fight because they disagree about the probability of victory (p) or because they disagree about the costs of fighting (a or b). One might wonder why the states cannot just tell each other about their capabilities and resolve, thus resolving the information problem. The basic answer is that states possess incentives to misrepresent or bluff.

Private information about the opponent's capabilities, morale, outside help, and quality of plans can all lead to different beliefs about the probability of winning. But if states revealed their private information, then it might undermine their advantage. For example, prior to the Six-Day War in 1967, Egypt and the other Arab states were confident in their ability to beat Israel. Israel was also confident in victory. Although they were greatly outnumbered in personnel, tanks, and aircraft, Israel felt that it had other advantages that numbers could not account for. In particular, Israel had a plan, called Focus (*Moked*), for achieving air supremacy right at the beginning of the war by destroying the air forces of Egypt, Syria, and Lebanon. The plan was initiated on the morning of 5 June, the first day of the war. Nearly all of Israel's 196 combat aircraft were committed to the airstrike, with only 12 being held back to patrol Israeli airspace. In a quick succession of attacking waves, Israeli aircraft struck fourteen Egyptian, one Iraqi, two Jordanian, and five Syrian air bases. The attack was a massive success for Israel, destroying 452 Arab aircraft for the loss of 19 Israeli aircraft (Hammel 1992; Oren 2002). Israel could not reveal this plan or other reasons why it was confident in victory because doing so would undermine its advantage.

Even if states do not have secret plans, strategies, or capabilities that give them reason to be confident of victory, they have incentives to pretend that they do. Thus, there are strong incentives for states to misrepresent their private information. Fearon (1995) argues that incentives to misrepresent were important in leading to the Russo-Japanese War in 1904 and World War I in 1914.

Commitment problems are also an important source of bargaining failures. Commitment problems arise when states cannot credibly promise to cooperate with agreements they reach. Thus, at least one side has an incentive to renege on its promise once the other side fulfills its part of the deal. Promises are credible only when it is in the self-interest of actors to carry them out.

Commitment problems are likely when bargaining over issues that affect future power, when power is changing exogenously, and when there are first strike advantages. When one side is rapidly growing in strength, its bargaining position is being improved continuously as time goes on. This creates a commitment problem because agreements that they agree to now will not be as good as they should be able to get in the future. This in turn creates powerful preventive motivations for opponents of such states to prevent such dramatic shifts. For example, Fearon argues that commitment problems were an important cause of World War I as "German leaders were willing to run serious risks of global conflict in 1914 . . . [because] they feared the consequences of further growth of Russian military power, which appeared to them to be on a dangerous upward trajectory" (1995, 407).

The final source of bargaining failures identified by Fearon (1995) is indivisible issues. Indivisible issues exist when circumstance dictates that the winner gets everything and the loser gets nothing. The bargaining model assumes that all issues are easily divided into increments. If issues are indivisible, there may not be possible divisions that lie within the bargaining space. For instance, Jerusalem is claimed as a capital and holy city by several groups and faiths. How can it be divided? Nonetheless, Fearon (1995) discounts indivisibilities as an important source of bargaining failure because virtually all issues are divisible in practice, especially if side payments are used to compensate the party that does not have direct control of the good.

A variety of bargaining models have further explained the outbreak of international conflict, looking at the influence of factors such as military mobilizations (Slantchev 2005), public commitment (Tarar and Leventoglu 2009), and international organizations (Chapman and Wolford 2010). Bargaining models have also been developed to explain war outcomes and termination. The basic logic is that war terminates when the information gap between the two sides closes enough to create a bargaining range (Filson and Werner 2002, 2004; Powell 2004; Slantchev 2003; Smith 1998; Smith and Stam 2004; Wagner 2000). Slantchev (2003) labels this the principle of convergence. We will explore how the bargaining model enables us to explain war termination in more detail in chapter 12. One of the greatest strengths of the bargaining model of war is that it allows explanations of war onset, duration, outcome, and termination all in one integrated theoretical model (Filson and Werner 2002).

Conclusion

The fundamental reason why wars and other forms of international conflict occur is that leaders decide to fight. Furthermore, decisions by leaders of states and other key people drive the ending of wars, the formation of alliances, and many other phenomena of interest related to international conflict and security. Rational choice theory focuses on explaining these decisions by assuming that decision makers are rational. Although there are many different notions of what it means to be rational, rational choice theory is based on the assumption of instrumental rationality. Instrumental rationality simply means that people make decisions according to their preferences.

Assuming instrumental rationality opens the way for formal, mathematical modeling, but it does not require it. Expected utility theory and game theory are formal modeling tools of rational choice. The specific examples of theories developed using expected utility theory (Bueno de Mesquita 1981) and game theory (Bueno de Mesquita and Lalman 1992) that we examined in this chapter are two of many theories of international conflict that scholars have developed using these tools—especially game theory. Game theory is especially useful because it focuses on explaining strategic interaction, which is an integral part of international politics. In particular, the bargaining model of war is an important game-theoretic framework that has become increasingly common for explaining international conflict.

Rational choice theory is useful in establishing the logic of theories by ensuring logical consistency. That is certainly one of the primary strengths of formal models such as game theory. One concern about game-theoretic models is that some have not been subjected to empirical testing. Regardless of the rigor of our logic, we still need to examine evidence. We now turn our attention to a variety of specific topics related to international conflict, beginning with contiguity and territory.

Key Concepts

Bargaining model of war

Bounded rationality

Dominant strategy

Expected utility theory

Game theory

Ideal point

Instrumental rationality

Nash equilibrium

Operational code

Poliheuristic theory

Procedural rationality

Prospect theory

Psychological approaches

Subgame perfect equilibrium

Unitary actor assumption

Part II

Causes of War and Militarized Disputes

4 Contiguity and Territory

Indian soldiers patrol the border with Pakistan in December 2013. India and Pakistan form one of the most conflict-prone contiguous dyads, and their relations are dominated by territorial issues over Kashmir.

Source: Associated Press

Why do countries engage in militarized disputes and wars against each other? This is an important question that is central to understanding international conflict and has dominated thinking about international relations over the years. The previous chapters—although shedding some light on the causes of international conflict—were focused more on preliminary topics such as the history of conflict and approaches to studying it. The next six chapters focus on prominent explanations of international conflict that try to answer this question.

Think of pairs of states that have fought one another repeatedly, such as France and Germany, Israel and Syria, India and Pakistan, Greece and Turkey, or Russia and China. What do these dyads have in common? Most obviously, these pairs of states are all neighbors. While some other dyads that do not share a border—such as the United States and the Soviet Union—have also engaged in many disputes, this clues us in to an important trend.

As discussed in the first chapter, Bremer (1992) found that contiguity is the most important factor for making dangerous dyads. **Contiguity** is the sharing of a border. So France and Germany are contiguous, but France and Russia are not. In this chapter, we examine the importance of contiguity as a cause of international conflict.

Borders are territorial in nature. The international system since the Peace of Westphalia in 1648 has been based on sovereign states. Sovereign states are territorial units, and (at least in principle) exercise control over life within their borders. Therefore contiguity is interrelated with territory, and so we will also look at the importance of territory as a cause of conflict.

Going back to the pairs of states highlighted above, we can see that not only are they contiguous, they have also had a history of disagreeing about territory. France and Germany disputed the border provinces of Alsace and Lorraine. India and Pakistan each claim to be the rightful owners of Kashmir. And territory has played a role in conflicts between each of the others as well.

In the next section, we examine the relationship between contiguity and international conflict and introduce the three primary reasons why contiguous states are so much more likely to fight. We then turn our attention to different ways to identify opportunity for international conflict. Finally, we examine the impact of territorial issues on international conflict.

Contiguity and International Conflict

In chapter 1, we saw that Bremer identified contiguity as the single most important factor causing dangerous dyads. In other words, contiguity is the factor that most increases the likelihood of war between pairs of states, or at least among the factors that Bremer examined.

In addition to the data sets we have discussed in other chapters, the Correlates of War (COW) project also collected data on contiguity (Stinnett, Tir, Schafer, Diehl, and Gochman 2002; see also Gochman 1991). The data set includes six categories of dyadic contiguity: contiguous by land (including a river border); contiguous by 12, 24, 150, or 400 miles or less of water; or noncontiguous. Water contiguity is based on whether a straight line of no more than a certain distance can be drawn between a point on the border of one state, across open water (uninterrupted by the territory of a third state), to the closest point on the homeland territory of another state. For example, France and Germany are contiguous by land. On the other hand, France and the United Kingdom are separated by the English Channel, which is only 21 miles wide at its narrowest point. Thus, France and the United Kingdom are contiguous by the 24, 150, or 400 miles of water categories, but not by the 12 miles of water or direct land contiguity categories.

Many scholars consider dyads that are contiguous by land or by less than 150 miles of water to be contiguous, and others to be noncontiguous. Coding states as contiguous if separated by less than 150 miles of water "reflected an estimate of the average distance that a sailing ship could cover in a single day during the 1816–1965 period. . . . If you were land (or river) contiguous, you did not need to pass through a third state to 'get at' another; and, if you were within 150 'line of sight' miles of water, you could 'get at' another state within a day without passing through a third state" (Gochman 1991, 94–5).

The contiguity data also contain information about colonial contiguity. For colonial contiguity, if two dependencies of two states are contiguous, or if one state is contiguous to a dependency of another, the data set reports a contiguity relationship between the two states. For example, the United Kingdom and China had direct land colonial contiguity through the former British colony of Hong Kong until 1997, when Hong Kong was transferred to Chinese control. The United States and United Kingdom have colonial contiguity through less than twelve miles of water between their colonies in the Caribbean Sea, the US Virgin Islands and the British Virgin Islands. The same six categories of land and water contiguity are used to classify colonial contiguity.

Given these data sets, we are able to systematically examine contiguity across the past two centuries. Table 4.1 shows the relationship between contiguity and militarized conflict. The table breaks down the dyads involved at the start of a conflict by whether or not they are contiguous. The top part of the table shows the trends for militarized intestate disputes over time. Over the entire time frame of the data, 1816–2001, 59.4 percent (1,586 of 2,671) of disputes began between contiguous adversaries. For the 1816–1945 time period (through World War II), 46.8 percent of disputes (449 of 959) began between contiguous adversaries. After World War II, that figure increases dramatically to 66.4 percent (1,137 of 1,712).

The pattern for interstate wars, shown in the bottom half of Table 4.1, is similar but even stronger. For the entire 1816–2001 time period,

Table 4.1 Contiguity and Militarized Conflict

CONTIGUOUS ADVERSARIES?	1816–1945	1946–2001	1816–2001
Militarized interstate disputes			
Yes	449 (46.8%)	1,137 (66.4%)	1,586 (59.4%)
No	510	575	1,085
Total	959	1,712	2,671
Interstate wars			
Yes	55 (60.4%)	29 (78.4%)	84 (65.6%)
No	36	8	44
Total	91	37	128

Source: Compiled by author.

w65.6 percent (84 of 128) of wars began between contiguous states. Prior to 1945, 60.4 percent (55 of 91) of wars were contiguous, while afterward, it rises to 78.4 percent (29 of 37).

Whether we focus on militarized disputes or wars, a high percentage of conflicts begin between contiguous states. Furthermore, this percentage is much higher after World War II than before. The explanation for that shift is not difficult to arrive at. The postwar period saw a major shift toward decolonization, as newly independent states emerged from former colonies from the late 1940s through the 1970s. There are many more states in the world after World War II than there were before. Thus, while major powers are much better able than minor powers to project power over distance, they represent a much smaller fraction of states in the world, and thus account for a smaller fraction of conflicts.

One might wonder whether the impact of contiguity on conflict is really that strong. After all, only about half of disputes and less than two-thirds of wars start between contiguous states. Of course, half of the time is just like flipping a coin—completely random. While a pair of states is either contiguous or not, however, that does not mean that the two categories are equally likely. In 2011, there were 195 states in the world, according to the Correlates of War state system membership data. How many neighbors does a state have? That of course depends on the geography of the state. The United States has only two neighbors by land (Canada and Mexico); if we add in contiguity with up to 150 miles of water, that adds three others (Cuba, Bahamas, and Russia). And what about a state with lot of neighbors? Russia shares a direct land border with fourteen different countries (Norway, Finland, Estonia, Latvia, Lithuania, Poland, Belarus, Ukraine, Georgia, Azerbaijan, Kazakhstan, Mongolia, China, and North Korea), plus water contiguity with two others (Japan and the United States). But even with such a large number of neighboring states, Russia is still a part of only 16 contiguous dyads compared to 179 noncontiguous dyads. Despite the fact that contiguous dyads comprise such a small percentage of all pairs of states, they still account for more than half of all conflicts. That is a very strong relationship.

Box 4.1

Case in Point: Falklands War

The Falkland Islands are located in the South Atlantic Ocean about 290 miles off the coast of Argentina. The long-running dispute between Britain and Argentina over the ownership of the Falkland Islands has its origins in the founding of an Argentine settlement on the islands in 1826. Most of the Argentine settlers were expelled by a US warship in 1831, and a British expedition took control of the territory in 1832. British sovereignty was declared in 1833, although Argentina has always disputed this. Territorial issues were key to the war even though the two countries are not neighbors.

On 26 March 1982, the Argentine military junta decided to invade the Falkland Islands. On 2 April, Argentina landed thousands of troops there. The British Royal Marines based on the islands put up some resistance before surrendering. The British government immediately cut diplomatic ties with Argentina and began assembling a large naval taskforce. The next day, Argentine troops seized the islands of South Georgia and the South Sandwich group following a short battle, prompting enthusiastic celebrations in Buenos Aires.

Any British military response was made more difficult by the more than 8,000-mile (13,000-km) distance from Britain to the Falklands. The British naval vanguard set sail for Ascension Island, which was the key British forward base in the South Atlantic, on 5 April. Given the vast distances, it took until 22 April for the British task force to reach Falklands waters. Three days later, a small British commando force retook South Georgia Island. Since their nearest air base, on Ascension Island, was 3,800 miles (6,100 km) from the Falklands, the British had to rely on aircraft carriers to provide air support to the task force. On 1 May, British aircraft began attacking the Port Stanley airfield. On 21 May, British troops landed near Port San Carlos and rapidly reinforced the bridgehead over the following days. After a week of ground fighting, Britain was able to take the Argentine positions Darwin and Goose Green on 28 May. The next day, British troops continued their advance, taking Douglas settlement, Teal Inlet, and Mount Kent, leaving Port Stanley surrounded.

On 12 June, British troops assaulted Argentine positions on Mount Longdon, Two Sisters, Mount Harriet, and Mount Tumbledown. On 14 June, the Argentine garrison at Port Stanley was defeated. And on 20 June, Britain formally declared an end to hostilities.

Argentina had 1,717 killed and wounded and 11,313 taken prisoner; naval losses were 1 cruiser (ARA *General Belgrano*), 1 submarine, and 7 other vessels sunk; 100 aircraft were lost. The United Kingdom had 1,033 killed and wounded and 115 taken prisoner; naval losses were 2 destroyers (HMS *Sheffield* and HMS *Coventry*), 2 frigates (HMS *Antelope* and HMS *Ardent*), and 3 other ships sunk; 34 aircraft were lost.

Now that we have explored the strong relationship between contiguity and international conflict, it is natural to wonder why contiguous states are so likely to fight, and there are a number of reasons. Research has identified three factors in particular as the most important. The first key characteristic of contiguity that makes it conflict prone is opportunity, which is related to proximity, the distance between states. The closer together two states are, the easier it is for them to reach one another militarily. Boulding (1962) developed the concept of a **loss of strength gradient** to reflect the importance of proximity. The loss of strength gradient reflects the idea that the amount of power that a state is able to project declines as distance increases. This is because simply transporting and supplying military forces over vast distances uses up a great deal of resources, leaving fewer resources available to actually fight or otherwise project power.

The importance of proximity is illustrated by the Falklands War. The Falkland Islands are about 8,000 miles away from the United Kingdom. Fighting a war at such a great distance poses such logistical problems that it is simply beyond the abilities of most states. Major powers such as Britain, however, have much greater power projection capabilities, without which they would have been unable to fight. In comparison, the Falklands are less than 300 miles off the coast of Argentina, yet even this much shorter distance created important difficulties. For example, the distance meant that Argentinian aircraft, which had to operate from land bases since they had no aircraft carriers, had an average of only about two minutes in the target area before they had to return to base to refuel.

Underlying all of this are the twin concepts of opportunity and willingness, which were introduced by Most and Starr (1989). **Opportunity** is essentially the possibility for militarized conflict between two states, while **willingness** refers to the desire by two states to engage in militarized conflict. In order for states in a dyad to become involved in an international conflict, they must have both the opportunity and willingness to fight. Thus, "opportunity and willingness should be viewed as jointly necessary conditions; neither alone is sufficient" (Most and Starr 1989, 40). Since contiguity provides opportunity, and opportunity is a necessary condition for international conflict, this leads to a strong relationship between contiguity and conflict.

The second key characteristic of contiguity is interactions. Increased interactions between proximate states provide increased opportunities for both cooperation and conflict. Neighbors are more likely to engage in trade of goods and services, have tourism between them, disagree about a border, or any number of things. It is easier to interact with those that are closer. For instance, although trade with China receives most of the attention, the number one trade partner of the United States is actually Canada.

The third key characteristic of contiguity is territory. The idea here is that contiguous states fight more often because they disagree about territory more often. This is important because disagreements about territory are highly contentious and conflict prone.

Map 4.1 Falklands War, 1982

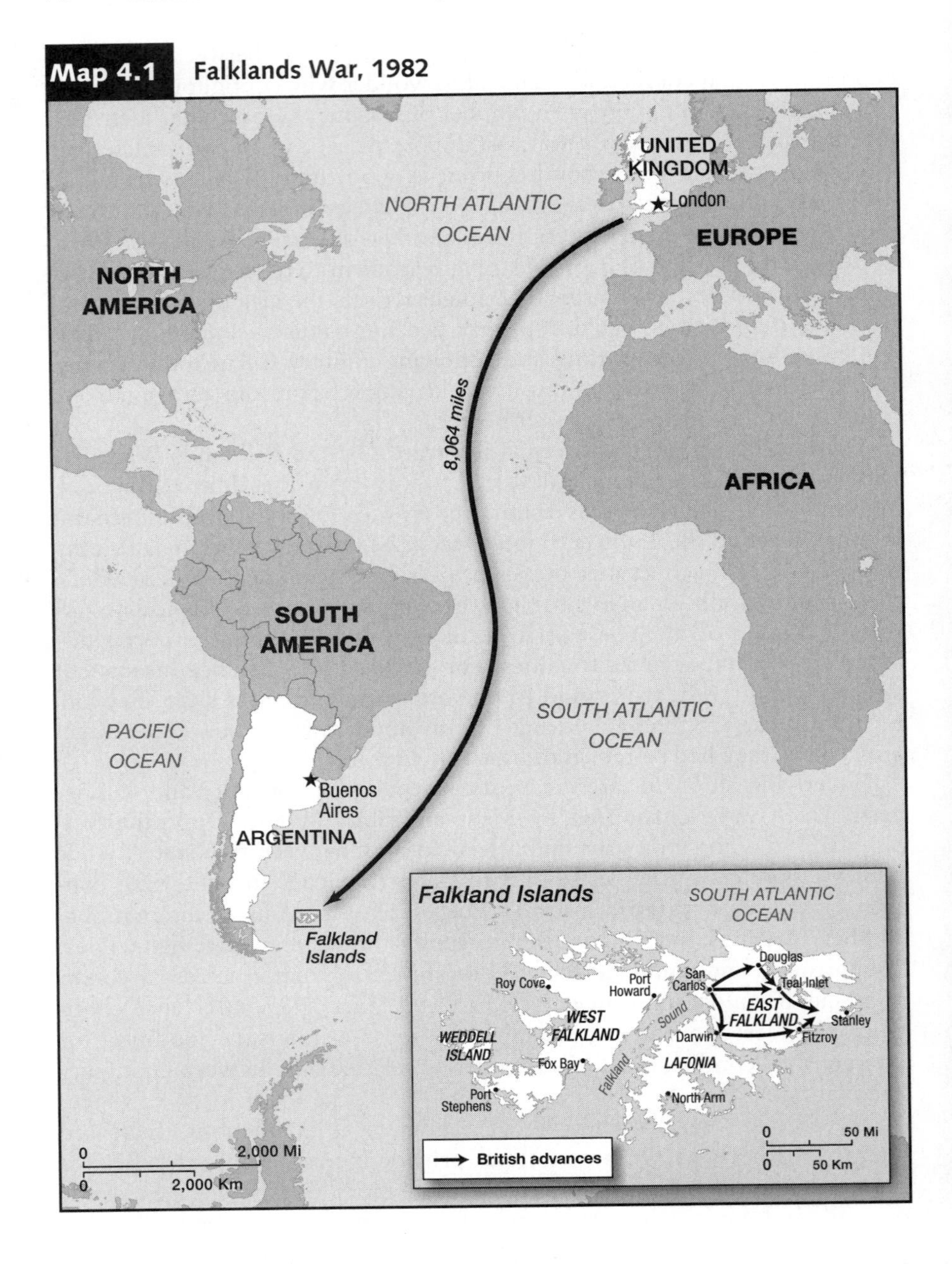

Opportunity, interactions, and territory have all been highlighted as key reasons for the importance of contiguity as a cause of wars and disputes, although interactions have received substantially less attention than the others. Some have speculated that increases in technologies such as drones and information may lead to a declining significance of contiguity in the future. These advanced technologies are mostly within the militaries of the strongest states, which are already the ones less constrained by proximity. Additionally, the influence of the other two characteristics of contiguity—interactions and territory—is not related to advances in technology. Thus, there is little reason to expect that the importance of contiguity as a cause of international conflict is likely to change significantly. Accordingly, we continue by examining attempts to identify opportunity for international conflict and then examining the importance of territorial issues.

Identifying Opportunity for Conflict

The first reason that contiguous states are more likely to fight is that contiguity creates the opportunity for conflict between states. Yet while contiguity makes conflict more likely, noncontiguous states also fight. So contiguity is not required to have the opportunity for conflict. What other factors provide states with the opportunity to fight each other? In other words, how can we identify opportunity for conflict? We will examine four of the most common concepts here.

Since opportunity is a necessary condition for conflict, the ideal measure of opportunity would include all dyads that have become engaged in international conflict. A measure of opportunity should determine those states that have the military capacity to reach one another. This capacity depends on the particular transportation assets available. In the nineteenth and early twentieth centuries, shipping and railroads were vital for power projection. Since the end of World War I, air transport has become increasingly important as well. In addition, terrain features such as mountains, deserts, and major forests could be considered. Power projection assets such as air and naval forces are essentially tools that states use to overcome the constraints of geography (Lemke 1995). Fortunately, there are concepts available that can serve as proxies for power projection capabilities, enabling the development of relatively simple measures of the opportunity for conflict.

An early attempt to identify those dyads having opportunity for conflict is the concept of **regional dyads**. Bueno de Mesquita (1981) divides the world into five different regions: Europe, Asia, the Middle East, the Americas, and Africa. "Membership in a region is held by any state physically located in the region and by any state that has taken a sustained and active interest in the affairs of one or more portions of the physical region" (Bueno de Mesquita 1981, 95). Bueno de Mesquita selects only dyads within the same region to analyze, arguing that "any analysis that heavily emphasizes these uninteresting cases [non-regional dyads] is itself bound to be uninteresting" (1981, 95).

Box 4.2

Concept in Focus: Understanding Cross Tabulations

A cross tabulation (or cross tab, for short) is a simple way of examining the relationship between two variables. A cross tab is a contingency table, showing the multivariate frequency distribution of statistical variables. To see how they are created, consider the following sample of ten dyads. For each observation, there are two variables: contiguous and war. Each variable can take either of two values: yes or no. In most applications, we would use numbers, with 1 for yes and 0 for no.

Dyad	Contiguous	War
1	No	No
2	Yes	No
3	No	No
4	No	No
5	Yes	Yes
6	No	No
7	Yes	No
8	Yes	Yes
9	No	No
10	No	Yes

The cross tabulation of the data above is shown below. Of the ten observations, contiguous is yes 4 times and no the other 6; war is yes 3 times and no the other 7 times. These numbers appear in the total column to the right (for contiguous) and the total row at the bottom (for war). There are four cells in the table that are at the intersection between the two variables. There are five cases where both war and contiguous are no; thus, a 5 appears in the appropriate cell of the table.

	War		
Contiguous	**No**	**Yes**	*Total*
No	5	1	6
Yes	2	2	4
Total	7	3	10

We can assess the statistical significance of the relationship between the two variables—the likelihood that it is due to chance—using a statistic called χ^2 (chi squared). For the example data here, χ^2 is 1.27, which is not statistically significant. Statistical significance, however, is quite difficult with only ten observations. If there were more than thirty observations with the same distribution of the variables, then the relationship would be statistically significant.

Cross tabulations of different conceptions of opportunity and militarized interstate disputes allow for easy examination of how well each measure captures recorded instances of international conflict. The cross tabulation of

Bueno de Mesquita's (1981) regional dyads and militarized interstate disputes (MIDs) is shown in Table 4.2. Each observation is a dyad-year, for a total of 656,870 dyad-years. Out of 3,002 dyad-years in which there was a militarized interstate dispute, 248 are considered by regional dyads to lack opportunity. Therefore, 33.2 percent of all dyad-years are classified as having opportunity by regional dyads, and 91.7 percent (2,754 of 3,002) of dispute-dyad-years are captured. Regional dyads account for many cases of international conflict, but also consider a significant portion of dyad-years to have opportunity.

It is clear, however, that regional dyads exclude some dyads that do have a positive probability of fighting. For example, although Italy and Austria-Hungary were each involved in the Boxer Rebellion in China in 1900, neither is considered to be a member of the Asian region. Regional dyads also do not exclude all dyads without opportunity for conflict (Lemke 1995). Consider Lesotho and Cameroon; each state is in Africa, but their chances of fighting are likely to be zero. Perhaps the most fundamental problem with the concept of regional dyads is that the defining characteristics (i.e., "a sustained and active interest" [Bueno de Mesquita 1981]) have more to do with a state's willingness, rather than opportunity, for conflict in a region.

The most commonly used concept for identifying opportunity for conflict is **politically relevant dyads.** Originally developed by Weede (1976, 1983), political relevance focuses on two factors: contiguity and major power presence. If a dyad has at least one of those conditions, then it is politically relevant. These factors are important for each of the remaining measures of opportunity.

Any pair of contiguous states is able to fight. States can also be contiguous through colonies: either between two colonies, or between one colony and another state's homeland territory. Colonial contiguity is important because the opportunities created by borders of homeland territory are created by colonial borders as well (Starr and Most 1976). Therefore, all contiguous dyads—either directly or indirectly through colonies—are considered to be politically relevant.

Major powers are, by definition, able to project power far beyond their boundaries. Major powers have the transportation capacity to move and supply military forces over vast distances. Thus, they not only form relevant dyads with their neighbors, but also with other states in the international

Table 4.2 Cross Tabulation of Regional Dyad Versus MID

	REGIONAL DYAD		
MID	0	1	TOTAL
0	438,215	215,653	653,868
1	248	2,754	3,002
Total	438,463	218,407	656,870

Source: Compiled by author.

system. For example, the Falklands War of 1982 was possible only because the United Kingdom was able to project its power to the South Atlantic.

In summary, the concept of political relevance focuses on contiguity and major power status because, as Maoz and Russett (1993, 627) state: "Over the period 1946–86 the international system averaged about 110 countries per year, which would give us roughly 265,000 dyad-years to study. But the vast majority are nearly irrelevant. The countries comprising them were too far apart and too weak militarily, with few serious interests potentially in conflict, for them plausibly to engage in any militarized diplomatic dispute."

Table 4.3 shows the cross tabulation of political relevance and MIDs. Out of 3,002 dyad-years in which there was a militarized interstate dispute, 357 are classified as irrelevant. Thus, political relevance captures 88.1 percent (2,645 of 3,002) of dispute-dyad-years while classifying only 13.9 percent of all dyad-years as relevant.

Politically relevant dyads classify many fewer dyad-years as relevant but also capture fewer instances of conflict. Analyzing the numbers differently, Maoz and Russett (1993) report that relevant dyads capture only 74 percent of militarized disputes. Dyads with the opportunity for conflict are those pairs of states that would have the ability to fight if they had the motivation to do so. But it is clear that many nonrelevant dyads do have the opportunity for conflict. One such example is the dyad of Israel and Iraq. They are not contiguous, nor are they major powers. Hence, this dyad is not considered to be politically relevant. Yet they have fought on opposite sides of several militarized interstate disputes. Since this and many other similar dyads have the opportunity to fight despite being classified as nonrelevant, it is clear that political relevance is limited as a measure of the opportunity for conflict.

Maoz (1996) offers a refinement of political relevance: the **politically relevant international environment (PRIE)**. The PRIE focuses on contiguity and major power status, which is similar to the previous definition. The standard definition of political relevance, however, classifies major powers as being relevant with all other states. But even within the group of major powers, there are significant differences between the power projection capabilities of states. While a limited number of states have truly global reach, other major powers can only project power regionally. Therefore, Maoz (1996) differentiates between global and regional powers. Regional powers can

Table 4.3 Cross Tabulation of Politically Relevant Versus MID

	POLITICALLY RELEVANT		
MID	0	1	TOTAL
0	565,377	88,491	653,868
1	357	2,645	3,002
Total	565,734	91,136	656,870

Source: Compiled by author.

fight all states within their region(s), whereas global powers are able to fight all states in the international system. A state's PRIE includes all contiguous states, as well as "regional powers of its own geographic region" and "major powers with global reach capacity" (Maoz 1996, 139). Clearly, the major change in this concept is the distinction between regional and global powers.

Table 4.4 lists the global and regional powers, the years of their power status, and their regions of activity. Since the politically relevant international environment and politically active dyads use different ranges of years, they are listed in separate columns. The baseline for his classification is the list of major powers coded by the Correlates of War project (Small and Singer 1982), which is used for classifying politically relevant dyads. For

Table 4.4 Global and Regional Powers

	YEARS				
STATE	*ACTIVE DYADS*	*PRIE*	*MAJOR POWER*	TYPE OF REACH CAPACITY	REGION
United States	1823–1897	1823–1898		Regional	Americas
	1898–2000	1899–1992	1898–2000	Global	World
Great Britain	1816–2000	1816–1992	1816–2000	Global	World
France	1816–1940	1816–1940	1816–1940	Global	World
	1945–2000	1944–1992	1945–2000	Global	World
Prussia/ Germany	1816–1866	1816–1866	1816–1918	Regional	Europe
	1867–1918	1867–1918		Global	World
	1925–1945	1925–1945	1925–1945	Global	World
	1991–2000		1991–2000	Regional	Europe
Austria-Hungary	1816–1918	1816–1918	1816–1918	Regional	Europe
Italy	1860–1943	1861–1943	1860–1943	Regional	Europe, Africa
Russia/ Soviet Union	1816–1917	1816–1917	1816–1917	Regional	Europe, Asia
	1922–1991	1922–1991	1922–2000	Global	World
	1992–2000	1992-		Regional	Europe, Asia
China	1950–1992	1949–1992	1950–2000	Regional	Asia
Japan	1895–1945	1895–1945	1895–1945	Regional	Asia
	1991–2000		1991–2000	Regional	Asia

example, while Prussia/Germany is considered to be a major power from 1816 to 1918 (and then again later), Maoz codes Prussia/Germany as a regional power in Europe from 1816 to 1866 and as a global power from 1867 to 1918. Note that the years for politically active dyads correspond to the major power years more closely than those of PRIE.

The cross tabulation of Maoz's (1996) politically relevant international environment (PRIE) and MIDs is shown in Table 4.5. The PRIE classifies 425 of 3,002 dispute-dyad-years to be irrelevant. Therefore, 11.4 percent of all dyad-years are classified as relevant by the PRIE, and 85.8 percent (2,577 of 3,002) of dispute-dyad-years are captured. Because of the distinction between global and regional powers, the PRIE considers fewer dyad-years to be relevant than the standard definition of political relevance, but it also accounts for fewer cases of international conflict. Thus, PRIE is still an imperfect measure of opportunity for conflict.

A fourth measure of opportunity for international conflict is **politically active dyads**, developed by Quackenbush (2006a). The factors used to determine which dyads are active are contiguity, power status, and alliances. Like the PRIE, politically active dyads differentiate between global powers and regional powers. Regional powers form active dyads with all states within the region(s) of the regional power, whereas global powers form active dyads with all states in the international system. Although Quackenbush (2006a) basically follows Maoz's (1996) classifications, he modified the dates to make them more consistent with the COW major power dates, as shown in Table 4.4. The key changes from Maoz's (1996) coding that Quackenbush makes are in the date of the American transition from a regional to a global power, the date of France's reentry as a major power following World War II, Germany and Japan's reentry as major powers in 1991, and the date of China's entry as a major power.

The criteria so far are similar to those for political relevance. Politically active dyads are distinguished by focusing on alliances. That alliances play an important role in creating the opportunity for conflict can be seen from Starr and Most's (1976, 611) argument that, "while a nation's borders are not readily manipulable by policy makers, its alliances are. It may be reasonable, therefore, to interpret alliance formations (and dissolutions) as attempts by policy makers to alter their border-related risks and opportunities—or, in other words, as efforts to 'create' or 'destroy' borders." By this same line of

Table 4.5 Cross Tabulation of PRIE Versus MID

	PRIE		
MID	0	1	TOTAL
0	581,794	72,074	653,868
1	425	2,577	3,002
Total	582,219	74,651	656,870

Source: Compiled by author.

reasoning, one can interpret alliances as ways of "creating" and "destroying" opportunity for international conflict.

The first way that alliances create opportunity is through alliances between minor powers. An alliance enables a state to militarily reach the neighbors of its ally. Consider Iraq and Israel, the example discussed previously. They have fought in several disputes, and Iraq contributed to the Arab effort in the Yom Kippur War of 1973. How was Iraq able to reach Israel? Through its alliance with Jordan and Syria, neighbors of Israel.

There is one other way that states can reach one another, forming an active dyad. Major powers have the transportation capacity to project power, and thus fight, over great distances. If allied with minor powers, the major powers can use their own transportation assets to transport the minor power's forces to the battlefield as well. Thus, if a state is allied with a major power, it potentially forms an active dyad with all states the major power does.[1] Since the minor power must rely on the major power's transportation assets to reach distant lands, however, there must be some willingness on the part of the major power to do so. This willingness comes in when the major power is in a dispute with another state. Therefore, a state allied with a major power forms an active dyad with all states that the major power is active and engaged in a militarized interstate dispute with (active dyads created this way cease to be active when the dispute is over).

In summary, a dyad is politically active if at least one of the following characteristics applies: (1) the members of the dyad are contiguous, either directly or through a colony; (2) one of the dyad members is a global power; (3) one of the dyad members is a regional power in the region of the other; (4) one of the dyad members is allied to a state that is contiguous to the other; (5) one of the dyad members is allied to a global power that is in a dispute with the other; or (6) one of the dyad members is allied to a regional power (in the region of the other) that is in a dispute with the other.

The cross tabulation of active and MID is shown in Table 4.6, and only 150 dispute-dyad-years are classified as nonactive. Thus, this measure

Table 4.6 Cross Tabulation of Politically Active Versus MID

	POLITICALLY ACTIVE		
MID	0	1	TOTAL
0	461,073	192,795	654,868
1	150	2,852	3,002
Total	461,223	195,647	656,870

Source: Compiled by author.

[1] This can be on either a regional basis (for alliances with regional powers) or a global basis (for alliances with global powers).

accounts for 95.0 percent of all dispute-dyad-years while classifying only 29.8 percent of all dyad-years as being active.

Clearly, politically active dyads more completely capture recorded instances of international conflict than do political relevance, the PRIE, or regional dyads. Furthermore, while active dyads classify significantly more dyad-years as having opportunity than either political relevance or the PRIE, it captures fewer dyad-years than do regional dyads.

Politically active dyads, however, still fail to capture 100 percent of all militarized interstate disputes. Therefore, it is clear that all of these concepts are imperfect indicators of the opportunity for international conflict. The simple response to the observation that dyads considered to lack opportunity have engaged in militarized interstate disputes would be to conclude that none of these concepts captures opportunity as a necessary condition, and therefore, all should be discarded. Such a response, however, ignores the reality that data are inevitably measured with error, as well as ignoring issues of reliability and validity (Braumoeller and Goertz 2000, 848).

Quackenbush (2006a) tests how well each of these measures captures opportunity as a necessary condition for international conflict by using the methodology developed by Braumoeller and Goertz (2000) for testing necessary condition hypotheses. The statement "opportunity is a necessary condition for international conflict" is not the hypothesis to be tested here. This is a theoretical statement that must be true, logically. Rather, the important hypotheses are each of the four measures of opportunity is a necessary condition for international conflict. In principle, these hypotheses state that the proportion of cases where Regional = 0 and MID = 1, Relevant = 0 and MID = 1, PRIE = 0 and MID = 1, or Active = 0 and MID = 1 should be zero. But in practice, the sample proportion of cases in that cell ($\hat{p}$) will depend on the error rate of the data.

Braumoeller and Goertz (2000) propose two tests to determine whether a given independent variable is necessary for the dependent variable. The first is the p_I-Test, which guards against Type I error (rejecting a proposition when it is true). The second test, called the p_{II}-Test, guards against Type II error (failing to reject the proposition when it is false). The results of the p_I-Test and p_{II}-Test for the four measures of opportunity—political relevance, PRIE, regional dyads, and political activeness—are shown in Table 4.7. The proportion of cases in the MID = 1 row that are in the opportunity = 0 cell ($\hat{p}$) is significantly lower ($p < 0.0001$) for political activeness (0.05) than for regional dyads (0.08), political relevance (0.12), or the politically relevant international environment (0.14). This results in a significantly lower p_I as well.

Ideally, the p_I for each definition would be compared to the error rate of the data. Unfortunately, however, the rate of measurement error in these data is unknown. Nonetheless, comparisons can still be made. When faced with unknown measurement error, Braumoeller and Goertz (2000, 852) argue that it is most reasonable to assume an error rate of 5 percent; in this case, only political activeness passes. But in order for regional dyads, relevant dyads, or PRIE to be accepted as identifying opportunity as a necessary

Table 4.7 An Empirical Test of Political Relevance Necessary Condition Hypotheses

H_0: PRESENCE OF . . .	NECESSARY FOR . . .	$K/N = \hat{p}$	P_I*	P_{II}**
Political relevance	MID	357/3002 = 0.12	0.109	0.00[2]
PRIE	MID	425/3002 = 0.14	0.131	0.00[2]
Regional dyad	MID	248/3002 = 0.08	0.074	0.00[2]
Political activeness	MID	150/3002 = 0.05	0.043[1]	0.00[2]

Source: Compiled by author.

*p_I refers to lower bound of 95 percent confidence interval, one-sided test.
**numbers for p_{II}-tests are probability of Type II error given N, H_1: $p = 0.5$, $\alpha = 0.05$.
[1]Pass assuming ≤ 5 percent error.
[2]Do not reject.

condition for international conflict, one would have to assume that the error rate of the data is at least 8, 11, or 14 percent.

Given the large sample size, all three pass the p_{II}-Test with ease. The first step of Braumoeller and Goertz's (2000) methodology for necessary conditions is now complete. The results indicate that one cannot reject the hypothesis that political activeness is necessary for international conflict, but one can reject the hypotheses that political relevance, the PRIE, and regional dyads are necessary.

Using the methodology developed by Braumoeller and Goertz (2000) has enabled fruitful comparison of the proposed concept of political activeness with measures of opportunity used previously. Assuming an error rate of the data of less than 5 percent, neither the standard definition of political relevance, the politically relevant international environment, nor regional dyads is adequate to identify opportunity as a necessary condition for international conflict. Active dyads, however, do identify opportunity as a necessary condition. Furthermore, this is a nontrivial necessary condition. Consequently, one can conclude that while regional dyads, politically relevant dyads, and the politically relevant international environment do not adequately identify opportunity for conflict, politically active dyads do.

Contiguity is an important cause of international conflict because it provides states with the opportunity to fight. Contiguous states, however, are not the only ones able to fight. Politically active dyads are the best way available to identify which pairs of countries are able to fight; power status and alliances are also important for providing opportunity.

Territorial Issues and International Conflict

Issues of disagreement between states have often been ignored in research. The prominent realist scholar Morgenthau (1967, 5) explicitly recommended "guard[ing] against two popular fallacies: the concern with motives and the concern with ideological preferences." According to realism, since power is central to international politics, other issues such as territory are irrelevant. The importance of issues, however, is highlighted by the process of war framework discussed in the first chapter. Recall that the process begins with underlying context, with the next step being a conflict of interest, which can then become militarized in future steps (or not). These conflicts of interest are what we mean by issues: something that states disagree about. Furthermore, the issues under contention can have an important effect on the likelihood that conflicts of interest become militarized and escalate to war (Diehl 1992).

To examine the importance of issues, we begin by reviewing different ways to identify issues of disagreement, especially territorial, between states. We then review research that has explored the relationship between territory and international conflict before examining the reasons why territorial issues are so conflict prone.

Identifying Issues of Disagreement Between States

One way to examine the impact of issues—such as territory—on international conflict is through the militarized interstate disputes data set. One of the variables included in the MID data identifies revisionist states. A revisionist state is a state that is seeking to revise the status quo in some way—in other words, a revisionist state wants some modification to the current relationship between the states. More than one state in a MID may be a revisionist actor; also, in some MIDs, no state is coded as revisionist. Along with this variable is another important variable, the revision type.

The revision type identifies what type of revision the revisionist state desires. There are five categories of revision types in the MID data: not applicable, territory, policy, regime/government, and other. The not applicable category covers cases in which the state is not considered revisionist; if the state is not revisionist, then it is not seeking any particular type of revision. The territorial category refers to an attempt by the revisionist state to gain control over a piece of land that it claims but does not effectively possess. For instance, there has been a series of disputes over the Spratly Islands among China, Taiwan, the Philippines, Malaysia, and Vietnam because the island group in the South China Sea is claimed by each state. Each of these disputes is over a territorial issue.

The policy category captures cases where the revisionist state seeks to change the foreign policy behavior of another state. For example, in the Kosovo War of 1999, the United States and its NATO allies sought to change Serbia's policy of ethnic cleansing in Kosovo. Thus, this is coded as a policy revision type. The regime/government category identifies the desire by the revisionist state to change the government of another state. An example is the invasion of Afghanistan by the United States and allies in 2001, where the invading forces sought to remove the Taliban regime from power. Finally, the other category is used when the revisionist state's objective was ambiguous.

Table 4.8 shows the relationship between territory and militarized conflict. The top portion of the table shows the relationship between territory and militarized disputes. Over the entire 1816–2001 time span of the data, 26.5 percent of disputes (708 of 2,671) involved territorial issues; the percentages for the pre- and post–World War II time periods are nearly identical. If each of the five different revision types was equally represented, then we would expect 20 percent of disputes to be over territory (and 20 percent over policy issues, etc.). The actual percentage is therefore stronger than that, but this is not a particularly strong relationship.

The real strength of **territorial issues** as a cause of conflict is revealed when we look at the relationship between territory and wars. Between 1816 and 2001, fully 47.7 percent of wars (61 of 128) involved territorial issues. This percentage is slightly smaller (45.1 percent) prior to 1945, and slightly larger (54.1 percent) after. While territorial issues account for less than a third of all militarized disputes, they account for about half of all wars. Thus, territorial issues are particularly likely to lead to war.

An important limitation of identifying territorial (or other) issues through the MID data is that we only observe the issue under contention when states are actively in a militarized dispute. However, issues are relevant during times of peace as well, as highlighted by the process of war. Accordingly, it is important to be able to identify issues more broadly than only during actual conflicts. Fortunately, there are a couple of options for doing so.

Huth has collected data on territorial claims by states against other states, identifying 348 cases of disputed territory between states between 1919 and 1995 (Huth 1996; Huth and Allee 2002). Territorial claims can arise in one of five different ways. The first type of territorial claim is where "at least one government does not accept the definition of where the boundary line of its border with another country is currently located, whereas the neighboring government takes the position that the existing boundary line is the legal border between the two countries based on a previously signed

Table 4.8 Territorial Issues and Militarized Conflict

TERRITORIAL ISSUE?	1816–1945	1946–2001	1816–2001
Militarized interstate disputes			
Yes	253 (26.4%)	455 (26.6%)	708 (26.5%)
No	706	1,257	1,963
Total	959	1,712	2,671
Interstate wars			
Yes	41 (45.1%)	20 (54.1%)	61 (47.7%)
No	50	17	67
Total	91	37	128

Source: Compiled by author.

treaty or document" (Huth 1996, 19). For instance, China has disputed territory with both India and Russia by arguing that nineteenth-century agreements regarding the borders were illegitimate. Such cases account for about 19 percent of territorial claims.

The second type is one in which "there is no treaty or set of historical documents clearly establishing a boundary line, and, as a result, bordering countries present opposing definitions of where the boundary line should be drawn" (Huth 1996, 20). This is the most common type of territorial claim, representing about 39 percent of all cases. For example, Saudi Arabia has had territorial disagreements with several of its neighbors (Iraq, Kuwait, UK/Qatar, and UK/United Arab Emirates) because historical documentation on the location of their borders was limited and vague.

The third type of territorial disagreement is when "one country occupies the national territory of another and refuses to relinquish control over the territory despite demands by that country to withdraw" (Huth 1996, 21). For instance, following the Six-Day War, Israel occupied territory of Egypt (the Sinai Peninsula) and Syria (the Golan Heights). Although the Sinai was returned to Egypt in 1982, Israel still occupies the Golan Heights. About 18 percent of territorial claims are similar cases.

The fourth type occurs when "one government does not recognize the sovereignty of another country over some portion of territory within the borders of that country" (Huth 1996, 21). The disagreement between India and Pakistan over Kashmir is such a case; similar cases account for about 15 percent of all territorial claims.

The final type of territorial claim is where "one government does not recognize the independence and sovereignty of another country (or colonial territory) and seeks to annex some or all of the territory of that country" (Huth 1996, 22). For example, Arab states such as Egypt, Syria, and Jordan refused to recognize Israel after Israel declared independence in 1948; similarly, West Germany refused to recognize East Germany as an independent country until 1972. Although these examples are among the most prominent of territorial disputes, such cases account for only 12 percent of territorial claims, making them the least common type.

Table 4.9 summarizes the territorial claims by region and shows the militarized confrontations (disputes and wars) that have emerged from them. Europe has had 95 territorial claims between 1919 and 1995, more than any other region. The Middle East had the second most territorial claims, 89, but by far the most militarized disputes (130) and wars (15) stemming from territorial claims of any region. Africa and the Americas had the fewest territorial claims, with 48 and 51 respectively.

Huth (1996) examines the causes of territorial disagreements between states. In other words, what factors lead states to disagree about territory? He finds that the issues at stake, international context, and domestic context all matter greatly. Relating to issues, when there is strategically or economically valuable territory near the border between two countries, a territorial claim between them is much more likely. If a state has previously lost territory to its opponent, then it becomes significantly more likely to initiate

Table 4.9 Military Confrontations Over Disputed Territory, 1919–1995

REGION	TERRITORIAL CLAIMS	NUMBER OF MIDS	NUMBER OF WARS
Europe	95	56	9
Middle East	89	130	15
Africa	48	27	3
Asia	65	109	12
Americas	51	52	1
Total	348	374	40

Note: The totals listed for militarized disputes include only those initiated by challenger states while the totals for wars include all military confrontations in which both challenger and target states resorted to the large-scale use of force.

Source: Paul K. Huth and Todd L. Allee, *The Democratic Peace and Territorial Conflict in the Twentieth Century* (Cambridge: Cambridge University Press, 2002), 30.

a territorial claim. On the other hand, if states have already reached some prior agreement clearly defining their borders, or they are allies, then they become less likely to have a territorial disagreement.

Paul Hensel and Sara Mitchell have led the Issues Correlates of War (ICOW) project, which seeks to collect broader data on disagreements between states about various issues. So far, they have collected data on territorial, maritime, and river claims (Hensel, Mitchell, Sowers, and Thyne 2008). In the future, they may collect data on regime, identity, or other types of claims. These data are important because they enable us to better account for the impact of various issues beyond just territory on the likelihood of international conflict.

The ICOW project identifies 122 distinct territorial disagreements in the Western Hemisphere and Western Europe between 1816 and 2001. There have been 772 attempts to settle territorial claims, with bilateral negotiations being the most often technique used to attempt settlement (59.7 percent of the time), followed by third-party activities (23.3 percent) and militarized conflict (17.0 percent). Table 4.10 shows how successful the different settlement attempts have been. Binding third-party arbitration or adjudication is by far the most likely to succeed, about 77 percent of the time. Bilateral negotiations and nonbinding third-party mediation are successful at a much lower rate, less than 20 percent of the time. But by far the least likely attempts to work are militarized conflict, which succeed less than 10 percent of the time (Hensel 2012).

Exploring the Relationship Between Territory and International Conflict

Substantial research has shown that territorial issues significantly increase the likelihood of international conflict and are consistently more conflict

Table 4.10 Ending of Territorial Claims

	DID SETTLEMENT ATTEMPT END MOST/ALL OF CLAIM?		
SETTLEMENT TECHNIQUE	*No*	*Yes (%)*	*Total*
Militarized conflict	185	20 (9.8)	205
Bilateral negotiations	312	68 (17.9)	380
Third party: Nonbinding	119	29 (19.6)	148
Third party: Binding	9	30 (76.9)	39
Total	625	147 (19.0)	772

Source: Paul R. Hensel, "Territory: Geography, Contentious Issues, and World Politics." In *What Do We Know About War?*, 2nd ed., ed. John A. Vasquez (Lanham, MD: Rowman and Littlefield, 2012), 19.

prone than other issues. In an initial look, Senese (1996) finds that, even while controlling for contiguity, territorial issues increase the likelihood that militarized disputes will escalate to higher levels of severity. In subsequent work, Senese and Vasquez (2003) find that territorial claims increase the likelihood of dispute initiation, and territory further raises the probability of escalation to war. These strong effects of territory hold even while controlling for selection effects.

Senese (2005) further explores the relationship between territory and contiguity, and finds that while territorial issues make both dispute onset and escalation to war more likely, contiguity only makes dispute onset more likely, with no significant effect on war escalation. These results make sense from an opportunity and willingness framework (Most and Starr 1989): Territory affects willingness, which matters at all stages of the steps to war; on the other hand, contiguity affects opportunity, and once states are already fighting, opportunity is largely irrelevant.

Huth (1996) finds that disputed territory is much more likely to lead to militarized conflict when it is strategically located, there are ethnic ties between the two states, the state challenging the territory is stronger than the opponent, there is a stalemate in negotiations, or there is a history of militarized disputes between the countries. If the disputed territory is economically valuable, there are alliances, or the challenger is a democracy, however, then militarized conflict becomes less likely.

What Makes Territorial Issues So Conflict Prone?

We have seen that one of the key reasons for the conflictual nature of contiguity is that contiguous states are more likely than other states to have disagreements about territory. Furthermore, we have seen that territorial issues are more likely to lead to militarized conflict than are other types of issues, and that disputes over territory are the most likely to escalate to war.

But what is it about territory that makes disagreements about it so prone to conflict?

Vasquez (1995) has argued that a key answer to this question is territoriality. Territoriality is "the tendency for humans to occupy and, if necessary, defend territory" (283). According to Vasquez (1993), this territoriality is an instinct inherited through evolution and is similar to the territoriality of many other species of animals. This begins to provide an explanation of why territorial issues are so conflict prone, although most scholars do not focus on these biological aspects of territoriality.

Huth (2000) identifies four other key reasons that territorial disputes are so conflict prone. First, states place a high utility on controlling disputed territory. There are different reasons why territory can be highly valued. Some territory is valued for strategic reasons. For example, Israel considers controlling the Golan Heights on their border with Syria to be very important because it provides ideal terrain to defend invasion; the flat, open terrain behind the heights is much harder to defend. Other territory is highly valuable for economic reasons, such as being endowed with important natural resources. Finally, territory also has symbolic value to states, particularly when there is historical, cultural, or religious significance to it. Thus, it is more likely that the perceived benefits of fighting over territorial issues will outweigh the costs for states. Because states place such a high utility on controlling territory, the presence of contested territorial issues between states lowers the utility of the status quo.

Second, military force is very effective in occupying and controlling territory (Huth 2000). Indeed, acquiring territory is what armies are designed to do. How does a state with a policy disagreement with another country use force to resolve the disagreement? States can bomb or launch some other kind of quick strike trying to convince the opponent to change their policy. For instance, Operation Desert Fox was a four-day series of airstrikes by the United States against Iraq in December 1998 in an attempt to coerce Iraq into cooperating with UN weapons inspections. Such methods, however, are very indirect. Although air strikes sometimes succeed in coercing changes in policy (such as the Kosovo War, in which NATO eventually succeeded in getting Serbia to end its ethnic cleansing campaign), larger scale ground invasions are more likely to succeed. For example, the Soviet Union invaded Hungary in November 1956 to crush Hungarian policies of nationalist, anti-Soviet reform. The use of military force is a direct way of dealing with territorial issues, while it is only an indirect way of dealing with other types of issues. Accordingly, military courses of action are more attractive to state leaders in dealing with territorial issues than with other kinds of issues in which force is less effective. The effectiveness of military force in occupying and controlling territory also raises the relative utility of conflict and lowers the relative utility of the status quo.

Third, it is easier for foreign policy leaders to mobilize domestic support behind territorial claims. This salience of territorial issues to domestic populations generates audience costs, which are negative repercussions that leaders face for honoring a commitment or failing to follow through on a threat. Audience costs such as these reduce the utility of backing down (Fearon 1994; Schultz 2001). Thus, when territory is in dispute it is harder for states to back down (Huth 2000).

Fourth, Huth (2000) argues that territorial issues are more conflict prone because territorial challengers are likely to be nondemocratic. This argument relies on the assumption that democracies are more peaceful than nondemocracies, and in particular are less likely to initiate conflict. This assumption is questionable, however, as we will see in our examination of the democratic peace in chapter 7.

Conclusion

Contiguity, as found by Bremer (1992), is the number one factor making dyads more dangerous, and it has been consistently found to increase the likelihood of international conflict because it provides the opportunity for states to fight one another. Although several different concepts have been developed to identify opportunity for international conflict, politically active dyads provide the best measure currently available.

Territory is an important cause because leaders have proven to be particularly willing to fight over territory. A large amount of research shows that territorial issues are more conflict prone than other types of issues. The evidence indicates that while territorial issues make both dispute onset and escalation to war more likely, contiguity only makes dispute onset more likely, with no significant effect on war escalation. This supports the idea that territory affects willingness, while contiguity affects opportunity. Territorial issues are so conflict prone not only because of the importance that people attach to territory, but also because military options are seen as more directly addressing territorial, as opposed to policy or regime, issues.

Key Concepts

Contiguity

Loss of strength gradient

Opportunity

Politically active dyads

Politically relevant dyads

Politically relevant international environment (PRIE)

Regional dyads

Territorial issues

Willingness

5 Power

A Soviet matchbox label with a fist smashing an American aircraft illustrates how the Cold War power struggle between the Soviet Union and United States pervaded society from the end of World War II until its symbolic end with the fall of the Berlin Wall in 1989.

Source: © Blue Lantern Studio/Corbis

To many, power is the key to understanding international politics. Accordingly, power has been a central element of international relations theories for many years. This focus can be traced back as far as Thucydides, who wrote *History of the Peloponnesian War,* detailing the war between Athens and Sparta in ancient Greece. Thucydides argues that power is the central element to explaining the actions of Athens and Sparta. Similarly, the classical realist Morgenthau (1967, 25) argued that "international politics . . . is a struggle for power."

In this chapter, we examine the impact of power on international conflict. We begin by considering the important questions of how to define and measure power. We then move on to discuss realist theory, focusing in particular on theories regarding the impact of polarity on stability and the offense-defense balance. We then examine power transition theory, an important nonrealist theory.

Defining and Measuring Power

Before examining the effect that power has on international conflict, it is important to discuss what we mean by power. One difficulty in doing so is that power has been viewed in many different ways. There are two primary ways to define **power,** as the ability to control or as resources (Rothgeb 1993). We discuss each in turn.

Power as the Ability to Control

The most common way that people define power in international relations is the **relational definition of power.** In this view, given two states (or people or organizations) A and B, "A has power over B to the extent that [A] can get B to do something that B would not otherwise do" (Dahl 1957, 202–3). In other words, power is all about the ability to control or influence others. Power exists only when a state exercises control or influence, so it can only be measured after the outcome is determined. The most powerful state is the one that prevails in a dispute.

There are several different ways that states seek to influence other states, or exercise power in the relational sense (Bueno de Mesquita 2006, 236–48). First, states can utilize persuasion, where A tries to convince B that B really should do, out of its own free will and interests, what A wants it to do. This is the art of campaigning on behalf of one's cause. Such diplomacy is at the heart of international politics. To persuade nation B, nation A might appeal to sentiments, values, ethics, morals, and principles; it might try to bring out facts that B has overlooked; or it could point to rewards and punishments. Most actors continually exercise persuasion in international relations. While this is the cheapest method of exercising power, it is not cost free. The most persuasive "talk" involves costs to oneself. This is known as a costly signal, which is a more credible signal. "Cheap talk" is often ignored because it could easily be a bluff.

Second, states can use rewards, where A offers B rewards for changing its actions in accordance with what A wants. In other words, A offers some benefit if B acts in a particular way. Such rewards are known as carrots. There are a variety of different types of rewards, especially political, economic, and military. Rewards rely on alteration of B's cost-benefit calculus by tipping the balance for B in favor of some outcome favorable to A.

Third, states can rely on punishments. With punishment, A threatens B with punishment if B does not do what A wants. In contrast to the carrot of reward, punishment is the stick. This represents a clear escalation in tension between the parties. Punishments are actions that have unpleasant consequences for the punished party. There is a fine line between rewards and punishments: Withholding a reward can be a punishment, while abstaining from a punitive action is a reward. Although separate in the interests of explication, persuasion and rewards are often combined.

The final way for states to exercise power is force, where A takes direct action to force B to do what A wants against B's will. Nations seek to control the behavior of another by coercing it through the use of superior military power. Force is frequently used when elites believe that there is no other way to change the other party's behavior. There is a critical difference between force as punishment and force as compulsion. The former is used to change B's costs and benefits of action, while the latter is used to take away B's ability to choose. There is a profound difference between the first three methods and the last: While the first three allow B to choose to act as A desires, with force, A tries to take away B's ability to choose.

The relational definition of power has several problems. First, how do we know what someone would "otherwise not do"? This puts a burden on establishing counterfactual history (establishing what would have happened in an alternative sequence of events) that analysts are often unable to meet. Second, how can we even be sure that a change in behavior has actually occurred? International politics is rife with secrecy, uncertainty, and incomplete information; changes in the behavior of other states are not always readily apparent. Third, whose power should be credited with that change in behavior? The general examples of A influencing B are nice to illustrate the ideas, but there are more than two countries in the real world. If states A, C, and D are all independently trying to influence B, and B changes its behavior, how do we know whose power was responsible for the change?

The relational definition of power can also lead to nontransitive cycles that are problematic. To illustrate, consider the US bombing of Libya in retaliation for Libyan support of terrorist groups in April 1986. For the mission, the United States used F-111 fighter-bombers based in Britain, but the aircraft were denied flyover rights by France.[1] French reluctance was due to fear of becoming a target of Libyan terrorists or a Libyan oil embargo in retaliation for supporting the US strike. What can be learned from the bombing of Tripoli about the power of the parties involved? A relational view of power would judge that the United States was more powerful than Libya because terrorism declined after the bombing. Libya changed its behavior because of American actions. (But is this causation or just correlation? The difference in Libyan behavior is not necessarily proof that it was a result of the American bombing, but it is plausible that it was.)

Second, one could conclude that France was more powerful than the United States. The French government's denial of flyover rights made the United States fly all the way around Europe and enter the Mediterranean through the straits of Gibraltar, which was something it would not have done otherwise. Finally, based on the relational definition of power, Libya was more powerful than France because Libya was able to get France to deny flyover rights, which it would otherwise not have done.

This power cycle that results demonstrates a fundamental problem with the relational definition of power. I would argue that in 1986, the United States was more powerful than France, which in turn was more powerful than Libya. While the United States did not get everything it wanted (such as permission to fly over France), one could also point out that it accomplished its primary objective: bombing Libya.

The root of the difficulty with viewing power as control is that causes within definitions are problematic. If the cause is true by definition, then we cannot use that concept to explain. For example, if we determine that Prussia was stronger than Austria in 1866 because it won the Seven Weeks War, we cannot in turn explain Prussia's victory by arguing that it was more powerful. We might be interested in examining the hypothesis: If A is more powerful than B, it will be more likely to get B to do what it wants. With the relational definition of power, however, this is true by definition, and thus the hypothesis is meaningless.

Power as Resources

The other primary way to view power is the **material basis of power**. According to this view, "power represents nothing more than specific assets or material resources that are available to a state" (Mearsheimer 2001, 57). This is a very different way of viewing power than the relational definition. While there is much debate about which approach to power is better, viewing it as resources has several advantages.

First, the balance of power is not a reliable predictor of military success. This is, of course, contrary to the claim of the relational definition. This contradiction occurs because power is not the only thing that determines

---

[1] Spain also denied the United States flyover rights. For simplicity, I focus on France in the discussion of the case, although Spanish reasoning was similar.

outcomes. We will consider this more closely when we examine war outcomes in chapter 11, but nonmaterial factors—such as strategy, intelligence, resolve, weather, or disease—sometimes provide one combatant with a decisive advantage.

There are additional reasons not to conflate power and outcomes. When focusing on outcomes it becomes almost impossible to assess the balance of power before a conflict, since the balance can be determined only after we see which side wins. This is problematic because it undermines our ability to measure power in any reliable and consistent way. In addition, the relational definition of power approach leads to implausible conclusions. After Napoleon decided to invade Russia in 1812, Russia beat France even though France was more powerful. Similarly, the United States failed to defeat North Vietnam in the Vietnam War, but by any other way of assessing power, the United States was stronger.

One of the most often-studied aspects of international relations is how power affects political outcomes. But there is little to say about the matter if power and outcomes are indistinguishable because there would be no difference between cause and effect. By viewing power simply as the resources available to the state, we can examine the relationship between power and outcomes because the two are then completely distinct.

I argue that viewing power as resources provides a much better way of accounting for power in international politics because focusing on the material basis of power avoids the problems associated with the relational definition. Interestingly, most studies that adopt the relational view in their conceptual definition of power turn to the resources view when they measure power. It would be more consistent and straightforward if we adopted the material basis of power in the first place.

Elements of Power

Highly related to questions of how to define power is consideration of how one can measure power. First, there are different types of power that are important in international politics, especially military power and economic power. This is particularly important when adopting the power as resources view because one must identify which resources are important elements of national power.

Scholars have identified a variety of different sources of power, including both tangible and intangible assets. Tangible assets are ones that can be counted. Intangible assets affect power but are much harder to measure precisely. Morgenthau (1967) devotes an entire chapter to discussing elements of national power. We discuss each of the nine elements that he identified in turn.

The first factor Morgenthau identifies is geography. Being surrounded by natural barriers such as rivers, mountains, or oceans is particularly important. This provides a passive source of power by making countries safer from attack. Despite facing a series of European enemies such as France under Napoleon and Germany under Hitler, Great Britain has not been invaded since 1066. Geography is the most stable element of national power that Morgenthau identifies.

Secondly, Morgenthau identifies natural resources as an important element. The key natural resources for international power are food and raw materials. Raw materials are important for manufacturing, equipping, and transporting military equipment. Of particular importance are energy sources (coal and oil) and metals (iron, aluminum, uranium, etc.). If an enemy can cut off your food (or oil, iron, etc.) supplies, it is difficult to challenge that enemy. Therefore, countries that are self-sufficient in food and natural resources have an advantage over those who must import crucial supplies. The United States and Russia have been very strong because they are blessed with large supplies of natural resources. Middle Eastern states such as Iran and Iraq grew to much greater importance in international politics because of their control of large quantities of oil.

Industrial capacity is the next element of power according to Morgenthau. Along with industrial capacity, the country's economy in general is a crucial source of power. Industrial capacity is important for the development and manufacture of weapons and technology. Great Britain and then the United States grew to be the most powerful countries in the world on the back of their massive industrial capacities. Similarly, the rise of China as a world power in recent decades has been driven by its tremendous economic growth. Brazil and India are also large states that seem poised for greater power in international relations following continued industrial and economic development.

Next, Morgenthau identifies military preparedness as a crucial element of power. This refers to a variety of characteristics, especially the technology, leadership, quantity, and quality of the country's armed forces. Military power is the only way to survive the ultimate challenge of war. All else being equal, large military forces can overwhelm small ones. Better technology and leadership, however, are key ways that small militaries can overcome much larger opponents.

The fifth element of power is population. Important here is not only the size of the population but also population growth trends. Having a large population is essential to having a large military. A well-educated population is also important because it is necessary for technological and economic development. States whose population is growing at a faster rate than others will become increasingly powerful over time, while countries with low growth rates will see a relative decline in power.

Morgenthau argues that national character has a "permanent and often decisive influence upon the weight a nation is able to put into the scales of international politics" (1967, 122). For example, he argues that German national character is marked by discipline and thoroughness, American character by individual initiative and inventiveness, British by common sense, and Russian character is said to be persistent. Yet do such inherent nationwide characteristics truly exist? There are many reasons to doubt that they do. As such, analysts have largely discarded national character as an element of power.

Another intangible characteristic that Morgenthau identifies as an element of power is national morale. By national morale, he means whether or not the citizens of a state are optimistic or support their government's military and foreign policy efforts. This refers to the will, or resolve, of the population when facing difficulties in international politics. For instance, the British people continued to support the war effort in World War II even after France had been conquered and Britain alone actively opposed Germany.

Morgenthau argues that the quality of diplomacy is the most important element of power. Morgenthau (1967, 135) defines quality of diplomacy as "the art of bringing the different elements of national power to bear with maximum effect upon those points in the international situation which concern the national interest." He argues that the British have had consistently high quality diplomacy while Germany's diplomacy has mostly been low quality. Rational choice theory highlights the importance of decision making and diplomacy, although many would consider them as something separate from power, rather than an element of it.

The final element of power that Morgenthau identifies is quality of government. The quality of government means the government's ability to balance resources and policy, as well as the ability to garner public support both domestically and internationally. As with diplomacy, government quality is an essential element for translating material resources into tangible policy achievements. Many scholars argue that democracies provide better government quality than nondemocracies, an insight that we will return to in chapter 7.

Measuring Power

The elements of power reviewed in the previous section provide ideas for what factors determine state power, but they are not sufficiently precise to provide an actual measure of power. In order to actually measure power, we need much more precision. There are two primary measures of power used in studies of international relations.

The most common measurement of power used in the study of international relations is the **Composite Indicator of National Capabilities (CINC)**, collected by the Correlates of War Project (Singer 1988; Singer, Bremer, and Stuckey 1972). This measure is available for each state annually from 1816 through 2007. It is an index that draws data on six individual indicators of power: military expenditures, military manpower, iron and steel production, commercial fuel consumption, urban population, and total population. Each of these indicators focuses on an important element of power; the composite indicator combines these together to provide a more complete measure of state power.

Military expenditures and military manpower focus on the military dimension of power. The number of military personnel is of obvious importance because a larger military is a stronger one, all else being equal. All else is never equal, however, because there is a wide disparity in the quality and preparedness of different states' militaries. Military expenditures provide a way of getting at this disparity because technology and training are expensive. Therefore, this provides a useful indicator of quality.

Iron and steel production and commercial fuel consumption focus on the economic dimension of power. Singer and colleagues (1972) focus solely on iron and steel production, rather than including additional commodities such as aircraft and electronics, because it provides a measure of industrialization that is appropriate for comparisons across the entire time period of the data. Commercial fuel consumption includes a variety of different sources such as coal, hydroelectric, and nuclear power, and is important because industry requires energy of some kind to operate.

Urban population and total population focus on the demographic dimension of power. Total population is important because large populations are necessary to sustain large militaries and economies. Urban population is a useful indicator of the education level and economic development of a country because as states develop from an agrarian economy to an industrial economy, there is a large shift from rural to urban population. Along with increasing the level of economic development, a larger relative urban population increases the level of technology available to a country and provides people who are more readily mobilized for a potential war effort.

The other common measure of power used in studies of international politics is simply a state's gross domestic product (GDP). Measuring state power this way focuses entirely on the economic dimension of power. Population is accounted for indirectly, however, because it takes a large population to have a large GDP. While small countries like Luxembourg or the United Arab Emirates can be very wealthy, with quite large GDP per capita, their total GDP is dwarfed by large states. Military strength is also indirectly accounted for because military forces and technology are expensive. Rich countries can always develop military capability even if their current military strength is low, whereas poor countries are always limited in their ability to field strong military forces.

Mearsheimer (2001) argues that CINC score is a good measure of states' current power, while GDP is a good measure of states' latent, or potential, power. The United States on the eve of World War II had a very large economy but a small, weak army. During the interwar period, the United States had about 250,000 military personnel, and even in 1939 had only 334,000. By 1942, the first full year of American participation in the war, personnel strength had grown to 3.9 million and continued to grow to over 12 million by 1945. Similarly, US military expenditures grew from only $980,000 in 1939 to an astounding $90 million in 1945. These tremendous rates of growth were only possible because of the massive size of the American economy, the largest in the world.

Figure 5.1 shows the power of the United States, Britain, and Russia/Soviet Union (using CINC scores) from 1816 to 2007. We can see that at the beginning of the time period, Britain was much stronger than the others while the United States was the weakest. American power, however, was growing throughout the century while Britain started a downward trend about 1850. The United States passed Russia in power around 1880, and then surpassed Britain about 1900.

Britain continued to decline throughout the twentieth century, while the United States had massive peaks of power during the world wars, declining

Figure 5.1 US, British, and Russian Power, 1816–2007

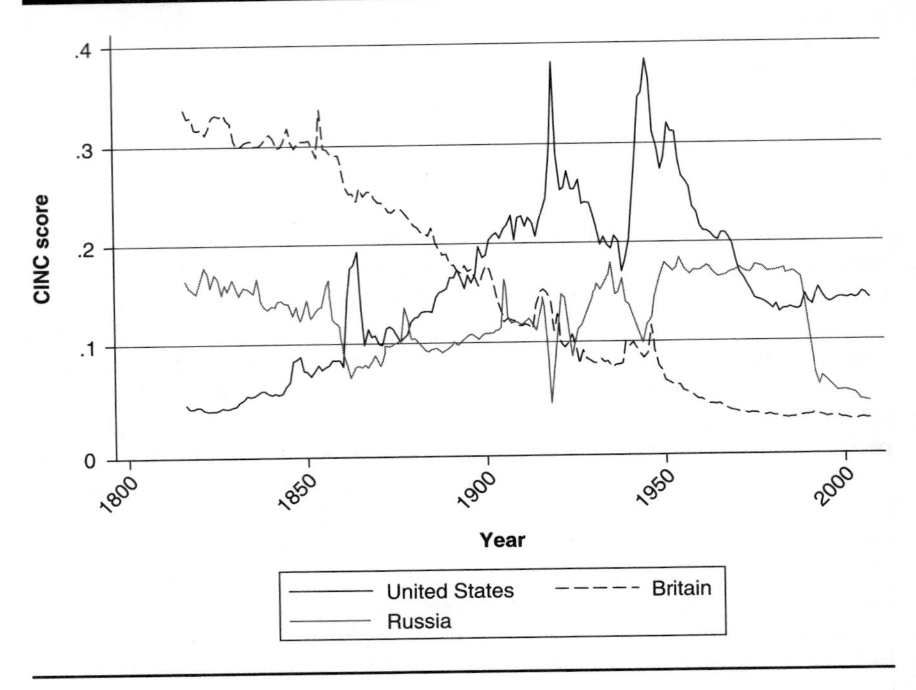

Source: Compiled by author.

to a new stable level by about 1975. Russian power grew and surpassed that of the United States during the Cold War—however, this is as measured by CINC; the GDP of the United States remained much higher than Soviet GDP throughout. Thus, using CINC scores to measure power can yield different conclusions about the relative power of states than using GDP to measure power. Following the breakup of the Soviet Union, Russian power plummeted to a level just larger than Britain's.

Polarity and Stability

We begin our examination of power as a cause of international conflict by looking at the relationship between polarity and the stability of the international system. **Polarity** refers to the number of power centers in the international system at a given time. This topic has been closely associated with realism, of which there have been different strands developed over the years. See Box 5.1 for a brief explanation.

The strongest realist argument about the impact of systemic polarity on stability has been made by Waltz, the father of neorealism. Neorealism is

Box 5.1

Concept in Focus: Realism

Realism is a traditional approach to studying international relations that focuses on the importance of power and the structure of the international system and traces its intellectual roots to writers such as Thucydides and Machiavelli. There are a variety of different variations of realism, and we will discuss the three most common variants here.

The most basic type of realism is classical realism. Hans Morgenthau is considered the father of realism and his book *Politics Among Nations,* first published in 1948, is the ultimate statement of classical realism. In particular, Morgenthau established six principles of realism, which laid out his central ideas that the key to understanding "international politics is the concept of interest defined in terms of power" (1967, 5), that states seek to maximize power, and that leaders of states as well as scholars should concern themselves with reality, not moral principles.

Classical realism (Morgenthau)	• Lust for power is inherent in human nature • States seek to maximize power
Neorealism, or defensive realism (Waltz)	• Anarchy and structure of the international system lead states to maximize security
Offensive relism (Mearsheimer)	• Anarchy and structure of the international system lead states to seek to maximize power • The ultimate goal is hegemony

Kenneth Waltz introduced a major new variant, called neorealism, structural realism, or defensive realism. Waltz's major work was *Theory of International Politics,* published in 1979. He argues that anarchy and the structure of the international system lead states to seek to maximize security. The argument that states maximize security, rather than power, makes defensive realism unique from other variants.

The third most common variant is offensive realism, developed in particular by John Mearsheimer in his book *The Tragedy of Great Power Politics,* published in 2001. Mearsheimer argues that anarchy and the structure of the international system lead states to seek to maximize power, with hegemony as their ultimate goal.

Realist theories and ideas—such as balance of power—have been prevalent not only in academic studies, but also in the rhetoric of politicians speaking about foreign policy. Hence, it is important to be familiar with the primary variants of realism.

based on several basic assumptions (Waltz 1979). First, neorealists assume that international politics is anarchic. **Anarchy** is the absence of government. While different states certainly have governments, there is no world government that rules over them. Thus, states are unable to appeal to a higher authority to resolve conflicts or enforce agreements between them. According to realists, this creates an international system in which might makes right, as each state must rely on itself to defend its interests.

The second basic assumption is that states, as rational unitary entities, are the central actors in international politics. While there are other actors—such as international organizations, nongovernmental organizations, and multinational corporations—that are involved in international politics, neorealists argue that none of them are nearly as important as states.

Third, Waltz assumes that states seek to maximize their security. This marks a decisive break from classical realism, where Morgenthau assumes that states seek to maximize their power. However, power maximization can lead to a phenomenon known as the **security dilemma**, which occurs when one state's attempts to increase its security decrease the security of others. Thus, when states attempt to maximize their own power, they form a threat to other states who then ally against them, making the state less secure.

Finally, Waltz assumes that states seek to increase their power as long as this does not place their security at risk. In contrast, Mearsheimer (2001) argues that the structure of the international system and anarchy lead states to maximize power since that is the only way to make themselves more secure. This is the central difference between offensive and defensive realism. Their logical incompatibility, however, also reveals that at least one of their theories is wrong since it is not possible for structure to simultaneously make all states maximize security and make all states maximize power.

Waltz argues that these assumptions lead to several different hypotheses. First, the distribution of power tends to be balanced. This is connected with **balance of power theory**, where parity between states is expected to lead to peace while preponderance is expected to lead to conflict. While Morgenthau (1967) and other classical realists argue that states must actively take steps to ensure a balance of power is formed, Waltz (1979) instead argues that system structure ensures that a balance will emerge regardless of states' efforts.

Waltz also argues that states mimic each other's behavior. They do so because the important determinants of state behavior are assumed to be the structure of the international system and the state's level of power, rather than any individual idiosyncrasies of states such as their form of government or leaders' preferences. Finally, neorealism argues that bipolarity is more stable than multipolarity. This is the hypothesis we will focus on here, so we turn to a more detailed exploration.

Bipolarity and Multipolarity

We are going to focus on the expected and observed differences between bipolarity and multipolarity. The international system at different points in time can be classified according to polarity. A bipolar system has two poles, whereas a multipolar system has more than two poles.

Polarity can be classified in two primary ways (Garnham 1985). Power polarity refers to the distribution of power between states. Cluster polarity refers to the distribution of power between alliances. We can combine these different types to identify four categories of systems, summarized in Figure 5.2. First, a system can be power bipolar and cluster bipolar, where there are two superpowers that are much stronger than any other states and the system is broken into two opposing alliance blocs. The Cold War was such a system, featuring the United States along with its NATO and other allies against the Soviet Union and its allies in the Warsaw Pact and elsewhere.

A system can be power multipolar and cluster bipolar, which occurs when there are multiple great powers but they are aligned into two opposing alliance blocs. The international system before World War I was arguably like this, as Britain, France, and Russia were aligned together in the Triple Entente while Germany, Austria-Hungary, and Italy were grouped together in the Triple Alliance.

A system could also be power bipolar but cluster multipolar. For such a system to occur, there would have to be two superpowers (because it is power bipolar), but there would have to be a third (or even a fourth) bloc of countries that are aligned together but separately from the two superpowers. For example, China's move away from the Soviet Union and the rise of the Group of 77 in the 1970s created additional power blocs that were not aligned with either of the superpowers during the Cold War.

Finally, a system could be both power and cluster multipolar. This occurs when there are multiple great powers that also form multiple blocs of power. For example, in post-Napoleonic Europe during the 1800s, Britain, France, Russia, Austria, and Prussia were all great powers that largely were not aligned together. It is important to be familiar with these different approaches to polarity. Most people, however, focus on power polarity: the number of great powers in the international system.

Why are bipolar systems expected to be more stable than multipolar systems? Waltz (1964) argues that, essentially, multipolarity allows

Figure 5.2 Categories of Polarity Systems

Power Bipolar + Cluster Bipolar	Power Multipolar + Cluster Bipolar	Power Bipolar + Cluster Multipolar	Power Multipolar + Cluster Multipolar
• Two superpowers, much stronger than any other state • Two opposing alliance systems centered on the superpowers	• Multiple great powers • Aligned into two opposing blocs	• Two superpowers • One (or more) bloc of nations aligned together and separate from them	• Multiple great powers • Multiple blocs of power centered on the great powers

recklessness and errors by failing to stop reckless or challenging behavior. First of all, spirals of hostility and the security dilemma are expected to be more dangerous under multipolarity. In a multipolar system, there are more opportunities for spirals of hostility and the security dilemma to operate because any state could potentially be an enemy of another state. But in a system with two major powers, it is clear who the enemy will be because there is only one possibility.

The next reason that neorealism expects bipolar systems to be more stable than multipolar ones is due to alliance problems. When there are many big states, there are many opportunities for alliances to form, and allies are very important because there are several states that are approximately equal in power. The importance of allies in multipolar systems can lead to either of two phenomena that cause problems and result in war—buck-passing and chain-ganging.

Buck-passing occurs when states do not take action because they expect allies to take action. Normally, if an opposing state becomes aggressive, states and their allies should step in to stop it. This is necessary if they want to maintain their security. Stepping in to stop a threatening state, however, is costly, so sometimes states will pass the buck and hope that others will deal with it. But if everyone passes the buck, no one will stand up to stop the aggressive state. For instance, Christensen and Snyder (1990) argue that buck-passing was an important cause of World War II. In particular, no one stepped in to stop Germany as it began rearming and challenging the status quo under Adolf Hitler in the 1930s. The United States stayed out because it was a European problem, not theirs. Britain felt it was a continental problem. France argued that it made more sense for the Soviet Union to stop Germany since it was so much bigger. By the time Britain and France had finally had enough and decided to stand up to Hitler in 1939, it was too late to avert war.

The other alliance problem is **chain-ganging**, where states are tightly connected by alliances. Because alliances are so important in a multipolar system, states may know or come to believe that they have unconditional support from their allies. States fear that if they do not back up their allies now, their allies will not back them up in the future. Christensen and Snyder (1990) argue that chain-ganging was a major cause of World War I. The war began because Austria-Hungary and Serbia got into a dispute over the assassination of the Archduke Franz Ferdinand in Sarajevo. But Russia felt it had to back Serbia against Austria-Hungary, Germany felt it had to back Austria-Hungary against Russia, France felt it had to back Russia against Germany, and England felt it had to back France against Germany. Thus, what might have been a small war between Austria-Hungary and Serbia grew to involve every major power because of tightly connected alliances.

Ultimately, the core of the argument concerning bipolarity and multipolarity concerns uncertainty. In a multipolar system, states have to worry about many possible enemies. Such a system has many possible threats, and there are many possible alliances that could form and conflicts that could occur. As a result, there is a lot of uncertainty in a multipolar world. Because of this uncertainty, states might make mistakes, such as misestimating the power of one or more of the other states.

This uncertainty is well illustrated by the European alliances prior to World War I, which were led by the Triple Alliance of Germany, Austria-Hungary, and Italy and the Triple Entente of France, Russia, and Great Britain. It is not as straightforward as that, however, because Italy was also allied with France and Russia, so a network of alliance ties existed that connected each of the European major powers with all of the others (not to mention family ties between royalty in the different countries). Ultimately, Italy joined the Allies in the war rather than fight alongside its Triple Alliance partners. All of this makes it evident that there was considerable uncertainty about how the great powers would respond to hostilities between Austria-Hungary and Russia over the future of Serbia. The combinations of states that eventually formed both the Allies and Central Powers in World War I did not arise from obviously distinct blocs prior to the war. The sides that appeared to be polar opposites after the war began seemed beforehand to be intricately intertwined.

By contrast, in a bipolar world the two major countries know who the main enemy is and can focus nearly all of their attention on that enemy. They can be certain about what their relative power is and judge the nature of their interactions accurately. For example, in the Cold War, two-thirds of US intelligence resources went toward monitoring the Soviet Union. The total breakdown was 66 percent toward the USSR, 25 percent toward China, 7 percent for the Middle East, 2 percent monitoring Latin America, and only 1 percent for the rest of the world (Bennett 2000, 65). While that is not a complete focus on the Soviet Union, it is highly skewed in a way that is not practical in a multipolar system.

The ambiguity of alliance commitments prior to World War I is sharply contrasted by the structure of European major power alliance commitments during the Cold War. The opposing sides during the Cold War years were clear. The United States led the North Atlantic Treaty Organization (NATO), and the Soviet Union led the Warsaw Pact. No alliance commitments ran from one bloc to the other. There was clear demarcation between east and west, communist and noncommunist, so uncertainty was minimized.

Uncertainty and Stability

The argument that bipolar systems have less uncertainty than multipolar systems certainly appears valid. Bueno de Mesquita (1978, 2010), however, argues that there is a considerable logical leap from the association of uncertainty with multipolarity to the association of multipolarity with instability and bipolarity with stability. Indeed, Deutsch and Singer (1964) argue that multipolar systems are more stable than bipolar systems precisely because multipolarity produces uncertainty. It is also possible that there is no relationship at all between polarity and stability.

There are several problems with the argument that because bipolar systems encompass less uncertainty than multipolar systems they yield greater stability. First, this argument is not implied logically by the four key assumptions of neorealism: (1) international politics is anarchic, (2) states are the central actors, (3) states seek to maximize their security, and (4) states seek to increase their power so long as it does not jeopardize their security. These assumptions say nothing at all about uncertainty or how uncertainty affects stability. Thus, additional assumptions are needed to establish a logical connection between polarity and stability.

In particular, we would need to make some assumption about how states respond to uncertainty. Uncertainty may prompt states to behave cautiously; in formal terms, this is known as being risk averse. If uncertainty promotes caution and certainty encourages opportunism, then bipolarity in fact encourages instability. This is what Deutsch and Singer (1964) implicitly assume. However, uncertainty might make states somewhat reckless; this is known as being risk acceptant. If uncertainty promotes risk-seeking behavior while certainty makes states cautious, then bipolarity does foster stability. This is what Waltz (1964) implicitly assumes. Furthermore, it is also possible that different states respond in different ways to uncertainty; some leaders may be risk acceptant, some may be risk averse, while others may be risk neutral (meaning their behavior is not influenced by risk). If this is true, then there is no real relationship between uncertainty and stability, so the purported relationship between polarity and stability does not exist.

Neorealism is also plagued by a variety of empirical problems. For example, the Cold War bipolar system was not especially stable in length of time compared with some of the multipolar systems in the past (Bueno de Mesquita 2010, 128–31). Thus, in addition to running into logical difficulty, the claim that bipolar systems are more stable than multipolar ones lacks empirical support as well. In addition, arguments that a balance of power is more peaceful are contradicted empirically. For example, why has the United States not attacked Canada sometime in the past century? It certainly is not because power is balanced, since the United States has been much stronger throughout. There are certainly good reasons that we could use to explain the absence of war between the United States and Canada, but they are not explainable by realism. Furthermore, World Wars I and II, along with most other major wars in history, were fought under conditions of parity. Simply put, as we saw when we examined Bremer's (1992) findings regarding dangerous dyads in chapter 1, a balance of power does not produce peace.

Offense-Defense Balance

The offense-defense balance is another topic relevant to the effect of power on international conflict that has received a great deal of attention from realist and nonrealist scholars. Is war more likely when conquest is easy? Could peace be strengthened by making conquest more difficult? What are the causes of offense dominance? How can these causes be controlled? These are some of the questions examined by offense-defense theory.

Offense-defense theory grew in the 1970s (Jervis 1978; Quester 1977), and continued developing in the 1980s and beyond (Glaser and Kaufmann 1998; Jones 1995; Snyder 1984; Van Evera 1984). The basic argument of offense-defense theory is that war is far more likely when conquest is easy. Furthermore, shifts in the offense-defense balance are expected to have a large effect on the risk of war. Although different authors disagree on specifics here and there, we will focus on the argument presented by Van Evera (1998) for the sake of simplicity.

The basic idea of what Van Evera and others mean by the offense-defense balance is well illustrated in the world of sports. Pretty much every team sport has elements of both offense, where you are trying to score, and defense, where you are trying to prevent the opponent from scoring. Additionally, in different eras offense or defense tends to have the upper hand throughout the sport. For example, in major league baseball, there have been different eras in which hitting (offense) or pitching (defense) had the upper hand. In the "year of the pitcher" in 1968, pitching was dominant; batting averages and scoring were historically low. In response, baseball authorities changed the rules by lowering the pitching mound in 1969, shifting the balance away from pitching.

Van Evera argues that the **offense-defense balance** is an important cause of war because offense dominance (or the perception of offense dominance) leads to a variety of different effects that raise the risk of war. Perceptions are important because offense-defense theorists argue that leaders sometimes misjudge the true nature of the balance between offense and defense.

First of all, offense dominance leads to opportunistic expansionism. When conquest is easy, aggression is attractive because it costs less to attempt and succeeds more often. In addition, Van Evera argues that we are likely to see defensive expansionism. When conquest is easy, states' borders are less secure. Therefore, they are more expansionist even if their motivations are defensive in nature. Furthermore, Van Evera argues that expansionism is met with fierce resistance by other states. When the offense dominates, states are cursed with neighbors made aggressive by both temptation and fear. Thus, everyone faces a greater risk of attack and is willing to defend themselves strongly.

Another key effect is that moving first is more rewarding when the offense is dominant. The incentive to strike first is larger because a successful surprise attack provides larger rewards and averts greater dangers. By moving first, states can take the offensive and dictate the course of events. States that wait for others to move first are forced into the disadvantaged position of being on the defensive. Thus, windows of opportunity to gain an advantage over other states are more dangerous because the incentive to capitalize on them is greatly increased.

Van Evera argues that when the offense is dominant, faits accompli are more common and more dangerous. A fait accompli is something that, once done, is not easily undone; states sometimes try to make a surprise move to suddenly change the status quo, thinking that once done, it will be difficult for other states to do anything other than accept the change. Van Evera argues that when conquest is easy, states adopt more dangerous diplomatic tactics such as these. A fait accompli, however, is a dangerous step that increases the probability of war: It promises a greater chance of political victory, but it also raises a greater risk of violence.

In addition, Van Evera argues that states negotiate less and reach fewer agreements when the offense is dominant. States have less faith in agreements because others break them more often. Accordingly, states will bargain harder and make concessions more grudgingly. States will also insist on better agreed procedures to enable verification and ensure compliance with agreements. Compliance, however, is harder to verify because secrecy is more common. An information advantage confers more rewards when conquest is easy because it allows attackers to gain an even greater advantage. Therefore, not only is secrecy more common, it is more dangerous.

Jervis (1978) argues that there is an important interaction between the offense-defense balance and the security dilemma. As discussed previously, the security dilemma occurs when one state's attempts to increase its security decrease the security of others. Jervis argues that the security dilemma exists when offensive postures are not distinguishable from defensive ones. For example, tanks are useful for both attacking and defending. If a state builds up a number of tanks, how can we determine whether they are building them for defensive or offensive purposes? Further, Jervis argues that the security dilemma is much more dangerous when the offense has the advantage. Figure 5.3 summarizes the argument.

When the offense has the advantage and offensive postures are not distinguishable from defensive ones, this creates a doubly dangerous situation. If, however, offensive and defensive postures are distinguishable (and offense

Figure 5.3 The Offense-Defense Balance and the Security Dilemmas

	Offense has the advantage	**Defense has the advantage**
Offensive posture not distinguishable from defensive one	Doubly dangerous.	Security dilemma, but security requirements may be compatible.
Offensive posture distinguishable from defensive one	No security dilemma, but aggression possible. Status-quo states can follow different policy than aggressors.	Doubly stable.

Source: Adapted from Robert Jervis, "Cooperation Under the Security Dilemma." *World Politics* 30 (1978), 211.

has the advantage), then there is no security dilemma but aggression is possible. In this case, Jervis argues that status-quo states can follow different policies than aggressors. If defense is advantaged and offensive postures are not distinguishable from defensive ones then the security dilemma exists, but security requirements between states may be compatible. Finally, if defense has the advantage and offensive and defensive postures are distinguishable, then this creates a situation that is doubly stable.

Determinants of the Offense-Defense Balance

What are the causes of offense and defense dominance? In sports, shifts in the offense-defense balance are usually driven by rules changes. For instance, American football had a dramatic shift toward the offense when forward passes were first allowed. Similarly, changes in rules about being offside have led to shifts in the offense-defense balance in both soccer and hockey over the years. In international relations, however, there is no rule book, there is no sanctioning body that institutes particular changes. Thus, shifts in the offense-defense balance must be driven by other factors.

Military factors are the primary driving force of changes in the offense-defense balance in international relations. In particular, shifts in military technology are argued to have a dramatic effect. For example, the introduction of the machine gun and improvements in artillery in the late 1800s and early 1900s led to massive increases in firepower that made attacking prepared defenses much more difficult. Other military factors that can have important effects on the offense-defense balance include shifts in military doctrine as well as force posture and deployments.

Van Evera argues that geography is another important factor affecting the balance between offense and defense. Conquest is harder when geography insulates states from invasion or effective embargo. Britain and the United States have been very difficult to attack because they are protected by large bodies of water, whereas attacking Germany has historically been much easier because of the wide open terrain of northern central Europe.

The third major driving force of the offense-defense balance is social and political order. Popular regimes are more able to develop greater offensive power, but they are also better at self-defense. Van Evera argues that overall, a world of popular regimes favored the offense prior to 1800 because they were able to raise larger, more loyal armies useful for attacking other countries. He argues that popular regimes switched to favor the defense after 1800 because increases in small arms technology make guerilla warfare more attractive, and vulnerable unpopular regimes form a threat to their neighbors.

The final category of factors that Van Evera argues shapes the balance is diplomatic factors. In particular, collective security systems, defensive alliances, and balancing behavior by neutral states are all types of diplomatic arrangements that strengthen the defense. All three impede conquest by adding allies to the defending side. If these factors are present in the international system, then the balance shifts toward the defense; if they are absent, the balance shifts toward the offense. For example, in the late 1800s, Germany created a network of defensive alliances and Britain actively worked as a balancer in Europe, creating a situation where diplomatic factors favored the defense in the period.

Empirically Testing Offense-Defense Theory

As with any theory, it is important to test offense-defense theory to determine if it is supported by the historical record. Van Evera (1998) conducts a series of case studies to do just that. He focuses on European great power relations from the 1700s to the 1900s, breaking history down into a series of time periods when important shifts occurred. His empirical observations are summarized in Table 5.1. Prior to 1792, offense and defense were roughly balanced and there was a medium amount of warfare.

Van Evera contends that between 1792 and 1815, the offense was strong militarily. This offense dominance arose in large part as a result of France's adoption of the popular mass army following the French Revolution. In addition, while in reality diplomatic factors were mixed, leaders perceived that aggressors were favored. Given this high level of offense dominance, a high level of warfare is predicted. And there certainly was a high level of warfare as the French Revolutionary and Napoleonic Wars, involving France, Britain, Austria, Prussia, Russia, and others, lasted from 1792 to 1815 with only brief pauses.

The period following the Napoleonic Wars, lasting from 1815 until 1856, was one of relative harmony in Europe. Van Evera argues that arms and diplomacy both favored defenders, particularly because mass armies disappeared and Britain remained active on the continent as a balancer. Thus, a low level of warfare is predicted, and there was only one great power war: the Crimean War (1854–1856), right at the end of the period.

The balance shifted back to the offense from 1856 to 1871. Van Evera asserts that there was no real advantage either way militarily, but diplomatic realities favored aggressors because Britain and Russia shifted to isolationism. Therefore, he argues, offense was favored overall, so there should have been a high level of warfare. This period was marked by a series of great power wars, particularly related to German unification and Italian unification. Because the wars were all short and involved only two or three great powers, however, Van Evera codes it as only a medium amount of warfare.

German unification in 1871 ushered in a new era, which lasted until 1890. In this period, German chancellor Otto von Bismarck created a network of defensive alliances and Britain had renewed activity as a balancer in Europe. Combined with changes in military technology that led to great increases in firepower, Van Evera maintains that the defense was favored. Furthermore, there was a low amount of warfare with no wars between great powers occurring during the period.

Van Evera contends that an important shift occurred around the turn of the twentieth century, from 1890 to 1919, as continued technological

Table 5.1 The Offense-Defense Balance Among Great Powers, 1700s–Present

ERA	MILITARY REALITIES FAVORED	MILITARY REALITIES WERE THOUGHT TO FAVOR	DIPLOMATIC REALITIES FAVORED	DIPLOMATIC REALITIES WERE THOUGHT TO FAVOR	IN AGGREGATE MILITARY AND DIPLOMATIC REALITIES FAVORED	IN AGGREGATE MILITARY AND DIPLOMATIC REALITIES WERE THOUGHT TO FAVOR	AMOUNT OF WARFARE AMONG GREAT POWERS
Pre-1792	Defs.	Defs.	Med.	Med.	Med.	Med.	Medium
1792–1815	Aggrs.	Aggrs.	Med.	Aggrs.	Aggrs.	Aggrs.	High
1816–1856	Defs.	Defs.	Defs.	Defs.	Defs.	Defs.	Low
1856–1871	Med.	Med.	Aggrs.	Aggrs.	Aggrs.	Aggrs.	Medium
1871–1890	Defs.	Med.	Defs.	Defs.	Defs.	Defs.	Low
1890–1918	Defs.	Aggrs.	Aggrs.	Aggrs.	Defs.	Aggrs.	High
1919–1945	Aggrs.	Mixed	Aggrs.	Aggrs.	Aggrs.	Aggrs.	High
1945–1990s	Defs.	Med.	Defs.	Defs.	Defs.	Defs.	Low

Aggrs.: The factor favors aggressors.
Defs.: The factor favors defenders.
Med.: A medium value: things are somewhere in between, cut both ways.
Mixed: Some national elites saw defense dominance, some saw offense dominance.

Source: Stephen Van Evera, "Offense, Defense, and the Causes of War," *International Security* 22 (1998), 24.

developments shifted military realities further and further in favor of the defense. However, he argues that everyone believed that the offense was dominant, while diplomatic realities also favored the offense. This is known as the **cult of the offensive** and is said to be a major cause of World War I (Snyder 1984; Van Evera 1984). With a perception of offense dominance, a high level of warfare is predicted, which was also observed in reality.

The key to the cult of the offensive argument is the observation that all of the great powers on the eve of World War I embraced offensive doctrines and strategies (Snyder 1984). Germany's plan for coming war was called the Schlieffen Plan, named after the former chief of the German General Staff, Alfred von Schlieffen. The plan called for German armies to sweep through Belgium in a grand arc, advancing through northeastern France on to Paris itself. France's main plan for war was Plan XVII, which called for an offensive into Alsace and Lorraine along the border with Germany. In addition, on the eastern front, Russia launched offensives against both Germany and Austria-Hungary.

After World War I, the offense once again gained the upper hand from 1919 to 1945. This shift came with the development of new technologies. In particular, the tank and other armored fighting vehicles allowed armies to have mobility, firepower, and protection while engaging prepared defenses. In addition, the introduction of wireless radios allowed armies to maintain command and control of these armored forces even after they broke through enemy lines. Van Evera contends that although military realities favored the offense, perceptions of the offense-defense balance were not so clear cut because some countries, in particular France, believed that defense was dominant. Nonetheless, given overall offense dominance, a high level of warfare is predicted. And with World War II and a number of other wars during this period, a high level of warfare was realized.

The final period that Van Evera examines is the Cold War from 1945 to the 1990s. He argues that military realities favored the defense in this period because of nuclear weapons, although leaders believed the offense and defense to be more balanced. In addition, diplomatic realities favored the defense as the United States served as balancer. Overall, the defense was dominant so a low amount of great power warfare is predicted, which was also the case.

In summary, Van Evera's analysis indicates strong empirical support for the basic predictions of offense-defense theory. Important criticisms of the theory, however, remain. First, empirical support for the theory rests importantly on the cult of the offensive. Without the cult of the offensive, World War I marks a key contradiction of the theory. Yet did such a cult truly exist? Gray (2012, 96) argues that "the historical record shows that, far from ignoring it, European armies had been debating the tactical crisis caused by rapidly evolving firepower for more than sixty years." Germany, France, and others adopted offensive war plans, but not because they believed that the offense was dominant. Rather, they believed that offensive action was the only way to seize the initiative and overcome the difficulties posed by massive increases in firepower.

Ultimately, the offensives failed in their intent. But "the armies of Europe were not collectively stupid. They were confronted with a tactical challenge to which there was no clever tactical solution ready at hand in 1914" (Gray 2012, 97). Such tactical solutions were developed during the course of the war, and by 1918 offenses succeeded despite the massive firepower of opposing defenses. Biddle (2001, 2004) claims that technology does not dictate an automatic balance between offense and defense. Rather, the key is how forces are used, regardless of the prevailing technology of the time. (We will examine Biddle's argument about force employment in more detail in chapter 10.) In addition, Levy (1984) questions whether the offense-defense balance is useful as an element of theories of conflict because it is an ambiguous concept that eludes effective measurement.

Power Transition Theory

Another important theory of international politics is power transition theory, developed by Organski (1968) as a counter to balance of power theory, and further developed by Kugler, Lemke, and others (Kugler and Lemke 1996; Organski and Kugler 1980; Tammen et al. 2000). Power transition theory shares with realism a focus on the importance of power in international affairs.

Power transition theory starts with assumptions that are quite different from those in realism. First, the international system is assumed to be a hierarchy rather than anarchy. The hierarchy is illustrated as a pyramid of power, as shown in Figure 5.4. At the top of the hierarchy is the dominant state, the strongest state in the world. The United States has been the dominant state since World War II. Before that, the United Kingdom was dominant. Below that are the great powers, very strong states such as Russia, China, France, and Germany, who may someday challenge to become the dominant state. In the next tier are middle powers, which are strong regionally but not at the same level as the great powers. Modern examples include Brazil, India, and Iran. Finally, at the lowest level are the small powers, which is the largest group of countries in the world but have little power. Small powers are countries such as Belgium, Cameroon, and Thailand that have little power or influence in the international system.

Power transition theory also assumes that some states are satisfied, while others are dissatisfied. The dominant state establishes the rules and norms of the international system. These define the status quo, so these are what states may be satisfied or dissatisfied with. The United States, as the dominant state in the system since World War II, set up a variety of international rules and norms in the international system. For instance, the Bretton Woods system was the currency regime set up after World War II; the General Agreement on Tariffs and Trade (GATT) and World Trade Organization (WTO) are trade regimes since World War II that focus on removing barriers to trade and increasing free trade in the world; further, the United States established strong norms in international relations for the promotion of democratic institutions and the protection of human rights.

Figure 5.4 The Hierarchy of Power According to Power Transition Theory

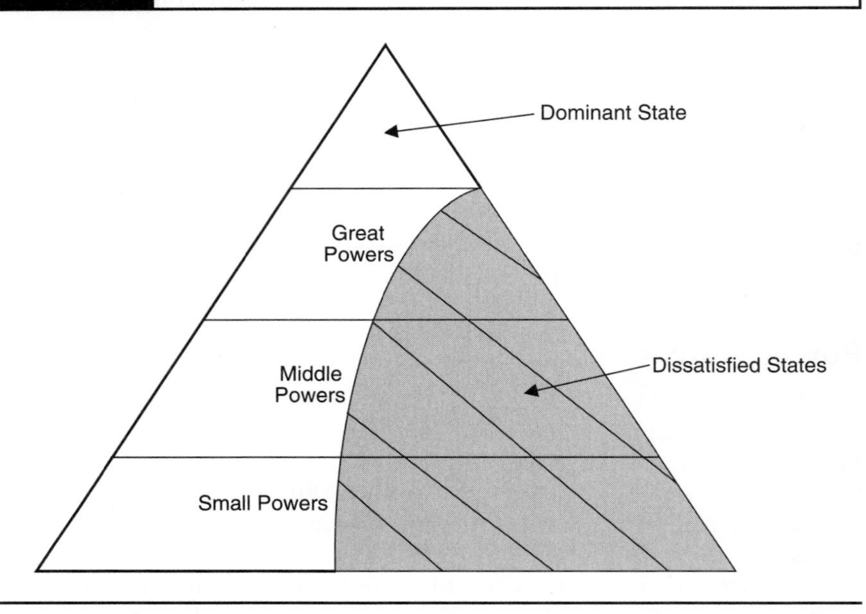

Source: Adapted from Ronald L. Tammen, Jacek Kugler, Douglas Lemke, Allan C. Stam, III, Carole Alsharabati, Mark Andrew Abdollahian, Brian Efird, and A. F. K. Organski, *Power Transitions: Strategies for the 21st Century* (New York: Chatham House, 2000), 10.

The Cold War, which lasted from 1945 to 1989, was a long period of tension marked by the Soviet challenge to the position of the United States as the dominant state. It is not difficult to imagine that many of the rules and norms promoted by the United States would have been discarded had the Soviet Union won in the competition to become the world hegemon. Furthermore, the status quo established by the United States marked a break from the past. The international rules and norms prior to World War II were much different. They were part of the system set up by Great Britain, the previous dominant state. In particular, Britain led the way in securing colonies around the world, while the United States fostered a norm of decolonization.

The pyramid of power shown in Figure 5.4 contains both satisfied and dissatisfied states. The dominant state is always satisfied, while the proportion of dissatisfied states increases as we move down the pyramid because less powerful states have little influence on the status quo. Power transition theory argues that satisfied states join the dominant state in a winning coalition to enforce the rules and norms of the system.

The core argument of power transition theory is illustrated in Figure 5.5. States grow at different rates of development because they began industrializing

Figure 5.5 The Power Transition

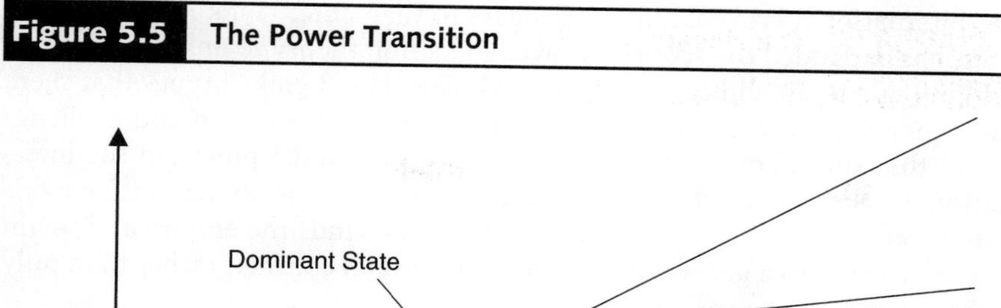

at different times and they have different capacities for growth based on different sizes and geographies (Organski 1968; Organski and Kugler 1980). Great Britain rose to be the dominant state in the 1800s because it was the first country in the world to begin industrialization. Germany, the United States, and others industrialized later but grew to challenge Britain's position as the dominant state.

Conflict is most likely to occur when a dissatisfied challenger moves up the pyramid enough to challenge or pass the dominant state. Figure 5.5 shows the power of the dominant state and a rising challenger over time. The time period begins with a large gap in power between the two. The rising challenger, however, grows more quickly than the dominant state and eventually the lines cross as the rising challenger becomes stronger than the former dominant state. Power transition theory argues that conflict is most likely around the point of transition, when the two states are roughly equal in power. Conflict is only likely, however, if the rising challenger is dissatisfied. A satisfied challenger would eventually become the new dominant state, but there would be no reason for war since the two generally agree on the functioning of the international system.

Recent research has focused on some important issues for power transition theory, particularly relating to regional powers and satisfaction. Power transition theory was developed to explain large, systemic wars such as World War I and World War II. The focus is on the dominant power in the international system, and challengers to that state. Lemke (2002), however, has expanded the scope of power transition theory to analyze relations within regions. In addition to the global hierarchy, Lemke argues that there are regional hierarchies, with a dominant power in the region and challengers to that state. For example, while Brazil is a middle power in the international system as a whole, it is the dominant state in its region. Lemke's extension of power transition theory greatly expands the empirical domain of the theory into a general theory of international conflict, rather than only a theory of great power conflict.

Satisfaction is a core component of the power transition argument because power transitions are only dangerous if the rising state is dissatisfied. Thus, the transition between Great Britain and the United States was peaceful because the United States was satisfied. However, it can be difficult to determine whether or not a state is satisfied. An important avenue of current research is attempting to measure satisfaction with the status quo systematically (Benson 2007; Kim 1989; Lemke and Reed 1996, 2001; Lemke and Werner 1996), which is important because it eliminates the problem of saying that a particular state was satisfied simply because there was no war.

Empirically Testing Power Transition Theory

As with other theories, it is important to examine whether there is empirical support for power transition theory. The theory points to several power transitions in history. In the early 1900s, the United Kingdom was the dominant state and Germany was the rising challenger. There was significant friction between the two and then Germany overtook Britain in power between 1906 and 1914. Of course in 1914, World War I began, providing strong support for the theory. After the war, the United Kingdom remained the dominant state, and in the 1930s, Germany once again grew to become a dissatisfied challenger. After Hitler took power in 1933, Germany once again passed Britain, and in 1939, World War II began.

In the first half of the twentieth century, there was also a transition in power between the United Kingdom and the United States. This was a peaceful transition, however, because the United States was satisfied. Indeed, the United States remained generally isolationist, not taking the role of the dominant state until World War II. Power transition theory explains the subsequent Cold War as the tensions between the United States as the dominant state and the Soviet Union as the dissatisfied rising challenger. Yet World War III never occurred because there was no overtaking. Although the Soviets narrowed the gap, the United States maintained a much larger GDP throughout.

Statistical evidence supports the theory as well. Organski and Kugler (1980) test power transition theory by examining great power relations from 1860 to 1975. They look at twenty-year periods for each great power dyad. Their central analysis focuses on contenders in the central system, and is shown in Table 5.2. They have a total of twenty observations. In the four cases in which power was unequal (because one great power was much stronger than the other), there were no wars, as the theory predicts.

Table 5.2 Power Transition Theory and Major Wars

WAR?	UNEQUAL	EQUAL, NO OVERTAKING	EQUAL AND OVERTAKING
No	4 (100%)	6 (100%)	5 (50%)
Yes	0	0	5 (50%)

Source: A. F. K. Organski and Jacek Kugler, *The War Ledger* (Chicago: University of Chicago Press, 1980), 52.

In the six cases where the two sides were approximately equal in power but there was no overtaking, there were also no wars. Finally, in the ten cases in which there was parity combined with overtaking (as one state became stronger than the other), there were five wars. Collectively, these observations provide strong support for power transition theory. Conflict is significantly more likely when there is a transition in power.

In looking to the future, power transition theory points to China as the rising challenger to the US position atop the power pyramid. Tammen et al. (2000) projected more than a decade ago that China would pass the United States in GDP around 2015. More recent forecasts expect China to pass the United States by 2020 (Rapoza 2011). Although the timing of the transition is of course uncertain, it seems highly likely to occur in the relatively near future. Of course, if power transition theory is correct, then this transition could be a very dangerous time. Accordingly, Tammen et al. (2000) suggest strategies for peace to deal with the coming transition. In particular, it is vital to manage Chinese dissatisfaction because a transition with a satisfied China should not lead to war.

Conclusion

Power has been a common focus of studies of international conflict. There has been a great deal of debate, however, about how to define it. The relational definition of power has a variety of limitations that are largely offset by focusing on the material basis of power. That is precisely what the primary ways of measuring power—CINC score and GDP—do.

Realism has been the primary power-centric approach to studying international politics. Although there have been many theories within the umbrella of realism, we focused on theories regarding the impact of polarity on stability and the offense-defense balance. Neorealists argue that bipolar systems are more stable than multipolar ones, but the theory has logical and empirical difficulties. Offense-defense theory seems to have better empirical support, but it also has met criticism. In conclusion, significant limitations exist in realist theory.

Power transition theory is an important nonrealist theory based on power. The logic and evidence supporting power transition theory indicates that significant support exists for the theory.

Power has been the central factor that many have used in attempts to explain international relations in general and international conflict in particular. Much of this focus on power comes from realism, but realist theory suffers from logical inconsistency and empirical inaccuracy. Although many analyses of the causes of war take a realist, power-centric approach, we do not take that view here because power is simply one of many different factors that shape international conflict.

Key Concepts

Anarchy

Balance of power theory

Buck-passing

Chain-ganging

Composite Indicator of National Capabilities (CINC)

Cult of the offensive

Material basis of power

Offense-defense balance

Polarity

Power

Power transition theory

Realism

Relational definition of power

Security dilemma

6 Alliances

Joseph Stalin, Franklin Roosevelt, and Winston Churchill met at Tehran, Iran, in November 1943 to coordinate Allied strategy for defeating Germany during World War II. The alliance between the Soviet Union, United States, and United Kingdom would not survive the war.

Source: Copyright Bettmann/Corbis / AP Images

International relations often contain a mix of conflict and cooperation. Although there are a variety of ways that states in the international system can cooperate with one another, one of the most prominent ways is through alliances. Alliances have been a continual feature of international relations for at least the past four hundred years. The longest active international alliance is that between Britain and Portugal, which was originally formed with the Treaty of London on 16 June 1373 and has been renewed at least ten times since (Gibler 2009, 4). There are many important questions we can ask about alliances. Why do states form them? What are the potential drawbacks? What do alliances tell us about the similarity of interests between states? Finally, what impact do alliances have on international conflict?

In this chapter, we seek to answer each of these questions. We begin by looking at the different possible types of alliances before considering the question of why states form alliances. We then examine some drawbacks of alliances as well as explore how they can be used to measure shared interests between states. Finally, we examine the relationship between alliances and conflict, including how alliances can make conflict both less likely and more likely.

Types of Alliances

We begin our exploration of alliances by being explicit by what we mean by the term. An **alliance** is a formal written agreement, "signed by official representatives of at least two independent states, that include promises to aid a partner in the event of military conflict, to remain neutral in the event of conflict, to refrain from military conflict with one another, or to consult/cooperate in the event of international crises that create a potential for military conflict" (Leeds et al. 2002). While some scholars argue that alliances do not have to be formal agreements (e.g., Walt 1987), we follow the mainstream and use precise language, reserving the term *alliance* for formal agreements, with *alignment* being the broader term for states that cooperate due to common interests (Snyder 1997; Leeds et al. 2002).

Military alliances typically spell out what is to be done in the event a member is attacked. They are anticipatory, tied to expectations concerning what might happen at some point in the future. Alliances are rarely pertinent in the case of a civil war or internal coup d'état; while the leadership in one country might decide to step in to protect the leader of an ally if there is a civil war or a coup, such action typically would not be in response to an alliance commitment. Furthermore, alliances are not binding commitments; sovereign states and their leaders can and do renege on promises. Rather, they are signals of intent designed in part to deter prospective foes and in part to facilitate preparedness by coordinating the actions of the signatories under specific contingencies (Leeds 2003).

Alliances come in a handful of varieties. First of all, we can categorize them based on how many states are involved. Bilateral alliances, such as the one between Britain and Portugal, are made between two states, whereas multilateral ones involve more than two states. A prominent multilateral alliance in the modern world is the North Atlantic Treaty Organization (NATO), formed in 1949 by the United States, Canada, and ten Western European countries. NATO has expanded several times, and now has twenty-eight member countries across Europe. By joining NATO, each member is making a promise to every other member that it will defend that member in the event that it is attacked by another country. We can also categorize alliances by the type(s) of commitment they contain. To explore the different types of alliance commitments, we now turn our attention to the two primary data sets of alliances. These data sets enable us to identify alliances and their types, which in turn allow us to examine the relationship between alliances and other factors, such as international conflict.

Correlates of War Typology

The oldest alliance data set is collected by the Correlates of War (COW) project and identifies 414 alliances covering the time period from 1816 to 2012 (Gibler 2009; Gibler and Sarkees 2004; Singer and Small 1966). The COW alliance data have traditionally used a typology with three basic types of agreement: mutual defense pacts, neutrality/nonaggression agreements, and consultation agreements. **Mutual defense agreements** involve a promise by each signatory that in the event a member is attacked by an outsider, every other member will come to its military defense. There are many examples of mutual defense pacts, including the NATO and the ANZUS alliance between Australia, New Zealand, and the United States.

Neutrality and nonaggression agreements involve a promise by each signatory that in the event one is attacked, the others will not join against it on the side of the attacker, or they agree to refrain from attacking each other. For example, on 23 August 1939 Germany and the Soviet Union signed a nonaggression pact. In Article I of the treaty, the two sides agreed to "obligate themselves to desist from any act of violence, any aggressive action, and any attack on each other" (Gibler 2009, 330). Despite the agreement, Germany attacked the Soviets less than two years later.

Consultation agreements (ententes) involve an agreement, or literally, an understanding, between member states to consult with one another on a course of action in the event that any one of them is attacked. Probably the most famous entente in history is the Triple Entente, the alliance between Russia, France, and Britain that decided to wage war against Germany following the Austro-Serbian crisis in 1914.

Of the 414 alliances in the COW data, 324 (78.3 percent) are bilateral. Looking at the types of commitments made within alliances, we see that 205 (49.5 percent) are mutual defense pacts, 145 (35.0 percent) are neutrality/nonaggression agreements, and 64 (15.5 percent) are consultation agreements. The latest version of the COW alliance data details the multiple types of commitments contained in an alliance, if any. Neutrality/nonaggression and consultation agreements were more common in the twentieth century than the nineteenth. In both centuries, defense pacts were common, representing the most serious of agreements between states.

Alliance Treaty Obligations and Provisions

A newer and more detailed data set is collected by the Alliance Treaty Obligations and Provisions (ATOP) project (Leeds et al. 2002). These data cover the years from 1815 to 2003, and are available online at http://atop.rice.edu/home. The ATOP data differ from the COW alliance data because they have additional obligation categories and they record all of the obligations in an alliance, not just the most serious one. The ATOP data also record the specific conditions that pertain to the alliance obligations, such as being directed at certain states or regions, dealing with a specific ongoing conflict, and so on. Finally, the ATOP data record asymmetric commitments—where the states make different promises to each other—when they are present.

The ATOP data record five different types of alliance obligations. Defensive alliances include obligations to provide active assistance to alliance partners in the event of an attack on a fellow alliance member, as discussed above. The data also have a category for **offensive alliances**, which involve "promises of active military cooperation in instances other than the defense of one's own sovereignty and territorial integrity" (Leeds et al. 2002, 241). An example of a purely offensive alliance is the 1866 alliance between Prussia and Italy, in which Italy pledged to also attack Austria if Prussia attacked within three months. The Seven Weeks War, in which Prussia and Italy attacked and defeated Austria, immediately followed.

Unlike the traditional categories of the COW data, the ATOP data distinguish between neutrality and nonaggression agreements. **Neutrality agreements** obligate members to stay out of wars that another member is involved in, refraining from aiding the ally's opponent(s). Neutrality provisions are much less common than nonaggression provisions and have been rarely used on their own since World War II. **Nonaggression agreements** involve a promise by each state to not attack the other(s) in the alliance. The Soviet Union signed a series of nonaggression pacts in the 1970s with states around the world. Some of them had both nonaggression and consultation provisions, including alliances with Egypt in 1971, Iraq in 1972, Somalia in 1974, and Angola in 1976. The Soviets also signed alliances with nonaggression provisions only with Sweden, Norway, Denmark, Iceland, and Turkey in 1975.

As with the COW alliance data, consultation agreements involve an agreement between member states to consult or cooperate with one another on a course of action in the event of international crises. For example, following the Korean War, the United States and South Korea signed an alliance on 1 October 1953 that is a consultation agreement. Article II of the treaty states, "The Parties will consult together whenever, in the opinion of either of them, the political independence or security of either of the Parties is threatened by external armed attack" (Gibler 2009, 395).

While the traditional categories in the COW alliance data are mutually exclusive, in the ATOP data, they are not. The ATOP project identifies 648 alliances formed between 1815 and 2003. Of these, 545 (84.1 percent) are bilateral. Looking at the types of commitments made within alliances, we see that 263 (40.6 percent) contain defense obligations, 80 (12.4 percent) contain offense obligations, 112 (17.3 percent) have neutrality provisions, 315 (48.6 percent) contain nonaggression provisions, and 354 (54.6 percent) contain consultation agreements. Thus, consultation agreements and nonaggression agreements are the most common, while offense obligations are the least common.

Why Do States Form Alliances?

Now that we have seen what alliances are and how we can identify them, we turn our attention toward answering the important question of why states form alliances. From a basic rational choice perspective, we can infer that states form alliances because the expected benefits outweigh the costs (Snyder 1997). But what are the key benefits that motivate states to form alliances? Answers to this question vary. At the most basic level, some scholars argue that alliance formation is all about security, while others argue that states form alliances to make a favorable tradeoff between autonomy and security. We examine theories of alliance formation, focusing on autonomy and security, and then turn our attention to empirical studies of alliance formation.

Security and Autonomy

Arguments that states are solely motivated by security when choosing whether to form alliances are closely associated with realism. To realists, alliances are designed to increase national security by aggregating the capabilities of allies together, reducing the need for each state to buy and produce as many weapons. An alliance is a quick way of accumulating security, which may be especially important in a time of threat. To those who believe in the balance of power, alliances are critical mechanisms by which balance can be maintained (Morgenthau 1967; Waltz 1979).

Realists have debated whether states are more likely to engage in balancing or bandwagoning when attempting to maximize their security (Snyder 1997). **Balancing** entails joining the weaker side in order to stop a strong state or coalition from becoming hegemonic, whereas **bandwagoning** involves joining with the stronger side to share in the spoils of dominance. Whereas balancing and bandwagoning are usually framed solely in terms of power, Walt (1987) argues that they are better conceived in terms of threat.

Balance of threat theory, developed by Walt (1987), focuses on the level of threat states pose to each other, rather than simply the amount of power. The threat posed by one state or coalition to another is driven by aggregate capabilities, geographic proximity, offensive power, and aggressive intentions. Walt argues that the balance of threat is the key to alliance formation and that other factors such as ideological similarity, foreign aid, or transnational penetration matter very little if at all.

A more recent, strategic argument is that alliances are formed by states seeking a favorable tradeoff between security and autonomy (Morrow 1991, 2000). This view recognizes the reality that alliances do more than simply provide a state with security. There are many states that are involved in alliances that they do not need in order to maintain their security. In particular, consider asymmetric alliances between one or more powerful states and one or more weaker states. Such alliances may prove costly if strong states like the United States are called on to help their far-flung allies. Many alliances are not explained very well from the realist perspective because of its exclusive focus on security.

It is useful at this point to be more explicit about what we mean by security and autonomy. At any given time, states are likely to be satisfied with some aspects of the status quo while being dissatisfied with other aspects. By **security**, we mean a state's ability to keep aspects of the status quo that it likes from changing. This might include keeping one's borders the same, maintaining a current trade regime, or maintaining certain configurations of friends and enemies. By **autonomy**, we mean the ability to pursue favorable shifts in the status quo in areas in which a state would like to see change. Such shifts might mean developing a new global trade pact, adjusting international borders, or changing international law in some way (Morrow 1991; Palmer and Morgan 2006). Some states have more autonomy than security, or vice versa. So states may be able to reach favorable trade-offs so that each state can get more of what it needs. A large state with lots of security can offer security to a small state that needs protection. In return, the small state can offer to back the large state when the large state pushes for some desired change in the status quo.

Consider a hypothetical example of an alliance between the United States and Kuwait. As the most powerful country in the modern world, the United States has a lot of security. Kuwait, on the other hand, does not have as much security. Thus, the United States might promise to protect Kuwait in exchange for basing rights or a Kuwaiti commitment to keep oil prices low. The United States would gain autonomy because it would increase its ability to pursue policies in the Middle East—because it now has a base and an ally—or increase its ability to spend money on goods other than oil. Kuwait gains security from attack because it has the support of a much more powerful country. Note that Kuwait gives up some autonomy in agreeing to this because it cannot do whatever it wants as freely with US troops sitting on Kuwaiti soil than if it had no links.

Palmer and Morgan (2006) argue that focusing on these two goods, which they label maintenance and change rather than security and autonomy, enables one to explain much more than just alliance formation, especially dealing with foreign aid, military spending, and the impact of power on international conflict. For example, the Marshall Plan was a major foreign aid effort undertaken by the United States beginning in 1948 to rebuild Europe following the massive destruction of World War II. In order to receive aid, recipient states had to agree to American provisions regarding financial stability, a free market economy, and worker-management relations. Foreign aid is a way that states can pursue change-seeking behavior. As Palmer and Morgan (2006, 57) state, "The Marshall Plan's purposes were to remake Europe in ways desired by the United States." Although it was offered to them as well, the communist states of Eastern Europe rejected the Marshall Plan aid because they were unwilling to agree to American demands associated with it.

There are many ways that states can reach tradeoffs between security and autonomy. The United States or other strong countries may need other countries to vote with it during a United Nations or World Trade Organization meeting to obtain international support and legitimacy for some policy that it favors that would change the status quo, such as promoting stronger human rights enforcement or pursuing lower tariff barriers. The United States might trade security through economic assistance, military assistance, or trade guarantees to get that additional autonomy from states that would then vote for policies that the United States favors.

Our focus here has been on theoretical explanations of why states form alliances, particularly security and autonomy. We now turn our attention to empirical findings regarding alliance formation and duration. Not only does empirical testing allow us to gauge the validity of the security and autonomy argument, it also enables us to see specific observable factors related to security and autonomy that drive states to form alliances.

Empirical Findings

Lai and Reiter (2000) conducted a general empirical analysis of the factors that explain alliance formation, focusing on regime type, culture, and threat. Their dependent variable codes whether the dyad has an alliance (of any kind) in a given year, and their analysis was replicated by Leeds et al. (2002) using the ATOP data (through 1944) and by Gibler and Sarkees (2004) using a newer version of the Correlates of War alliance data.

Box 6.1

Concept in Focus: Reading Tables of Regression Results

Quantitative studies of international conflict often use some type of regression analysis to test hypotheses. There are a variety of different types of regression. Determination of the appropriate type to use is largely driven by the dependent variable (the thing we want to explain). Ordinary least squares, or OLS, is the most basic regression technique and is appropriate for continuous dependent variables, such as levels of trade or economic development. If the dependent variable is dichotomous (such as *war,* where a value of 1 means there is a war and 0 means there is not one), then either a logit or probit model is appropriate.

Bivariate regression is the simplest and only involves two variables: the independent and dependent. This is limited, however, because we rarely find situations in which only one factor is relevant to explain the dependent variable. Multivariate regression is much more useful because it involves multiple independent variables to explain the dependent variable. Putting multiple variables into the same model allows us to assess the influence of each variable individually while controlling for the effect of others. In a laboratory experiment, we can control the environment so that we can keep everything constant except for the one factor that we change. Because studies of international relations generally focus on observations of actual historical events rather than experiments, researchers are not able to control the environment like they can in a laboratory setting. For example, the models of alliance formation in Table 6.1 have twelve independent variables. Thus, we can evaluate the effect of joint democracy, joint language, distance, and so on individually while controlling for the effects of the other eleven variables (i.e., keeping them constant). This allows us to test hypotheses using what some call "quasi-experiments" (Achen 1986).

There are two key pieces of information that you should focus on when reading a table of regression results: the sign of the coefficients and their statistical significance. By the sign of the coefficient, we simply mean whether it is positive or negative. A positive coefficient means that as the independent variable gets larger (or smaller), the dependent variable does too. A negative coefficient means that the two variables move in opposite directions: As the independent variable gets larger, the dependent variable gets smaller, or vice versa.

By **statistical significance**, we mean the probability that this coefficient is the result of chance. By comparing the size of the coefficient with the standard error, each estimated coefficient has a particular p-value (the probability that the relationship is due to chance) associated with it. Often, a p-level of 0.05 is considered the minimum for statistical significance, meaning that we are more than 95 percent confident that the independent variable is actually correlated with the dependent variable, rather than appearing to be related because of chance. Higher p-levels of 0.01 or 0.001 give us more confidence that we are observing a real relationship.

Studies should also go beyond statistical significance to assess the **substantive significance** of their results. This is often done by calculating the effect of changes in the independent variable on the predicted probability of various outcomes. No matter how statistically significant a variable is, if changes in an independent variable only produce irrelevant changes in the dependent variable, than it is not very substantively significant.

Table 6.1 shows the results from Gibler and Sarkees (2004) for the COW alliance data (1816–1992) and the ATOP data (1816–1944). They examine two different variables capturing the influence of regime type. Joint democracy is positive and significant, indicating that dyads where both states are democracies are more likely to form alliances than other dyads. For polity difference—which measures how similar the two countries' regime types are (larger values meaning a greater difference)—the results vary between the two models. In the model using COW alliance data, polity difference has a negative and significant effect, indicating that the more dissimilar the two states' regimes are, the less likely they are to ally. In contrast, the model using the ATOP data shows the opposite result. This difference, however, is due to the different time frame, not the different alliance data, because a positive and significant effect also results when the COW data are examined from 1816 to 1944 (Gibler and Sarkees 2004). Thus, prior to World War II, more dissimilar states were more likely to ally, but since then, they have become less likely to ally.

Table 6.1 Probit Results for Dyadic Alliance Formation

VARIABLE	COW ALLIANCES	ATOP ALLIANCES
Regime type		
Joint democracy	0.176 (0.031)***	0.152 (0.067)*
Polity difference	-0.010 (0.002)***	0.012 (0.004)***
Culture		
Joint religion	0.309 (0.021)***	0.029 (0.038)
Joint language	0.369 (0.035)***	0.144 (0.060)**
Joint ethnicity	-0.040 (0.042)	0.187 (0.061)**
Threat		
Conflict relations	-0.108 (0.040)***	0.036 (0.072)
Joint enemy	0.106 (0.027)***	0.428 (0.053)***
Amount of threat	0.024 (0.002)***	0.010 (0.005)**
Other controls		
Distance	-0.012 (0.000)***	-0.009 (0.001)***
Major power	-0.018 (0.033)	0.202 (0.041)***
Learning	0.151 (0.017)***	0.019 (0.030)
Ally lag variable	4.040 (0.023)***	3.892 (0.047)***
Constant	-2.321 (0.033)***	-2.630 (0.055)***
Wald χ^2	47,406.51	8,964.97
N	411,013	93,321

Notes: $^{*}p < 0.05$, $^{**}p < 0.01$, $^{***}p < 0.001$

Source: Adapted from Douglas M. Gibler and Meredith Reid Sarkees, "Measuring Alliances: The Correlates of War Formal Interstate Alliance Dataset, 1816–2000." *Journal of Peace Research* 41 (2004), 220. Model with COW alliances covers 1816–1992; model with ATOP alliances covers 1816–1944.

Gibler and Sarkees (2004) also examine three variables comparing the culture of the states in the dyad. Both models indicate that dyads that share a joint language are significantly more likely to ally than other dyads. The effect of joint religion and joint ethnicity, however, varies in the two models. For the model using the COW alliance data, joint religion makes an alliance more likely while joint ethnicity appears to have no effect. In contrast, joint religion has no effect while joint ethnicity makes the dyad more likely to ally in the model with the ATOP data. Nonetheless, both models show that similar regime types and cultures make states more likely to form alliances. These findings contradict realist arguments (e.g., Walt 1987) that ideological similarity does not matter for explaining alliance formation.

The third group of variables that they examine describes the level of external threat facing the states in the dyad. A recent history of conflict relations between the two states, determined by whether they have engaged on opposite sides of a militarized dispute in the last decade, has the expected negative impact on alliance formation when using the COW data. The model using the ATOP data, however, reveals no effect; this is consistent for the COW data in the pre–World War II time frame (Gibler and Sarkees 2004). The joint enemy variable identifies cases in which the two states have each fought militarized interstate disputes against a common enemy within the past decade. The effect of joint enemy is found to have a consistently positive effect. Finally, the amount of threat—the number of disputes the states have been involved in within the past ten years—also has a consistently positive effect. Thus, states facing greater security threats are more likely to form alliances, as we would expect from theories of alliance formation.

For their other control variables, Gibler and Sarkees (2004) find mostly unsurprising results. First, they find that distance has a negative effect, meaning that pairs of states that are farther apart are less likely to ally. For the effect of major power status, their results depend on the measure of alliances used. The model with the COW data indicates that major power status has no effect on alliance formation, whereas the model with the ATOP data indicates that major powers are more likely to form alliances.

Lai and Reiter (2000) introduced a learning variable, indicating whether states had learned a previous lesson favoring involvement in alliances or a previous lesson favoring neutrality. The results from the COW alliance data indicate that positive lessons about alliances make alliance formation more likely, whereas the results from the ATOP data indicate that learning has no effect. Finally, the ally lag variable is strongly positive, indicating that dyads that were allied last year are likely to still be allied this year.

Maoz (2000) also examines the origins of alliances, using a different research design focusing on individual states in relation to their politically relevant international environments—the set of countries a state has the opportunity for conflict with. He finds that the size of a state's politically relevant international environment (PRIE) has a consistently positive effect on alliance commitments, indicating that the more states in a state's strategic reference group, the more likely the state is to form alliances.

The proportion of democracies in the PRIE also increases the likelihood of forming alliances. Maoz also finds that power matters, as the stronger a state is relative to its PRIE, the more likely it is to form alliances. Finally, he finds that the more alliances that other states in the PRIE are in, the more likely a state is to form alliances.

Alliance Maintenance. Bennett (1997b) examines factors that affect alliance duration, focusing on four different explanatory models. He finds the most empirical support for the security and autonomy model of alliance formation and duration. In particular, he finds that alliances where members have large changes in capabilities are likely to end sooner, and alliances with greater symmetry in power between the members are also likely to be shorter than asymmetric alliances. However, Bennett (1997b) finds little support for the capability-aggregation model of alliances, which argues that states form alliances purely in pursuit of security. Of the three variables following from this explanation, one is significant in the expected direction (security improvement due to alliance), one is significant in the opposite direction than expected (change in security), and one has no effect (mutual threats).

With regard to other factors, Bennett (1997b) finds that domestic politics have a significant effect on the duration of an alliance, as the more democracies are in the alliance, the longer it lasts. Changes in alliance members' regime types, however, appear to have no effect on its duration. Furthermore, he finds no support for arguments that institutionalization of alliances drive how long they last. On the contrary, he finds that the longer an alliance has been in existence, the more likely it is to terminate.

Maoz (2000) also examines the management of alliances (alliance duration), using a similar research design to his analysis of alliance formation. He finds that the size of a state's politically relevant international environment (PRIE) has a negative impact on the probability of alliance termination. In addition, the level of democratization in the PRIE has a negative impact on the probability of alliance termination. Finally, he finds that the stronger a state is relative to its PRIE, the more likely is alliance termination.

In summary, a variety of different factors have been found to affect alliance formation and maintenance. Some of these factors, such as amount of threat, joint enemy, and distance, support ideas that states form alliances in order to enhance their security. Factors relating to regime type and culture, however, also have important effects on alliance formation, and these are not explained by a realist perspective. Overall, empirical analyses provide much better support for the argument that states form alliances to gain a favorable tradeoff between security and autonomy than the argument that alliances are all about security.

Drawbacks of Alliances

Politicians and scholars have pointed to several drawbacks of alliances over the years. We will examine several of these potential drawbacks, focusing on entrapment, chain-ganging, unreliability, and the potential for alliances to lead to conflict.

Entrapment is the idea that alliances can harm a country by dragging it into the wars of other countries. Concerns about entrapment dominated US foreign policy through World War II. In his farewell address, George Washington (1796) argued that "it is our true policy to steer clear of permanent alliances with any portion of the foreign world." Many people believe that Washington called them entangling alliances, but that was Thomas Jefferson (1801), who quite similarly argued that the United States should pursue "peace, commerce, and honest friendship with all nations, [but] entangling alliances with none" in his inaugural address.

These statements by Washington and Jefferson were the beginnings of the long-standing American policy of isolationism. Although the United States began to emerge from isolationism somewhat by involvement in the Spanish-American War of 1898 and the last year of World War I, it retreated back into isolationism, illustrated primarily by the US refusal to join the League of Nations. Not until after World War II did the United States fully emerge from this shell of isolationism and take an active role in alliances, as highlighted by NATO. Nonetheless, this move away from isolationism was controversial within American politics (Fromkin 1970).

A related drawback of alliances is chain-ganging, which we discussed in the previous chapter. This is the problem in multipolar systems pointed to by Christensen and Snyder (1990) in which alliances are tight and states can get drawn into situations that they do not really want to be involved in. This is said to be what led to World War I, leading a dispute between Austria-Hungary and Serbia to become a large war involving all of the major powers of Europe. Chain-ganging could be thought of as entrapment on a larger scale.

A third problem with alliances is that empirical analyses have indicated that they are not particularly reliable. Sabrosky (1980) finds that from 1816 to 1965 only 27 percent of alliance members fought alongside their partners when they were attacked. Leeds, Long, and Mitchell (2000) demonstrate that Sabrosky underestimates alliance reliability because he does not account for the actual obligations and provisions included in alliances. Nonetheless, they also find that alliance commitments are not always honored. Theoretical analyses indicate that this trend occurs because challenging states tend to target unreliable alliances (e.g., Quackenbush 2006b; Smith 1996). Thus, wars usually remain bilateral (Gartner and Siverson 1996), a point we examine in more detail in chapter 11.

Finally, several scholars argue that another drawback of alliances is that they can lead to conflict. Vasquez (1993) and Senese and Vasquez (2008) argue that alliances, as a central feature of power politics, threaten other states and are an important part of the steps to war. We will examine this steps-to-war argument in more detail in our discussion of the escalation of disputes to war in chapter 9. Kimball (2006), however, argues that alliances are often formed in anticipation of a coming war. Therefore, the causal arrow is reversed, with the war causing the alliance rather than the alliance causing the war. We return to the subject of alliances as a cause of conflict for a closer examination later in the chapter.

Measuring Shared Interests

Alliances have also been used to measure shared interests between states. This can be done because states tend to form alliances with other countries that they share some common interests with, and they tend to not form alliances with countries that they share nothing in common with. To measure shared interests across states, Bueno de Mesquita (1975) evaluated the similarity in alliance portfolios. An **alliance portfolio** is defined as a nation's complete array of alliance commitments. Thus, in comparing alliance portfolios we are assessing the degree to which the full array of alliance commitments of one nation matches that of another nation.

Some basic examples will help to illustrate alliance portfolios, and how they help to identify shared interests between states. Table 6.2 shows the alliance portfolios of the Netherlands and Canada in 1951, two countries with very similar portfolios. In fact, their alliance portfolios are identical, as there are fourteen other countries (through NATO) that they both had defense pacts with, and neither state had an alliance with the other sixty-three countries in the data.

Table 6.3 shows an example of two countries with dissimilar alliance portfolios, the United States and Soviet Union in 1953. There are no countries that both the United States and Soviet Union were allied with. The United States had defense pacts with thirty-four countries and ententes with four (through NATO and other alliances), while the Soviet Union had no alliance with any of them. The Soviet Union had defense pacts with ten countries and a neutrality agreement with one (through the Warsaw Pact and other alliances), while the United States had no alliance with any of them. Finally, there were eleven additional countries in the data that neither side had an alliance with.

Table 6.2 Alliance Patterns for Netherlands and Canada, 1951

	NETHERLANDS			
CANADA	*DEFENSE*	*NEUTRALITY*	*ENTENTE*	*NO ALLIANCE*
Defense	14	0	0	0
Neutrality	0	0	0	0
Entente	0	0	0	0
No alliance	0	0	0	63

Source: Adapted from Bruce Bueno de Mesquita, "Measuring Systemic Polarity." *Journal of Conflict Resolution* 19 (1975), 197.

Table 6.3 Alliance Patterns for United States and Soviet Union, 1953

	UNITED STATES			
SOVIET UNION	*DEFENSE*	*NEUTRALITY*	*ENTENTE*	*NO ALLIANCE*
Defense	0	0	0	10
Neutrality	0	0	0	1
Entente	0	0	0	0
No alliance	34	0	4	11

Source: Adapted from Bruce Bueno de Mesquita, "Measuring Systemic Polarity." *Journal of Conflict Resolution* 19 (1975), 198.

Bueno de Mesquita (1975) introduced a measure called τ_b (tau-b) to identify the similarity in alliance portfolios. Signorino and Ritter (1999), however, demonstrate that τ_b is flawed as a measure, and they introduce a new measure, called the *S*-score (or simply *S*), which provides a much better measure of shared interests.

Although it is not necessary for our purposes to be able to calculate an *S*-score, it is important to understand what it represents. The *S*-score is an index that ranges from -1 to 1. When the score is -1, the alliance portfolios of nations A and B are as different as possible. When the score is 1, the alliance portfolios of nations A and B are as similar as possible. Scores falling in between these extremes reflect varying degrees of foreign policy similarity. Going back to the examples shown above, the *S*-score for the Netherlands and Canada in 1951 is 1, while the *S*-score for the United States and Soviet Union in 1953 is -0.354.

To see an example of how this can help us measure shared interests between states, we can examine the alliance portfolios of Britain, France, and Germany on the eve of World War I. Britain had defense pacts with Japan and Portugal, and ententes with France, Russia, and Spain. Germany had defense pacts with Austria-Hungary, Italy, and Romania. France had a defense pact with Russia, a neutrality pact with Italy, and ententes with Britain, Japan, and Spain.

Given these alliance portfolios, the *S*-score for Britain and Germany equals 0.24, while for Britain and France, it is 0.59, and for France and Germany, it is 0.40. We can easily see from these scores that Britain favored France over Germany and that France favored Britain over Germany. This helps to explain why Britain and France fought together in the war.

Signorino and Ritter (1999) argue that alliance portfolios are limited as a measure of shared interests because alliances are just one of a number of other behaviors that indicate shared interests between states. Gartzke (1998, 2000) and others have used United Nations (UN) General Assembly voting data to measure the affinity between states. This literature on UN voting (e.g., Gartzke 1998, 2000; Voeten 2000, 2004) assumes that higher levels of agreement between states in their UN General Assembly votes reflect higher levels of satisfaction with one another. This provides a measure of foreign policy similarity that captures a broader range of behaviors than alliance portfolios, although it is only available in the post–World War II time period.

Alliances and Conflict

Ultimately, our primary interest here is understanding how alliances affect international conflict, rather than an overall concern with the impact of alliances on international relations. The short answer is that the relationship between alliances and international conflict is complicated, as they can be a cause of peace but they can also be a cause of conflict. We consider each trend in turn.

Alliances as a Cause of Peace

We begin our assessment of the consequences of alliances for conflict within a dyad by analyzing politically active dyads from 1816 to 2000. The results of regression models with militarized interstate disputes as the dependent variable are shown in Table 6.4. In the left column, alliance is measured using the COW data, while the ATOP data are used to measure alliance for the model in the right column. In each case, the alliance variable is dichotomous, simply indicating whether the particular dyad in the given year is allied or not. In both models, the coefficient on the alliance variable is negative and highly significant. Thus, alliance commitment has a consistently pacifying effect on conflict initiation between allies. Maoz (2000) finds that alliances make the escalation of disputes between allies to war less likely as well.

Table 6.4 Logit Results for Prediction of Militarized Interstate Disputes

VARIABLE		COW ALLIANCES	ATOP ALLIANCES
Alliance	β	-0.5586***	-0.3324***
	Se_β	0.0503	0.0458
Joint democracy		-1.1095***	-1.0986***
		0.0724	0.0731
Contiguity		2.1439***	2.1204***
		0.0392	0.0393
Relative power		-0.3904**	-0.3238*
		0.1291	0.1291
Constant		-4.1997***	-4.2766***
		0.1115	0.1118
Wald χ^2		3,317.1***	3,236.7***
Log-likelihood		-13,073.9	-13,114.1
N		180,987	180,987

Notes: $^{*}p < 0.05$, $^{**}p < 0.01$, $^{***}p < 0.001$

Source: Compiled by author.

The control variables are factors highlighted by Bremer (1992) in his study of dangerous dyads. Joint democracy is negative and significant, indicating that jointly democratic dyads are less likely to fight than other pairs of states. We explore this in more detail in the next chapter. Contiguity is strongly positive and significant, indicating that contiguous states are far more likely to fight, as our discussion in chapter 4 led us to expect. Finally, relative power is negative and significant, indicating that as power becomes more imbalanced within the dyad, conflict is less likely. In chapter 5, we examined debates about whether a balance of power or a preponderance of power is more peaceful, and these results support the latter expectation.

The results show that the overall impact of an alliance on the likelihood that allied states fight each other is to make conflict less likely. There are different types of alliance agreement, however, as documented above. Because they entail different types of commitment, they may also have different effects on the likelihood of conflict between allies. Accordingly, the model in Table 6.5 breaks down the different types of alliance agreements according to the categories in the ATOP data.

Table 6.5 Logit Results for Prediction of Militarized Interstate Disputes

VARIABLE		ATOP ALLIANCES
Defense agreement	β	-0.3927***
	Se_{β}	0.1027
Offense agreement		0.03417
		0.1774
Neutrality agreement		0.4801***
		0.1403
Nonaggression agreement		-0.1733*
		0.0732
Consultation agreement		-0.0482
		0.1013
Joint democracy		-1.0632***
		0.0733
Contiguity		2.1212***
		0.0394
Relative power		-0.3793**
		0.1291
Constant		-4.2130***
		0.1116
Wald χ^2		3,315.7***
Log-likelihood		-13,074.6
N		180,987

Notes: *$p < 0.05$, **$p < 0.01$, ***$p < 0.001$

Source: Compiled by author.

Dyads with defense pacts are significantly less likely to fight each other, which is exactly as expected. On the other hand, offense agreements have no impact on the likelihood of conflict. This is not surprising, since offense agreements are specifically made with conflict in mind, rather than the deterrence of conflict (Leeds 2003).

Perhaps the most surprising finding is that neutrality agreements make conflict much more likely. Neutrality provisions are the least common type of provision, other than offensive provisions. It would seem that states include them in alliances only when they believe that their new allies might have reason to not remain neutral in a potential conflict. For example, India and Bangladesh signed an alliance in March 1972; Bangladesh had just become independent (from Pakistan) in December 1971 following the latest Indo-Pakistani war. Given this history of conflict, they could reasonably have been concerned about the potential for future conflict, and therefore agreed that each state "shall refrain from giving any assistance to any third party taking part in an armed conflict, against the other party" (Gibler 2009, 456). Despite the alliance, which continued until 1997, India and Bangladesh engaged in eight militarized disputes within twenty years (in 1976, 1980, 1981, 1983, 1986, 1987, 1995, and 1996).

Although neutrality agreements are problematic, nonaggression agreements make conflict less likely between allied states, as they are intended to do, although their pacifying effect is weaker than defense pacts. Nonaggression provisions are the most frequent other provision to go along with neutrality agreements, happening in about 62 percent of neutrality agreements (69 of 112). In these alliances in which both neutrality and nonaggression provisions are included, their effects cancel each other out somewhat. On the other hand, the nonaggression agreements without neutrality provisions, which are the strong majority (246 of 315, or 78 percent), are more likely to lead to peace within the dyad.

Like offensive alliances, consultation agreements are also found to have no effect on the likelihood of conflict. It is important to remember that these are not mutually exclusive categories, because all obligations contained within an alliance are included in the ATOP data, not only the most serious. Thus, the most pacifying alliance would be one with both defense and nonaggression provisions, but not a neutrality provision. There are forty-eight such alliances, including NATO and the alliance formed between Russia, Belarus, Kazakhstan, and other former Soviet republics following the breakup of the Soviet Union. Finally, the findings for democracy, contiguity, and power are all the same as before.

In summary, allied states are generally less likely to fight each other than other pairs of states. However, their effect depends on the specific type(s) of agreements included within the alliance. Defense and nonaggression agreements make conflict significantly less likely, but offense and consultation agreements have no effect, while neutrality provisions make conflict between the allies significantly more likely.

Alliances as a Cause of Conflict

Although alliances can reduce the likelihood of conflict between member countries, they also have the potential to make conflict more likely as states that form alliances then sometimes fight wars with other states (Gibler and Vasquez 1998; Senese and Vasquez 2008; Vasquez 1993). Gibler (2000) examines the question of why some alliances cause war and others cause peace and argues that the formation of an alliance sends signals to other countries, who use those signals to assess how much of a threat the alliance poses.

Most alliances are security alliances, and there are important cues that other states can focus on to assess the threat. The first is the power status of the member states. We have already seen that major powers are more likely to fight than minor powers. Thus, an alliance formed between major powers is more threatening than one formed between minor powers. The second signal Gibler (2000) identifies is success in war. Countries with a history of success in war are more likely to be willing to fight again in the future, while those with a history of failure are less likely to do so. Finally, the satisfaction of the states with the status quo sends an important signal to others, as dissatisfied states are more likely to seek to challenge the status quo, leading to conflict.

Gibler (1996) identifies a different category of alliance, the **territorial settlement treaty**. These treaties contain provisions settling territorial issues between states. Gibler identifies two types of territorial settlement treaties. Most of them (21 of 27) are status quo settlements, "where two or more states agreed to settle a disputed territory by formally accepting the *status quo* distribution of control in the region" (1996, 80). For example, newly independent Bangladesh signed a neutrality/nonaggression pact with India in 1972 in which they agreed to their shared border. Less frequently observed are exchange settlements (6 of 27), "where alliance members have actually exchanged territory through the use of an alliance" (Gibler 1996, 78). For example, in 1977 the United States and Panama signed a defense pact in which the Panama Canal was transferred to Panama and Panama agreed to the permanent neutrality and continued operation of the canal.

We already saw in chapter 4 how conflict prone territorial issues are, so settling them bodes very well for peace (Gibler 2012). Thus, Gibler (2000) argues that these territorial settlement treaties do not affect the war proneness of states because they send nonthreatening signals to other countries.

Table 6.6 shows the impact of territorial settlement treaties on the likelihood of war following the alliance. To be counted, the war must have occurred within five years of the alliance formation and must be related to the alliance. As we can see from the average alliance row in the table, there have been fifty-seven wars following 193 alliances. Thus, wars follow alliances nearly 30 percent (Pr[war] = 0.2953) of the time. There have been 27 territorial settlement treaties between 1815 and 1980, and if war follows them at the average rate, we would expect about 8 wars (which we can find by multiplying 27 by 0.2953). However, only 1 war was observed; furthermore, as the z-score indicates, this difference is statistically significant. While we would expect about 49 wars following the 166 alliances that were not territorial settlement treaties, we actually observe 56. Thus, Gibler's argument that territorial settlement treaties are more peaceful than security alliances is strongly supported.

Gibler (2000) moves beyond the simple distinction between territorial settlement treaties and security alliances to assess the threat posed by security alliances. To classify the combativeness of security alliances, Gibler develops a bellicosity score. To do so, he focuses on the signals discussed above: success in war, satisfaction, and major power status. Each of these characteristics can be absent (where none of the members has them), mixed (where some do but some do not), or present (where all of the members has the characteristic). Combined, the scale varies between 1 and 27, with larger scores indicating higher levels of bellicosity.

Table 6.7 shows the overall relationship between alliances and war. The top row shows all alliances, which have seen 57 wars following 193 alliances, which is about 30 percent of the time. This is the same result as shown in Table 6.6. Following territorial settlement treaties, Gibler (2000) expects decreased armament levels, decreased involvement in crises, and decreased involvement in alliances, resulting in a higher likelihood of peace. The actual result is 1 war in 27 cases, or 3.7 percent of the time. Thus, territorial settlement treaties are about 10 percent as likely to result in war as alliances on average.

For security alliances, Gibler's expectations depend on the bellicosity of the alliance. Low-bellicosity alliances (with a bellicosity score less than 3) are expected to be relatively peaceful. We see 6 wars following the 40 alliances in this category, or 15 percent of the time, half the rate of alliances on average. Medium-bellicosity alliances are expected to be situation dependent, and we see 31 wars in 91 cases, or about 34 percent of the time, which is

Table 6.6 Conditional Probabilities of War Involving an Ally

ALLIANCE CHARACTERISTIC	1815–1980 WAR COUNTS: ACTUAL	EXPECTED	N	Z	PR(Z)
Territorial settlement treaty					
Yes	1	(7.97)	27	-2.94**	0.0016
No	56	(49.03)	166	1.19	0.1170
Average alliance	57	(57.00)	193	Pr(war) = 0.2953	

** $p < 0.01$

Source: Adapted from Douglas M. Gibler, "Alliances: Why Some Cause War and Why Others Cause Peace." In *What Do We Know About War?*, ed. John A. Vasquez (Lanham, MD: Rowman and Littlefield, 2000), 155.

Table 6.7 The Overall Relationship Between Alliances and War (Predicted and Actual)

		ACTUAL RESULT		
TYPE OF ALLIANCE	PREDICTED RESULT	WARS	N	RATE
All alliances	Unknown	57	193	0.2953
Territorial settlement treaty	Peace	1	27	0.0370
Security alliances				
Low-bellicosity alliances (bellicosity score: 1–3)	No effect/peace	6	40	0.1500
Medium-bellicosity alliances (bellicosity score: 4–9)	Situation-dependent	31	91	0.3407
High-bellicosity alliances (bellicosity score: 12–27)	War	19	35	0.5429

Source: Adapted from Douglas M. Gibler, "Alliances: Why Some Cause War and Why Others Cause Peace." In *What Do We Know About War?*, ed. John A. Vasquez (Lanham, MD: Rowman and Littlefield, 2000), 160.

about average. Gibler expects high-bellicosity alliances to lead to increased armament levels, increased involvement in crises, and increased involvement in alliances, and thus be highly war prone. The actual result is 19 wars in 35 cases, which is over 54 percent of the time, nearly twice the average rate.

These findings provide strong support for Gibler's argument about the war proneness of alliances. Territorial settlement treaties are much more peaceful than security alliances. In addition, success in war, satisfaction with the status quo, and the major power status of allies are all important indicators of the bellicosity of alliances.

Conclusion

Alliances have been a primary way for states to cooperate with one another in the international system and have been our focus in this chapter. There are a variety of different possible types of alliances. When people think of alliances they often think about defense agreements. However, offense, neutrality, nonaggression, and consultation agreements are also primary types of provisions contained within alliances.

The question of why states form alliances is an important one. Although scholars traditionally argued that alliance formation is driven entirely by states' concerns about security, focusing on pursuing favorable tradeoffs between security and autonomy provides a better explanation of alliance formation and duration. Furthermore, empirical analyses support the security and autonomy explanation.

Although alliances are a frequently used instrument of foreign policy, they have some drawbacks. Nonetheless, alliances can be used to measure shared interests between states. The relationship between alliances and conflict is complicated, as alliances can make conflict both less likely and more likely. Allies are less likely to fight each other, although the formation of an alliance may form a threat to other states and lead to conflict.

Key Concepts

Alliance

Alliance portfolio

Autonomy

Balance of threat theory

Balancing

Bandwagoning

Consultation agreement (entente)

Entrapment

Mutual defense agreement

Neutrality agreement

Nonaggression agreement

Offensive alliance

Security

Statistical significance

Substantive significance

Territorial settlement treaty

7 Democratic Peace

An Iraqi woman casts her ballot in Fallujah on 20 June 2013. Democratic peace theory suggests that such elections have important consequences for international conflict.

Source: Associated Press

In his 1994 State of the Union address, President Bill Clinton said that "ultimately, the best strategy to ensure our security and to build a durable peace is to support the advance of democracy elsewhere. Democracies don't attack each other" (Clinton 1994). Similarly, President George W. Bush stated in a 2004 press conference that "the reason why I'm so strong on democracy is democracies don't go to war with each other. And the reason why is the people of most societies don't like war, and they understand what war means. . . . I've got great faith in democracies to promote peace. And that's why I'm such a strong believer that the way forward in the Middle East, the broader Middle East, is to promote democracy" (Bush 2004).

These comments by presidents from both major political parties in the United States are quite similar in nature and have been echoed by other leaders around the world. But are their views justified? Does democracy lead to peace? Before examining these questions further, we need to establish what we mean by democracy. A **democracy** is country with three basic characteristics: "(1) most citizens can vote, (2) the government comes to power in a free and fair election contested by two or more parties, and (3) the executive is either popularly elected (a presidential system) or is held responsible to an elected legislature (a parliamentary system)" (Dahl 1971; Russett and Oneal 2001, 44).

The short answer to the question of whether these views are justified is that substantial evidence indicates that pairs of democracies are less likely to fight each other than are other pairs of states. This is known as the **democratic peace**, and the findings are so robust that Levy (1988, 662) claimed that the democratic peace is "as close as anything we have to an empirical law in international relations." Although democracies rarely fight other democracies, they also regularly fight nondemocracies.

In the first section of the chapter, we examine empirical findings regarding the democratic peace. We then review the primary explanations of the democratic peace, which focus on either democratic norms or on the structure of democratic institutions. Democracy is just one aspect of liberalism; the Kantian triangle of peace focuses on broader conceptions of a liberal peace, and we examine it in the third main section of the chapter. Finally, we explore alternative explanations for peace between democracies, particularly those focusing on common interests and capitalism.

Empirical Findings Regarding the Democratic Peace

Studies about the effects of regime type have become increasingly prominent in research on international relations in the past several decades. At the center of these efforts is the democratic peace, which is a well-established empirical law. Theoretical explanations of the democratic peace were developed in response to early empirical findings. In keeping with this basic order of events, we examine empirical evidence regarding the relationship between democracy and international conflict before turning our attention to explanations.

Studies of the effects of democracy, and regime type more generally, depend on the ability to measure democracy. The measure of democracy that is most commonly used in studies of international relations comes from the Polity IV data set (Marshall, Jaggers, and Gurr 2010). These data focus on the competitiveness of political participation, the openness and competitiveness of executive recruitment, and constraints on the chief executive and are available online at www.systemicpeace.org/polity/polity4.htm. The primary variables used are the democracy and autocracy scores, which each range from 0 to 10. We can subtract the autocracy score from the democracy score, resulting in an overall scale (called *polity*), which ranges from -10 to 10. We can use a dichotomous measure by coding each state as a democracy or nondemocracy; when a state is democratic it is coded as a 1, otherwise 0. For the empirical analyses in this section, a state is considered a democracy if its polity score is greater than or equal to 5, or a nondemocracy otherwise. Sometimes other thresholds are used (typically 6 or 7), but the results do not change much.

Figure 7.1 uses Polity data to show trends in democracy over time. We can easily see that there has been tremendous growth in democracy in the

Figure 7.1 Growth of Democracy in the International System

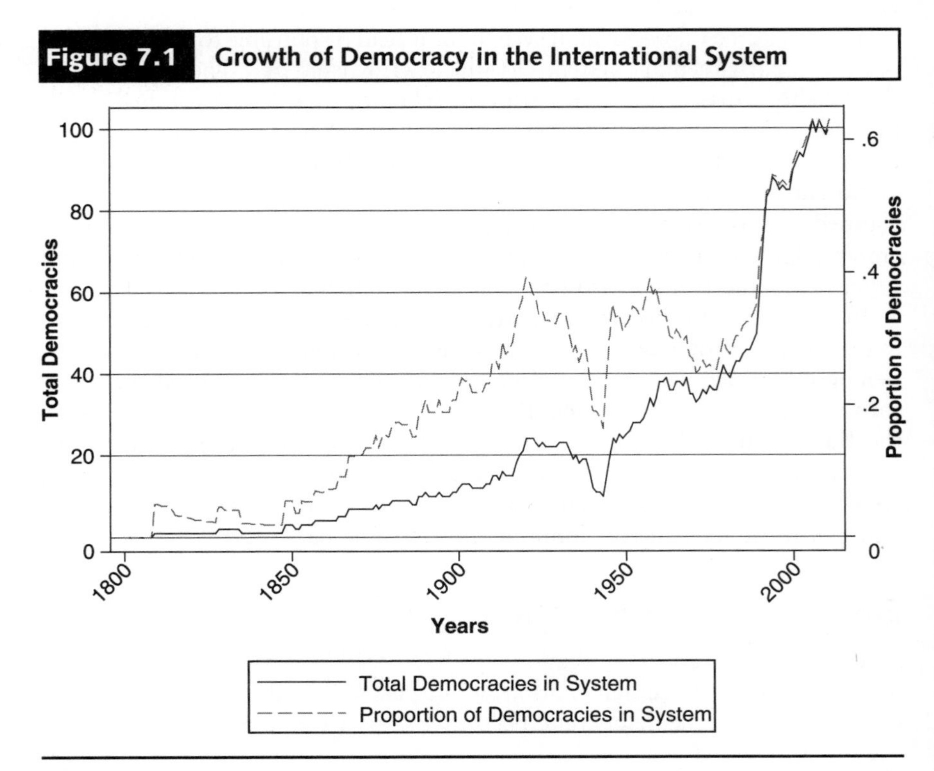

Source: Compiled by author.

international system over the past two hundred years. The solid line in the figure shows the total number of democracies in the system, while the dashed line indicates the proportion of democracies (the number of democracies divided by the total number of countries). From 1800 through 1932, there was a slow but steady growth in the number of democracies, peaking at a total of 23 countries or 33.8 percent of the system. Through World War II, there was a sharp decline in democracies, bottoming out at 10 democracies (16.4 percent of the system). Since then, the international system has experienced continual growth in democracy, which accelerated rapidly after 1990. In 2010, there were a total of 102 democracies in the world, representing 62.2 percent of all countries.

There are a number of different possible ways to examine the relationship between democracy and international conflict. In particular, we can examine (and form expectations about) whether democracies are more peaceful in general or are only peaceful in their relations with other democracies and also about whether democracies are more or less likely to initiate conflicts than nondemocracies. Accordingly, it is important to examine the logic of different democratic peace expectations in order to focus attention on appropriate empirical tests of the relationship between democracy and conflict.

The fundamental argument of the dyadic democratic peace is that pairs of democracies are less likely to fight than any other pairs of states. In other words, the conventional wisdom within the democratic peace literature is that democracies are peaceful *only* in their relationships with other democracies, not in general. Furthermore, substantial empirical evidence has supported this idea (e.g., Babst 1972; Chan 1997; Maoz and Abdolali 1989; Maoz and Russett 1993; Morgan and Campbell 1991; Oneal and Russett 1997). We can state this idea more formally as

$$\Pr(\text{fight} \mid \text{joint-D}, \mathbf{x}) < \Pr(\text{fight} \mid \text{not joint-D}, \mathbf{x}) \qquad \text{(Equation 7.1)}$$

where fight represents militarized interstates disputes and wars, joint-D represents a jointly democratic dyad, and **x** represents the set of other factors explaining international conflict.[1]

While most of the literature focuses on the dyadic peace, some argue that it is a monadic phenomenon. The idea of the **monadic democratic peace** is that democracies are more peaceful than other regimes in general, not just in their relationships with other democracies (Huth and Allee 2002; MacMillian 1998, 2003; Ray 2000; Rousseau, Gelpi, Reiter, and Huth 1996; Rummel 1979, 1983, 1985, 1995). The similar expectation for the monadic democratic peace argument is that democracies are less likely to fight than other states. More formally,

$$\Pr(\text{D fights} \mid \mathbf{x}) < \Pr(\sim\text{D fights} \mid \mathbf{x}) \qquad \text{(Equation 7.2)}$$

where D represents a democracy and ~D represents a nondemocracy. Note that in equation 7.1, the pacifying nature of democracy is contingent on the

[1] Thus, Pr(fight | joint-D, **x**) means the probability of a fight, given joint democracy and **x**.

opponent's regime type; democracies are only more peaceful when facing other democracies. In equation 7.2, however, the expectation is not contingent; democracies are expected to be more peaceful regardless of the opponent.

Strong empirical support for equation 7.2 would be the strongest possible support for the monadic proposition. Some, however, argue that the key to the monadic peace is that democracies are less likely to initiate conflicts than nondemocracies (e.g., Huth and Allee 2002):

$$\Pr(\text{D initiates} \mid \mathbf{x}) < \Pr(\sim\text{D initiates} \mid \mathbf{x}). \qquad \text{(Equation 7.3)}$$

Thus, even if equation 7.2 is not empirically supported, empirical support for equation 7.3 would provide some evidence in favor of the monadic democratic peace. To test this equation, we need to look not only at the effect of democracy on the frequency of international conflict (as we do in the next section), but also the effect of democracy on the initiation of international conflict (as we do in the section after that).

If democracy truly pacifies relations between states, then democracies should also be less likely to be targeted than nondemocracies. Thus, we would expect that

$$\Pr(\sim\text{D initiates} \mid \text{D}, \mathbf{x}) < \Pr(\sim\text{D initiates} \mid \sim\text{D}, \mathbf{x}). \qquad \text{(Equation 7.4)}$$

Together, equations 7.3 and 7.4 lead to explicit expectations for the probability of initiation for any pair of states, as follows:

$$\Pr(\text{D initiates} \mid \text{D}, \mathbf{x}) < \Pr(\text{D initiates} \mid \sim\text{D}, \mathbf{x}) <$$
$$\Pr(\sim\text{D initiates} \mid \text{D}, \mathbf{x}) < \Pr(\sim\text{D initiates} \mid \sim\text{D}, \mathbf{x}). \qquad \text{(Equation 7.5)}$$

Equation 7.5 shows the combined expectations that democracies are less likely to initiate conflict than nondemocracies and that democracies are less likely to be targeted than nondemocracies.

A clear logical implication of equation 7.5 is that dyads with at least one democracy are less likely to fight than jointly nondemocratic dyads. More formally,

$$\Pr(\text{fight} \mid \text{not joint-}{\sim}\text{D}, \mathbf{x}) < \Pr(\text{fight} \mid \text{joint-}{\sim}\text{D}, \mathbf{x}). \qquad \text{(Equation 7.6)}$$

Together, equations 7.1 and 7.6 nicely capture MacMillan's (2003, 233) monadic argument "that while liberal states are *especially* peace prone in relations with other liberal states, they are not *only* peace prone with other liberal states, but also more broadly." Our examination of empirical evidence regarding the relationship between democracy and international conflict in the following three sections is guided by the logic set forth in these equations. The first two sections focus on dyadic analyses to test equations 7.1, 7.4, 7.5, and 7.6, while the third section uses monadic analyses to test equations 7.2 and 7.3.

Frequency of Democratic Conflict

We begin our look at the empirical evidence regarding the democratic peace by examining the impact of regime type on the frequency of international conflict. These dyadic tests are obtained through analyses of politically active dyads from 1816 to 2000, and are based on the work of Quackenbush and Rudy (2009). As the discussion above indicates, there are different ways to test the democratic peace. We want to test both the dyadic and the monadic democratic peace arguments. We begin here by examining the impact of democracy on involvement in militarized disputes.

Table 7.1 displays the results of a logit model in which the dependent variable is the occurrence of a militarized interstate dispute within a (nondirected) dyad year. Model 1 tests the expectations expressed in equations 7.1 and 7.6 that conflict is less likely in jointly democratic dyads and most likely in jointly nondemocratic dyads, with mixed dyads in between. The coefficient for Both Democratic is negative and highly significant, indicating that when both states in a dyad are democratic, disputes are much less likely to occur than if neither is. On the contrary, One Democratic is positive and significant, which indicates that dyads containing exactly one democracy (i.e., mixed dyads) are significantly more likely to fight than jointly autocratic dyads. The effects of the control variables are all in the expected directions, although the effect of power parity is not statistically significant.

Table 7.1 Logit Results for Prediction of Militarized Interstate Dispute Occurrence

VARIABLE		MODEL 1
Both Democratic	β	-0.3527***
	Se_β	0.1279
One Democratic		0.4586***
		0.0776
S-Score		-0.2747***
		0.2330
ln (Distance)		-0.2469***
		0.0131
Power Parity		0.0477
		0.1514
Peace Years		-0.2979***
		0.0171
Constant		-1.1699***
		0.1909
Wald χ^2		2,094.7***
Log-likelihood		-8,631.5
N		163,920

Notes: *$p < 0.05$, **$p < 0.01$, ***$p < 0.001$

Unit of analysis is dyad-years. Peace years cubic spline variables not shown. Standard errors are robust standard errors adjusted for clustering within dyads.

Source: Compiled by author.

Although the direction and significance of the coefficients suggest the effect of democracy on dispute involvement, we can get a better idea of the substantive effect by examining the predicted probabilities. Setting the control variables to their means and just varying the regime type, we find that the probability of conflict within a jointly democratic dyad is 0.0034. As expected, the predicted probability for nondemocratic dyads is higher, at 0.0049, but the predicted probability that mixed dyads fight is 0.0077, the highest overall. Thus, while equation 7.1 is supported, these results are the opposite of the prediction made by the monadic democratic peace in equation 7.6; instead, the presence of a single democracy within a dyad significantly *increases* the likelihood of international conflict.[2]

While the results in model 1 strongly contradict the monadic democratic peace proposition, it is possible that this is driven by some peculiarities in the Polity data that we use to measure democracy. Casper and Tufis (2003) point out that different measures of democracy can produce very different results in various applications. Nonetheless, Quackenbush and Rudy (2009) examine three other measures of democracy and find very consistent results.

Democracy and Conflict Initiation

These dyadic results make it clear that whereas jointly democratic dyads are the most peaceful, mixed dyads of one democracy and one nondemocracy are the most conflict prone. Thus, when we focus on the frequency of international conflict, there is strong support for the dyadic democratic peace, but no support for the monadic democratic peace. There are other ways, however, to examine the relationship between democracy and conflict.

Supporters of the monadic democratic peace (e.g., Huth and Allee 2002; MacMillan 2003; Rioux 1998) have argued that while democracies may indeed fight as frequently as other states, they are less likely to initiate conflict. This argument is well summarized by equation 7.5. Therefore, it is important to examine the impact of democracy on militarized interstate dispute initiation. We do this through an analysis of directed dyad years, where the dependent variable is dispute initiation as shown in Table 7.2.

Model 2 begins to address the impact of regime type on conflict initiation by including separate variables for whether State A (the potential initiator) and State B (the potential target) are democratic. The effect of State B democratic is positive and highly significant, indicating that democracies are indeed more likely to be targeted than nondemocracies. Although the effect of State A democratic is negative, however, it does not come close to a reasonable level of significance ($p = 0.15$). Again, the expectations of the monadic democratic peace argument are not supported.

The control variables are all in line with expectations. As states' foreign policy positions become more similar (as reflected by the *S*-score in Table 7.2), as the distance between the states increases, or as the number

[2] Of course we cannot determine from this result whether democracies are the targets or initiators; we examine that below. There should, however, be an increased likelihood of peace when a democracy is in a dyad in order to meaningfully speak of a monadic democratic peace.

Table 7.2 Logit Results for Prediction of Militarized Interstate Dispute Initiation

VARIABLE		MODEL 2	MODEL 3
State A democratic	β	-0.1213	0.3499***
	Se_β	0.0842	0.0985
State B democratic		0.1821*	0.5954***
		0.0760	0.0905
State A democratic * State B democratic		—	-1.2622***
			0.1522
S-Score		-0.4162	-0.3162
		0.2363	0.2365
ln (Distance)		-0.2627***	-0.2729***
		0.0139	0.0162
Relative power		0.6359***	0.6326***
		0.0865	0.0860
Peace years		-0.2787***	-0.2758**
		0.0180	0.0178
Constant		-1.8822***	-2.0603***
		0.2158	0.2132
Wald χ^2		1,751.7***	1,902.8***
Log-likelihood		-10,443.9	-10,373.4
N		325,990	325,990

Notes: *$p < 0.05$, **$p < 0.01$, ***$p < 0.001$

Unit of analysis is directed-dyad years. Peace years cubic spline variables are included in the analysis but not shown in the table. Standard errors are robust standard errors adjusted for clustering within dyads.

Source: Compiled by author.

of peace years since the last dispute increases, each state is less likely to initiate conflict. Finally, the stronger that a state is relative to its potential adversary, the more likely it is to initiate a militarized interstate dispute. The results for these control variables are consistent across both of the models we examine here.

Although equation 7.5 indicates that the probability of initiation is contingent on the target's regime type, model 1 does not allow this. In order to do so, an interaction term, State A democratic * State B democratic, is included in model 3.[3] This allows us to account separately for monadic and dyadic effects of democracy. Once the interaction between regime types is

[3] An interaction term is one in which two variables in the model are multiplied by each other. This allows us to estimate whether each variable's effect on the dependent variable depends on the value of the other independent variable, rather than being independent of it. In model 3, the interaction term (State A democratic * State B democratic) allows us to determine whether the effect of effect of State A democratic depends on the value of State B democratic and vice versa.

controlled for, we find that not only are democratic states significantly more likely to be targeted by autocracies, they are also significantly more likely to initiate disputes against nondemocracies. Democracies, however, are significantly less likely to initiate disputes against other democracies, as indicated by the strong, highly significant, negative effect of the interaction term.

Together, these results strongly contradict the monadic peace expectations laid out in equation 7.5. The likelihood of initiation in a jointly democratic dyad ($p = 0.0019$) is reduced by 30 percent when compared to a dyad with no democracies ($p = 0.0027$).[4] Conflict initiation, however, is more likely in a mixed dyad than in a nondemocratic dyad: The probability that the democracy initiates versus the autocracy ($p = 0.0038$) is increased by 41 percent, and the probability that the nondemocracy initiates versus the democracy ($p = 0.0048$) is increased by 78 percent. Thus, contrary to the expectations of the monadic democratic peace argument, democracies are more likely to initiate disputes versus nondemocracies than nondemocracies are. Rather than equation 7.5, the true relationships between regime type and initiation are

$$\Pr(\text{D initiates} \mid \text{D}, \mathbf{x}) < \Pr(\sim\text{D initiates} \mid \sim\text{D}, \mathbf{x}) <$$
$$\Pr(\text{D initiates} \mid \sim\text{D}, \mathbf{x}) < \Pr(\sim\text{D initiates} \mid \text{D}, \mathbf{x}). \qquad \text{(Equation 7.7)}$$

Again, the results are consistent if other common measures of democracy are used (Quackenbush and Rudy 2009).

A Monadic Analysis

We conclude our empirical evaluation of the democratic peace by looking at the monadic level of analysis. By focusing directly on each state's conflict behavior individually, monadic analyses seemingly provide a useful way to test equation 7.2 regarding dispute involvement and equation 7.3 regarding dispute initiation. The expectation laid out by these equations is that the coefficients on democracy would be negative and significant, indicating that democracy makes dispute involvement or initiation less likely. The results, shown in Table 7.3, reveal once again that there is no empirical support for the monadic democratic peace.

In model 4, we estimate the impact of democracy on militarized interstate dispute (MID) involvement, controlling for power. The coefficient is positive but insignificant ($p = 0.938$). The effect is also insignificant using other measures of democracy (Quackenbush and Rudy 2009). Thus, rather than making conflict less likely, democracy has no effect on the likelihood of dispute involvement, contradicting equation 7.2.

Model 5 examines the relationship between democracy and MID initiation. Although the coefficient for democracy is negative, it again is not significant ($p = 0.980$). Thus, the expectation of equation 7.3—that democracies

---

[4]These predicted probabilities are calculated based on model 3 in Table 7.2. Only the democracy variables are changed; other variables are held at their means.

Table 7.3 Logit Results for Monadic Analyses

VARIABLE		MODEL 4: INVOLVEMENT	MODEL 5: INITIATION
Democracy	β	0.0109	-0.0042
	Se_{β}	0.1404	0.1647
ln (Power)		0.2860***	0.3479***
		0.0326	0.0347
Constant		0.7092***	0.2990
		0.2155	0.2072
Wald χ^2		77.5***	104.2***
Log-likelihood		-6,573.2	-4,821.6
N		11,654	11,654

Notes: *$p < 0.05$, **$p < 0.01$, ***$p < 0.001$
Unit of analysis is nation-state years. Standard errors are robust standard errors adjusted for clustering by state.

Source: Compiled by author.

are less likely to initiate disputes than other regime types—is not supported. As with the previous analyses, the results are fairly consistent if other common measures of democracy are used (Quackenbush and Rudy 2009).

These results are based on the monadic level of analysis. Although this would seem to be the appropriate level of analysis to test monadic democratic peace expectations, the (directed or nondirected) dyadic level of analysis is more appropriate because interstate conflict—by definition—can only occur (at least) at the dyadic level. As Most and Starr (1989, 76–8) state, when international conflict "is conceived as the outcome of the interactions of at least two parties, the attributes of all of those parties—not just one of them—must be considered in one's attempts to understand and explain when" conflicts will and will not occur. Decisions to fight are not made in a vacuum. Furthermore, they are not made by only one state, since it "takes two to tango." Thus, dyadic analyses are the most appropriate for studying international conflict (Bremer 1992) since they allow one to account for the international context in which conflict occurs.

Furthermore, recall that the monadic democratic peace expectations as expressed in equations 7.2 through 7.6 above are that democracies are less likely to fight or initiate given **x**, the set of other factors explaining international conflict. It is not possible, however, to control for important factors such as relative power and contiguity in a monadic analysis, because these factors require one to know information on two states. For example, France and Germany are contiguous, but France and India are not, but if our analysis is only looking at France individually, then the idea of contiguity does not make sense. Nonetheless, these results are important because they allow the most complete test of monadic democratic peace expectations, and they demonstrate that even if we focus on the monadic level of analysis, there is still no empirical support for the monadic democratic peace.

Democracy and International Conflict

Two key findings are generally considered to mark the cornerstone of the democratic peace: First, democracies almost never fight other democracies, and second, democracies regularly fight nondemocracies. As Maoz and Russett summarize, there appears to be "something in the internal makeup of democratic states that prevents them from fighting one another *despite the fact that they are not less conflict-prone than nondemocracies*" (1993, 624, emphasis in the original). Thus, the democratic peace is entirely a dyadic phenomenon; there is no empirical support for a monadic democratic peace (Quackenbush and Rudy 2009).

A variety of additional observations concerning the impact of democracy on international conflict have been made. While these are not an integral part of the democratic peace argument, they are nonetheless trends regarding the effect of democracy that should be accounted for in explanations of the democratic peace. We highlight six trends regarding the effect of democracy; the first three deal with the outbreak of conflict, and the second three deal with the evolution of war.

The first three trends relate to the relationship between democracy and the outbreak of conflict. First, when two democracies have a conflict of interest between them, they are much more likely to successfully negotiate peaceful settlements than are other pairs of states (Dixon 1994; Dixon and Senese 2002). Furthermore, although democratic dyads are generally associated with peace, they are more likely to fight with one another when they are in transition to democracy (Mansfield and Snyder 1995). Finally, democracies are more likely to initiate conflict against autocracies than are autocracies to initiate conflicts against them (Quackenbush and Rudy 2009).

The next three trends deal with the impact of democracy on the evolution of war itself. First, democracies are more likely than are other states to emerge victorious in war (Lake 1992; Reiter and Stam 1998; 2002), although this advantage declines the longer a war lasts (Bennett and Stam 1998). In addition, democracies are more likely to fight shorter wars than are autocracies (Bennett and Stam 1996). Finally, democracies also tend to experience fewer battle deaths in wars they fight (Horowitz, Simpson, and Stam 2011).

Regime type has many more effects related to international conflict than just peace between democracies. It is important to be able to explain each of these trends. A theory that is able to explain most or all of these effects of democracy is superior to one that is only able to explain peace between democracies.

Explaining the Democratic Peace

The empirical evidence regarding the democratic peace is very strong. Whereas much other research on international conflict started with the logic of the idea and then turned to examining empirical evidence, modern democratic peace research started with the evidence and then went in search of an explanation. The scholar credited with first observing that democracies are less likely to fight each other was Babst (1972). Small and Singer (1976, 67) reexamined the evidence and found that democracies "do not seem to fight against one another" even though they are regular participants and initiators in international wars. They conclude that the relationship is spurious, arguing that it is better explained by contiguity—the fact that "democracies have rarely been neighbors" (Small and Singer 1976, 67), rather than that they are democratic.

Others disagree, and therefore embarked on efforts to explain the democratic peace. The observation that democracies are less likely to fight one another seems to fit well with the expectation of political thinkers such as Immanuel Kant and Woodrow Wilson that democracies are more peaceful (Doyle 1986). Explanations of the democratic peace were initially grouped into two primary arguments: the normative, or cultural, explanation and the institutional explanation (Maoz and Russett 1993). A third explanation is based on selectorate theory. We review each in turn.

Normative Explanation

The **normative explanation** of the democratic peace is centered on the impact of democratic norms of behavior, which emphasize regulated political competition through peaceful means (Dixon 1994; Doyle 1986; Maoz and Russett 1993). When one party wins a democratic election, there is no need to eliminate the opponent, and it is perfectly accepted (and expected) that the loser will come back to challenge again. Thus, political conflicts in democracies are resolved by compromise rather than by the elimination of opponents.

The prevalence of democratic norms in domestic politics leads democracies to externalize these norms in their relations with other states. Therefore, relations between two democracies are characterized by democratic norms, and thus they compromise on negotiated settlements rather than fighting (Dixon 1994).

In contrast, nondemocracies exhibit norms in which political conflicts are more likely to be resolved through violence and coercion. Because nondemocracies also externalize their norms, they are less likely to reach negotiated settlements and more likely to engage in militarized conflict when conflicts of interest arise between them. Therefore, conflict between democracies is much less likely than between nondemocracies.

Explaining conflict between democracies and nondemocracies is where the normative argument runs into some difficulty. Maoz and Russett (1993) argue that if nondemocracies with nondemocratic norms face democratic states with democratic norms, they will be able to exploit the democracies. Thus, in order to guard against exploitation and ensure their own survival, democratic states employ nondemocratic norms in their relations with nondemocracies. Accordingly, mixed dyads of one democracy and one nondemocracy are expected to be as conflict prone as nondemocratic dyads.

This explanation faces several limitations. First of all, nondemocracies sometimes adopt democratic norms of conflict resolution. As the proportion of democracies in the system has increased, nondemocracies have adopted democratic norms of third-party conflict management, respect for human rights, cooperation through international organizations, and respect for territorial boundaries (Mitchell 2002; 2012). Given this pattern of nondemocracies adopting democratic norms of behavior, one could reasonably question why democracies would be expected to adopt nondemocratic norms, as Maoz and Russett (1993) claim they do. In addition, while the explanation implies that nondemocratic dyads should be at least as conflict prone as mixed ones, the empirical evidence indicates that mixed dyads are much more conflictual.

Institutional Explanations

The second type of explanation for the democratic peace focuses on the structure of democratic institutions. There are three primary approaches to doing so. The first two focus on institutional constraints or information, and are reviewed in turn. The third, known as selectorate theory, is examined in the following section.

The basic idea of the **institutional explanation** of the democratic peace is that democracies are characterized by institutional constraints—checks and balances between the executive and legislative branches of government (Morgan and Campbell 1991). Furthermore, democracies have large (as a percentage of their population) electorates that compel democratically elected leaders to seek popular support for their policies. Because war entails high costs, many of which are paid by the general public, wars are generally unpopular, particularly the longer they last (Gartner and Segura 1998; Mueller 1973). Leaders of democratic states are therefore compelled to resolve international conflicts more peacefully in order to avoid the costs of war and the increased probability of removal from office.

These democratic domestic imperatives cause democratic dyads to be peaceful in their relations with one another. Nondemocratic states, however, act without as many domestic constraints and are therefore able to pursue more aggressive and conflictual foreign policies, which causes nondemocracies to be rather conflict prone. Conflicts between democracies and nondemocracies are driven by the lack of structural constraints. Thus, the nondemocracy imposes on the democratic political system emergency conditions enabling the government to rally support, and conflict escalation becomes a distinct possibility.

The second type of institutional explanation focuses on the role of information (Schultz 1998, 2001). We saw the importance of incomplete information (along with incentives to misrepresent) as a cause of war in our discussion of the bargaining model of war in chapter 3. This explanation is based on the observation that democracies have greater audience costs than nondemocracies (Fearon 1994). **Audience costs** are political costs that leaders have to pay for making foreign threats and then backing down. Because of these potential costs, threats made by leaders of democracies to other states are more credible. Thus, "democracy facilitates peaceful conflict resolution by overcoming informational asymmetries that can cause bargaining to break down" (Schultz 1999, 233).

These information arguments, however, "are fundamentally claims about democratic states, rather than democratic dyads" (Schultz 1999, 243). Thus, if they are correct, there should be evidence of a monadic democratic peace. Since evidence suggests that democracies are only peaceful in their relations with other democracies, this argument is contradicted.

Selectorate Theory

The third primary explanation of the democratic peace comes from **selectorate theory** (Bueno de Mesquita et al. 1999, 2003). Selectorate theory is also an institutional explanation, but it is considered separately because the theory provides an alternative view of governance. At the heart of the theory is the assumption that leaders' primary preference is to stay in power. While they certainly have preferences for different policies, leaders cannot hope to shape policy without being in power.

The theory provides a formal, rational choice explanation of the impact of regime type on the behavior of states and is based on several parameters that are summarized in Table 7.4. The first is the total number of residents of a country, *N*. All other groups are subsets of this total. The selectorate (*S*) is the set of people who have a legal right to participate in the selection of the government leadership. These are the residents of the country who have at least a nominal say in choosing leaders; they can become members of a winning coalition.

The winning coalition (*W*) is the subset of the *S* without whose support the leader cannot be sustained in office. These are the members of the selectorate whose support is essential to keep the incumbent government in office at any given moment. In a democracy, it is essential to maintain the support of enough people to get reelected, and certainly not get impeached

Table 7.4 Selectorate Theory

TRAIT	DESCRIPTION
N	Total number of a country's residents
S	*Selectorate:* Set of people who have a legal right to participate in selecting the government's leadership
W	*Winning coalition:* A subset of *S*, without whose support the leader could not hold office
R	Available revenue
N-S	The disenfranchised
W/S	*Loyalty norm:* A large *W/S* is evident in democracies and is very small in nondemocracies

or removed through a vote of no confidence. In nondemocracies, elections (if any) are much less important, but the leader must still satisfy enough people to avoid a coup d'état, revolution, or other removal from power.

The disenfranchised (*N-S*) are people who have no legal right to participate in the selection of government leadership. In many countries, this includes children and criminals. At various times and places, women, nonlandowners, and minorities have been disenfranchised.

Much of selectorate theory's explanation is driven by the ratio between the winning coalition and the size of the selectorate (*W/S*), which is called the loyalty norm. *W/S* is large in states that people normally refer to as democracies. It is miniscule in places, such as North Korea, that operate on the basis of rigged elections. A leader needs *W* people out of *S* to support him or her. The larger *W* is relative to *S*, the more likely any individual is to be in the winning coalition. For small *W/S*, coalition members face a higher risk of being replaced in the winning coalition. Therefore, small *W/S* leads to greater loyalty to the incumbent.

Different political systems are characterized by different sizes of the winning coalition (*W*) and selectorate (*S*). A typical democracy has a large *S* relative to the polity size and a large *W* relative to *S*. While democracies have large *S* and large *W*, there are considerable variations within systems that we usually label as democracies (e.g., proportional representation parliamentary democracies have a smaller *W* than presidential democracies); this affects their policies. A typical one-party autocracy has a large *S* relative to polity size, but a small *W* relative to *S* because of rigged elections. Finally, a typical monarchy or military dictatorship has both small *S* and small *W*.

Leaders attempt to satisfy enough members of the selectorate to maintain a winning coalition by allocating the various goods that the leader controls. There are two primary tools available to leaders in their attempts to stay in office: private goods and public goods. Private goods are goods that benefit a single person or a well-defined group. Public goods are goods that are nonexcludable and indivisible, meaning than no one in a group can be excluded from the benefits of the good and the good cannot be divided into smaller parts.

The type of goods that leaders typically utilize to satisfy a winning coalition varies depending on the coalition size. When there is a small winning coalition, leaders can rely on private goods to stay in office. For example, leaders might provide "privileged access to government contracts, exploitation of a black market, or protection against prosecution" (Bueno de Mesquita 2010, 32) to their inner circle of supporters. However, there are simply not enough resources available to focus on private goods when the winning coalition is large. For a large winning coalition, leaders must rely on public goods to stay in office. "Examples of public goods include national defense, free speech and free assembly, public parks, equal protection under the law, and free access to education" (Bueno de Mesquita 2010, 33).

Bueno de Mesquita et al. (1999) lay out the implications of selectorate theory for the democratic peace. Regimes with large *W* focus on effective policies, including foreign policies, in order to provide the public goods required to stay in office. When leaders' policies do poorly, they either switch them or pour more resources into trying to make them succeed. Regimes with small *W* tolerate failed policies more because in small-coalition settings, loyalty to a leader depends more on receiving private rewards than public benefits. Therefore, they do not change course as quickly as large-coalition leaders. The likelihood that a leader survives in office despite failed policies increases as *W/S* decreases and decreases as *W/S* increases.

The maximum average private goods for members of the winning coalition is *R/W*, where *R* is the available revenue and *W* is the size of the winning coalition. This obviously decreases as *W* increases. Democratic leaders need policy success—including in war. They are likely to lose power without successes (Bueno de Mesquita and Siverson 1995; Goemans 2000), as we will examine in more detail in chapter 12. Therefore, democracies are likely to try harder to win. They devote more and more resources toward the war effort, and they are better at selecting only winnable wars to fight.

Because autocratic leaders rely more on private goods, victory is not as vital. They are still interested in winning, but they can secure their hold on power through private goods. If they put too many resources into the war effort, then that reduces the resource pool for private goods, and can therefore reduce their hold on office.

Thus, democracies are unattractive targets. Democracies are therefore less likely to fight other democracies because they prefer to avoid the difficult fight. Furthermore, democracies tend to overwhelm autocracies because they are willing to mobilize more of their resources for the war effort rather than reserving them to reward domestic backers. Wars between democracies and autocracies are generally short in duration and relatively low in cost to the democracy.

Bueno de Mesquita et al. (1999) argue that the selectorate explanation is consistent with each of the trends—that the democratic peace is dyadic, not monadic, that democracies are more likely to win wars, and so on—highlighted in the previous section summarizing the various effects of democracy on international conflict. Another advantage of selectorate theory over the normative and institutional explanations is that it explains more than just the democratic peace. In addition to peace between democracies, selectorate theory has been used to explain how regime type affects phenomena such as economic growth, the duration of leaders' tenure in office (Bueno de Mesquita et al. 2003), foreign aid (Bueno de Mesquita and Smith 2007, 2009), and corruption in both democracies and autocracies (Bueno de Mesquita and Smith 2011). Because of its broad scope, logical consistency, and empirical support, selectorate theory appears to provide the best explanation of the democratic peace.

The Kantian Triangle

An idea that is closely related to the democratic peace is the **Kantian triangle of peace**, a concept developed by Russett and Oneal (2001) that builds on the arguments of eighteenth-century German philosopher Immanuel Kant in his book *Perpetual Peace*. These ideas represent a broader liberal peace than a simple focus on democracy and echo ideas advanced by Woodrow Wilson for peace following World War I.

Box 7.1

Case in Point: Woodrow Wilson's Fourteen Points

Ideas of a liberal peace are well summarized in the Fourteen Points presented by Woodrow Wilson, the president of the United States from 1913 to 1921. Although World War I began in August 1914, the United States did not join the war until April 1917. President Wilson presented his plan for postwar peace in fourteen points in a speech to a joint session of Congress on 8 January 1918. His plan was later accepted by France and Italy, and mostly accepted by Britain, and formed the basis of the Allies' approach to postwar peace.

The fourteen points, summarized, were:

1. Diplomacy should proceed "in the public view," rather than in secret alliances.
2. There should be "freedom of navigation upon the seas," in both peace and war.
3. Trade barriers among nations should be reduced.
4. "Armaments will be reduced to the lowest point consistent with domestic safety."
5. Colonial claims should be adjusted in the interest of the inhabitants as well as of the colonial powers.
6. The Russian territory should be evacuated and its government welcomed to the society of nations.
7. Belgian territories in Germany should be restored.
8. All French territory, including Alsace-Lorraine, should be evacuated.
9. Italian boundaries should be readjusted along clearly recognizable lines of nationality.
10. There should be independence for various national groups in Austria-Hungary.
11. The Balkan nations should be restored as well as free access to the sea for Serbia.
12. There should be protection for minorities in Turkey and the free passage of the ships of all nations through the Dardanelles.
13. Poland should have independence, including access to the sea.
14. A league of nations should be established to protect "mutual guarantees of political independence and territorial integrity to great and small nations alike."

Wilson's fourteen points highlight economic interdependence (points 2 and 3), self-determination (points 5–13), disarmament (point 4), and international organizations (points 1 and 14) as the path to peace. While not addressed in the fourteen points, Wilson's speech asking Congress to declare war argued that the United States should join to "make the world safe for democracy." Together, Wilson's arguments highlight the major themes of liberalism in international relations theory.

Figure 7.2 Kantian Triangle of Peace

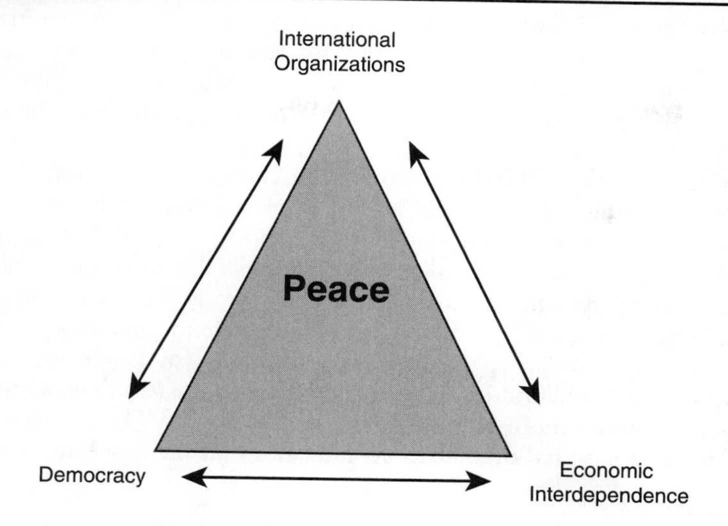

The Kantian triangle represents three key factors that make international peace more likely, as shown in Figure 7.2. The first leg of the triangle is democracy. As we have already seen so far in this chapter, there is substantial evidence that democracies rarely fight each other. Kant argued that democracies "are capable of achieving peace among themselves" even while they "remain in a state of war with" nondemocracies (Doyle 1986, 1162).

Economic interdependence is the second leg of the Kantian triangle of peace. Russett and Oneal (2001) argue that economically important trade and investment with other states limit the likelihood that a state will use force against its commercial partners because the cost of severing these economic ties is too great. Although some scholars have argued that international trade makes conflict more likely (e.g., Barbieri 1996), the majority of research indicates that trade makes conflict less likely (e.g., Dorussen and Ward 2010; Maoz 2009; Oneal and Russett 1997).

The final leg of the triangle is international organizations. An **international organization** is an organization that has members in multiple countries. Russett and Oneal focus on membership in intergovernmental organizations (IGOs), such as the United Nations (UN), Organization of Petroleum Exporting Countries (OPEC), and World Trade Organization (WTO), which have states as their members. There are also international nongovernmental organizations (INGOs), whose members are individuals or private organizations. Examples of INGOs include Amnesty International, International Committee of the Red Cross, and Greenpeace.

Russett and Oneal argue that international organizations make a direct contribution to preventing and resolving conflicts between countries.

International organizations do this in different ways. They may directly coerce and restrain those who break the peace, they can serve as agents of mediation and arbitration, or they can reduce uncertainty in negotiations by conveying information. Furthermore, Russett and Oneal expect that the more international organizations to which two states belong together, the less likely they will be to fight one another or even to threaten the use of military force.

Russett and Oneal (2001) argue that these three legs of the triangle work together to produce peace. Their central empirical results, drawn from an analysis of politically relevant dyads from 1886 to 1992 and using militarized interstate disputes as the dependent variable, are shown in Table 7.5. The negative coefficient for lower democracy means that as the less democratic state in a dyad becomes more democratic, the likelihood of conflict is reduced. Similarly, the negative coefficient for lower dependence means that the likelihood of conflict is reduced as the less dependent state in a dyad becomes more economically interdependent. The negative coefficient for international organizations indicates that the more international

Table 7.5 Logit Results for the Kantian Triangle of Peace

VARIABLE	COEFFICIENT (STANDARD ERROR)	SIGNIFICANCE LEVEL
Lower democracy	-0.0608 (0.0094)	< 0.001
Lower dependence	-52.9 (13.4)	< 0.001
International organizations	-0.0135 (0.0043)	< 0.001
Alliance	-0.539 (0.159)	< 0.001
Power ratio	-0.318 (0.043)	< 0.001
Noncontiguity	-0.989 (0.168)	< 0.001
ln (Distance)	-0.376 (0.065)	< 0.001
Only minor powers	-0.647 (0.178)	< 0.001
Constant	-0.128 (0.536)	< 0.41
Wald χ^2	228	< 0.001
N	39,988	

Source: Adapted from Bruce M. Russett and John R. Oneal, *Triangulating Peace* (New York: Norton, 2001), 316.

organizations that the two states are joint members of, the less likely they are to have a militarized dispute. The impact of their control variables dealing with contiguity, power, and alliances are all as we would expect from our discussions in chapters 4, 5, and 6 respectively.

Thus, Russett and Oneal find strong empirical support for their expectations regarding each leg of the Kantian triangle of peace. In addition, they argue that peace produces these factors because democracy is easier to sustain in a peaceful environment, trade is discouraged by international conflict and especially by war, and international organizations are most often formed when a certain level of peace seems probable. Russett and Oneal argue that since democracy, economic interdependence, and international organizations all lead to peace, and peace in turn leads to each of them, feedback loops are created. These feedback loops create the potential for what Russett and Oneal label virtuous circles of peace.

These findings regarding democracy, economic interdependence, and international organizations seem to bode very well for the future of world peace. As we saw in Figure 7.1, democracy has spread dramatically across the world in recent decades. Similarly, economic interdependence and membership in international organizations have grown dramatically since World War II. In many ways, Russett and Oneal's argument echoes that of Norman Angell, who published a book called *The Great Illusion* in 1911. Angell argued that the interdependence in Europe created a situation in which war could no longer serve states' economic interests; the idea that war could do so was a great illusion. Many people took Angell's argument to mean that war was not going to happen, much like Mueller (1989) argued that major war is obsolete in the modern world. Nonetheless, World War I began shortly after Angell's writing, and further wars have occurred in the century since, regardless of whether they served the states' economic interests.

Alternative Explanations

Despite the strong empirical finding that democracies are less likely to fight each other than are other pairs of states, the democratic peace argument is not universally accepted (Thompson and Tucker 1997). That democracies are less likely to fight each other is not really subject to debate. What is debated is whether democracy causes this peace or whether some other factor is responsible. Two of the strongest alternative explanations in particular deserve our attention here: The first is focused on common interests, and the second is focused on capitalism. We review each in turn.

Common Interests

Farber and Gowa (1995) reassess the democratic peace and find that it is period specific. In particular, they find that democratic dyads are only more peaceful than others during the Cold War period. Prior to World War I (1816–1913), they find that democratic dyads are significantly more likely to get into militarized disputes than other pairs of states. Thompson and

Tucker (1997) argue that Farber and Gowa's pre–World War I results are driven by the nineteenth-century rivalries between France and Britain and the United States and Britain. This counterargument is questionable, however, since these dyads are identified as rivalries because they fought a number of militarized disputes (as we will see in chapter 13), which is precisely Farber and Gowa's point.

Farber and Gowa (1997) and Gowa (1999) argue that it is common interests between states that lead to peaceful relations, rather than democracy. They find that prior to 1945, democracies tended to ally with one another less and fight each other more, whereas since 1945 democracies have tended to ally with one another more and fight each other less.

Their results indicate that common interests between states explain the observed conflict patterns better than common polities do. In the Cold War, democracies fought each other less because they all faced a common enemy, the Soviet Union. Furthermore, if the democratic peace was only a Cold War phenomenon, then we would expect democracy to no longer be significant in the post–Cold War world. Gowa (2011) finds that in the post–Cold War period, the relationship between joint democracy and peace no longer holds. Nonetheless, while common interests are certainly important, by themselves they appear to be insufficient to explain peace between democracies (Gartzke 1998).

Capitalist Peace

A different set of alternative explanations of the democratic peace, known as the **capitalist peace**, focuses on capitalism as the force driving peace between democracies, rather than democracy itself (Schneider and Gleditsch 2010; Weede 2003). Capitalism is an economic system that is characterized by private ownership of capital goods (such as factories, machinery, etc.) and the reliance on the free market to determine the price and quantity of goods and services. The importance of capitalism, and economic issues in general, for explaining peace between democracies was highlighted by several studies showing that the impact of democracy depends on the level of economic development (Hegre 2000; Mousseau 2000; Mousseau, Hegre, and Oneal 2003).

Gartzke (2007) focuses on economic development and financial and monetary integration as the key indicators of capitalism. Many studies have argued and found that international trade is associated with peace between states, and Russett and Oneal (2001) demonstrate that democracy still matters even when we account for trade. However, Gartzke, Li, and Boehmer (2001) argue that trade is not the only aspect of economic interdependence that should be accounted for. In particular, they argue that financial and monetary integration are even more important aspects of economic interdependence because they provide a key way for states to convey credible signals to others and thereby overcome potential bargaining problems. Gartzke (2007) finds that greater levels of financial openness significantly decrease the probability of conflict in a dyad. Furthermore, once financial openness and economic development are controlled for, democracy no longer has an effect on the likelihood of wars or fatal disputes. Thus, his results indicate that capitalism drives peace between democracies, not democracy.

Mousseau (2009, 2012) argues that the key element of capitalism is a contract-intensive society. "A contract-rich economy is one where most citizens normally contract with strangers in the market to obtain their incomes, goods, and services. In contract-poor economies, in contrast, citizens are more dependent upon favors reciprocated among friends and family" (Mousseau 2012, 194). Mousseau (2009) finds that contract-intensive economies are significantly less likely to engage in international conflict, and once they are accounted for, democracy no longer has a significant impact on conflict.

Pitting the capitalist peace against the democratic peace in empirical tests is difficult because there is a strong overlap between capitalism and democracy. Furthermore, the capitalist peace has faced criticism from a couple of different fronts. Russett (2010) argues that capitalism and democracy both make peace more likely, although capitalism has both monadic and dyadic effects, while democracy's effect is only dyadic. In addition, some evidence suggests that findings regarding the capitalist peace may result from statistical anomalies rather than reality (Choi 2011; Dafoe 2011).

Whether the democratic peace or the capitalist peace is correct has important implications for the real world. For example, power transition theory points to the potential for conflict between the United States and China in the next few decades, as discussed in chapter 5. Although China is not democratic, it has become increasingly capitalist in recent decades. Thus, the democratic peace would suggest that the risk of conflict is great because the dyad is mixed, with the United States democratic and China autocratic. On the contrary, if China continues to move toward capitalism in its economic system, the capitalist peace would indicate that the risk of conflict between the two is greatly reduced (Weede 2010).

Conclusion

Examination of the history of international conflict in the past two centuries clearly demonstrates that democracies are less likely to fight each other than are other pairs of states. However, although pairs of democracies are relatively peaceful, democracies are not less likely to fight in general. These observations mark the cornerstone of the democratic peace. The logical explanation of the relationship between democracy and conflict has been subject to debate.

The normative explanation argues that democratic norms of compromise and nonviolent conflict resolution explain peace between democracies. In contrast, institutional explanations focus on the constraints created by democratic institutions and the information revealed to other states because of audience costs to explain the democratic peace. However, these explanations lead one to expect there to be a monadic democratic peace as well;

since the historical record contradicts the idea that democracies are more peaceful in general, these explanations are problematic. Probably the best explanation of the democratic peace comes from selectorate theory, which argues that the size of the selectorate drives the democratic peace and also explains a variety of other effects of regime type.

The Kantian triangle of peace is the idea that democracy, economic interdependence, and international organizations all work together to produce peace, and there is evidence to support the idea that they do. Alternative explanations for why democracies have fought each other less often than other pairs of states have been developed. In particular, there is some evidence to suggest that this peace between democracies is driven by common interests between states or capitalism, rather than democracy. Nonetheless, these alternative explanations remain quite controversial as most scholars are convinced that joint democracy is a major driving force for peace.

Key Concepts

Audience costs

Capitalist peace

Democracy

Democratic peace

Economic interdependence

Institutional explanation

International organization

Kantian triangle of peace

Monadic democratic peace

Normative explanation

Selectorate theory

8 Deterrence

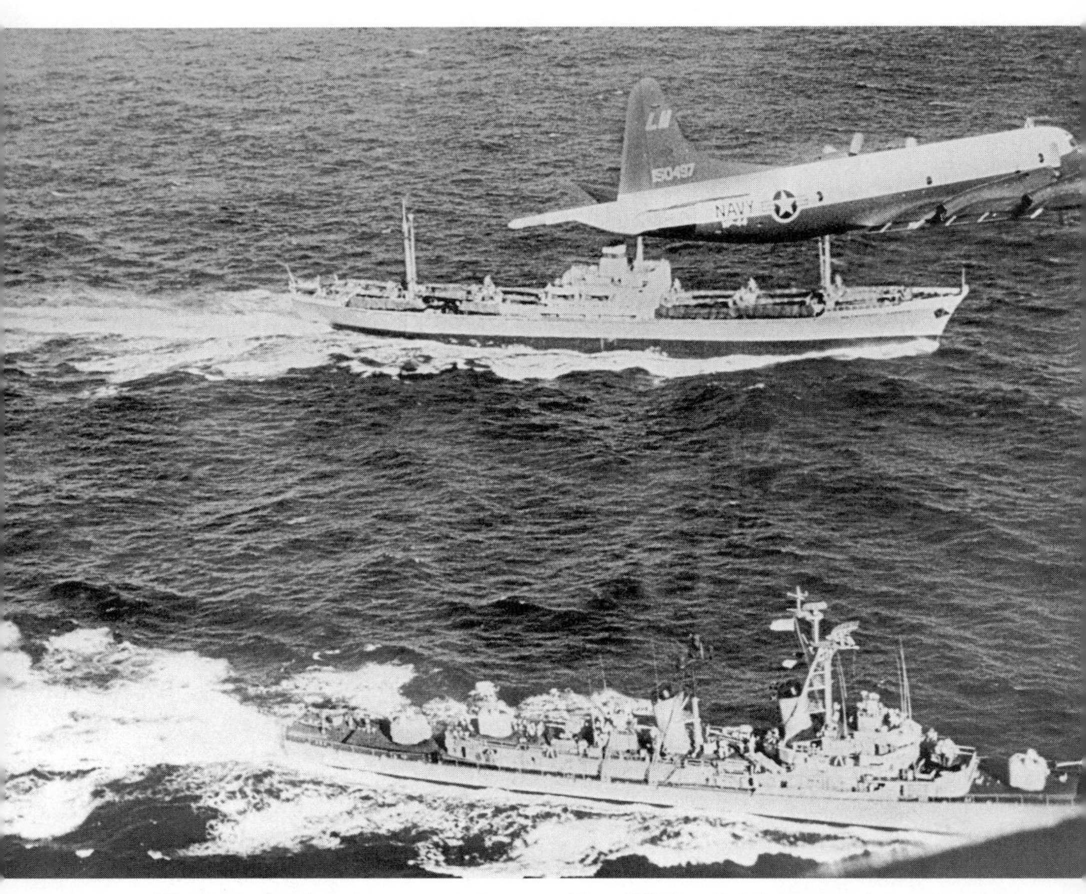

The Soviet freighter *Anosov,* rear, is escorted by a US Navy P-3 Orion patrol aircraft and the destroyer USS *Barry* while it leaves Cuba, probably loaded with missiles under the canvas cover seen on deck. The Cuban Missile Crisis in October 1962 was one of the most important examples of immediate deterrence.

Source: Getty Images

When people think about deterrence, they tend to think about nuclear weapons or crises. At first glance, each reaction seems perfectly justified. After all, although World War II resulted in more casualties than all other modern interstate wars put together, the atomic bombs dropped on Japan in August 1945 introduced an unprecedented level of destructive power. The bomb dropped on Hiroshima had the explosive power of thirteen thousand tons of TNT (13 kilotons) and killed 90,000 to 166,000 people; the bomb dropped on Nagasaki was 21 kilotons and killed between 60,000 and 80,000 people. While conventional bombings of some other cities—such as Dresden and Tokyo—led to similar numbers of casualties, those resulted from thousands of bombs dropped from hundreds of airplanes, rather than one bomb dropped from one plane.

Observers quickly realized that the enormous costs of nuclear weapons compared to conventional ones revolutionized warfare in many ways. Once thermonuclear weapons were introduced in 1952, the destructive power increased by orders of magnitude. Although warheads on modern intercontinental ballistic missiles (ICBMs) such as the Minuteman III are "only" 300 kilotons, warheads as large as 25 megatons (million tons of TNT) were deployed by the United States and former Soviet Union during the Cold War. In contrast, World War II consumed about three million tons of TNT in total (Rhodes 1986, 563). Given this massive destructive power, all-out nuclear war threatened the existence of life on earth and therefore made "winning" such a war virtually useless. Thus the advent of nuclear weapons meant that states would have to rely on threats of retaliation rather than using military force to dissuade others from taking certain actions. Armed forces now existed more to deter wars rather than to win them (Brodie 1946).

Probably the closest the world has come to nuclear war was the Cuban Missile Crisis between the United States and Soviet Union in October 1962. It is also a high-profile example of deterrence. President John F. Kennedy estimated the chances of nuclear war as "between one out of three and even." War, however, was successfully avoided when the Soviets backed down in the face of an American naval quarantine (a polite term for a blockade). As US Secretary of State Dean Rusk stated, "We're eyeball to eyeball and the other fellow just blinked." Although deterrence worked in the Cuban Missile Crisis, in other crises it has failed. For example, the July Crisis of 1914 began with a dispute between Austria-Hungary and Serbia over the assassination of the Archduke Franz Ferdinand, heir to the Austrian throne, in Sarajevo by Serbian nationalist Gavrilo Princip. All efforts by the European major powers to deter others from becoming involved (such as Germany's blank check to Austria-Hungary, Russia's partial mobilization, and many others also involving France and Britain) failed and the crisis boiled over into World War I.

While deterrence is certainly relevant to understanding nuclear weapons and crises in international relations, it is by no means limited to these areas. **Deterrence** is the use of a threat by one party in an attempt to convince another party to refrain from some action. Reliance on nuclear weapons is neither specified in this definition nor even implied. Indeed, deterrence is not even limited to international relations. When a mother tells her daughter, "Susie, don't do that or I'll send you to your room," she is practicing deterrence. Similarly, criminal justice systems assess fines, prison terms, and other punishments in an attempt to deter violations of the law.

In this chapter, we are focused on the dynamics of deterrence within international relations, and there are a variety of areas beyond nuclear weapons where deterrence applies. First of all, states are not only interested in deterring nuclear attacks against them, nor do they rely only on threats of nuclear retaliation to deter challenges. States also have strong incentives to deter conventional attacks on themselves and their allies. For example, in the July Crisis that led to World War I, each of the major powers tried to deter intervention by other major powers (Zagare 2011). Conventional deterrence has also played an important role in relations between Israel and Arab states (Shimshoni 1988; Sorokin 1994) as well as numerous other cases.

Increasing attention has also been devoted to the functioning of deterrence in areas beyond the traditional focus on interstate conflict. Terrorism has been an increasingly popular tactic used by various groups around the world. Most famously, the attacks by al-Qaida against the World Trade Center and other targets within the United States on 11 September 2001 led many to question why the United States was unable to deter the group—and by extension, whether terrorists can be deterred at all (e.g., Lebovic 2007; Trager and Zagorcheva 2005/06). The NATO conflict with Serbia over Kosovo in 1999 also illustrates the attempt to deter other states from engaging in ethnic cleansing (Quackenbush and Zagare 2006).

While in one light the Cuban Missile Crisis was a deterrence success and the July Crisis was a deterrence failure, they were both crises (recall that we defined crisis in chapter 2). Indeed, many studies of deterrence have focused their attention squarely on crises. For example, George and Smoke (1974) examine deterrence in American foreign policy through case studies of a series of Cold War crises: the Berlin blockade in 1948, the outbreak of the Korean War in 1950, the Taiwan Straits Crisis in 1954–1955, the Cuban Missile Crisis in 1962, and several others. This connection between deterrence and crises is quite common, particularly within the case-study literature on deterrence (e.g., Jervis 1982/83; Jervis, Lebow, and Stein 1985; Lebow 1981, 1984; Lebow and Stein 1990).

Deterrence, however, is relevant to many cases beyond crises, and focusing exclusively on crises creates a distorted, incomplete, and inaccurate understanding of deterrence (Achen and Snidal 1989; Fearon 2002). Revisiting the comparison of the Cuban Missile and July crises, one could easily consider both cases to be deterrence failures precisely because they were both crises. This point raises the distinction between general and immediate deterrence, as discussed in more detail in the next section.

While some have argued that deterrence is irrelevant in the post–Cold War world (e.g., Krauthammer 2002), deterrence is a general phenomenon that is not limited to any particular time or space. And once the focus is on every-day relations between states and general deterrence, rather than crises and extended deterrence, deterrence theory provides a general theory of international conflict (Powell 1985; Zagare 2007).

Types of Deterrence

Deterrence is the use of a threat (explicit or not) by one party in an attempt to convince another party to not take an undesired action. Deterrent threats have two purposes. The purpose of **direct deterrence** is to deter a direct attack on the defender. Conversely, the goal of **extended deterrence** is to deter attack on one's allies. But while direct deterrence and extended deterrence are usually considered separately, they are nonetheless clearly related (Snyder 1961, 276–7).

We can further categorize cases of direct deterrence by considering which states desire alterations to the status quo. Deterrence is not needed to prevent a state that is satisfied with the status quo from initiating a challenge; it is only relevant when states seek alterations in the status quo. If neither state in a dyad seeks to alter the status quo, then there is no deterrence. If one state seeks to alter the status quo but the other does not, there is **unilateral deterrence**. And finally, a situation where both sides of a dyad seek alterations to the status quo is one of **mutual deterrence**.

As developed by Morgan (1983), there are two basic types of deterrence situations: general and immediate. **Immediate deterrence** "concerns the relationship between opposing states where at least one side is seriously considering an attack while the other is mounting a threat of retaliation in order to prevent it" (Morgan 1983, 30). Immediate deterrence is all about crises and leaders' attempts to prevent the crisis from escalating to war. Thus, immediate deterrence is responsible for the common connection of deterrence with crises previously discussed. **General deterrence**, however, "relates to opponents who maintain armed forces to regulate their relationship even though neither is anywhere near mounting an attack" (Morgan 1983, 30). Thus, general deterrence has less to do with crisis decision making than with everyday decision making in somewhat conflictual or adversarial relationships.

Immediate deterrence is much narrower in scope than general deterrence. For example, consider a case of immediate deterrence such as the Cuban Missile Crisis. Studies of immediate deterrence seek to understand how escalation can be controlled within the context of a crisis. That is, once a state has already challenged the status quo, how can a defending state deter the challenger from taking further action and thus avoid all-out war? Successful immediate deterrence entails a challenger's backing down following the defender's threat to retaliate, whereas the failure of immediate deterrence results in the challenger's attacking despite the defender's retaliatory threat.

The need for immediate deterrence indicates that general deterrence has previously failed (Danilovic 2001). If general deterrence always succeeds, crises and wars do not occur. It would seem reasonable, then, to focus on the origins of international crises before examining management of those crises. Since general deterrence necessarily precedes immediate deterrence, its analysis is more important for a general understanding of international conflict than that of immediate deterrence (Quackenbush 2011b). Furthermore, as the literature on selection bias (e.g., Fearon 2002; Reed 2000) makes clear, examination of immediate deterrence without consideration of the origins of immediate deterrence cases (i.e., the failure of general deterrence) can lead to misleading empirical results.

Now that we have examined the basics of what we mean by deterrence, along with different types of deterrence that we can examine, it is important to consider how deterrence can be explained. How does it work? Answering this question is the role of deterrence theory, which we now turn to exploring.

Box 8.1

Case in Point: Cold War

Deterrence theory was initially developed early in the Cold War. The Cold War was a long-running conflict between the United States and the Soviet Union, who emerged from World War II as the strongest states in the world, which dominated international politics in the decades following the war. Although the United States and Soviet Union were allied together during the war, ideological and strategic differences between them began to emerge at the Potsdam Conference, before the war was over. These differences significantly influenced their actions toward one another, as well as toward other states who might align with one or the other. The Soviets dominated Eastern Europe, establishing communist regimes throughout the region. In turn, US influence was greater in Western Europe. Yet the United States deemed communism a threat to its interests and adopted a policy of containment, in which it sought to restrict communism's geographic expansion. Under the Marshall Plan, the United States authorized massive foreign aid for rebuilding throughout Western Europe, partly to help these war-torn nations recover and partly to contain the spread of communism.

Although there had already been a series of disagreements between the two sides, Cold War tensions greatly escalated with the Berlin Blockade from June 1948 through May 1949, during which the Soviet Union blockaded all Western transportation into and out of Berlin and the United States responded by airlifting supplies to Berlin. Militarization of the Cold War escalated during the Korean War from June 1950 through July 1953, when the American defense budget more than tripled as the United States increased its military preparedness in order to deter further

(Continued)
challenges by the Soviet Union or other communist states. Additional crises over the divided city Berlin—one half overseen by US forces, the other by Soviet—between 1958 and 1961 dominated Cold War discord until the Cuban Missile Crisis of October 1962.

Despite US efforts, communism spread geographically from the Soviet Union through Eastern Europe, China, North Korea, and elsewhere, and US concerns about the power of its ideological foe increased. During the 1950s and 1960s, US foreign policy was guided in part by domino theory, the belief that communism would spread from one country to the next like falling dominoes. To prevent this, the United States used its financial, diplomatic, and military powers to prop up states in hopes of preventing their fall. Belief in domino theory was the primary motivation for US involvement in the Vietnam War from 1965 through 1973. Ultimately, of course, the United States was unable to prevent the North Vietnamese takeover of South Vietnam in 1975. Yet the nations around Vietnam did not, in fact, fall like dominoes in reaction to their communist neighbor.

The 1970s saw a period of détente, during which Cold War tensions relaxed a great deal. Key events at this time include the Strategic Arms Limitation Talks (SALT I), the first US-Soviet arms control agreement reached in 1972; the Apollo-Soyuz joint space mission in July 1975; and the Helsinki Accords signed in August 1975, in which the United States, Soviet Union, and other states agreed to recognize the territorial integrity of other states and respect human rights.

The Soviet invasion of Afghanistan in December 1979, however, led to a rapid increase in tensions, including the US boycott of the 1980 Olympic Games in Moscow. After Ronald Reagan became president on an anti-détente platform in 1981, the United States began a significant military buildup. The 1980s marked a renewal of Cold War turmoil that only lessened after Mikhail Gorbachev became leader of the Soviet Union in March 1985. His reform policies of *perestroika* (restructuring) and *glasnost* (openness) sparked a chain of events leading to the fall of the Berlin Wall in November 1989, the dissolution of the Soviet Union in December 1991, and the end of the Cold War.

Throughout the Cold War, the United States and Soviet Union were engaged in mutual deterrence, as each had incentives to challenge the status quo and had incentives to deter the other from doing so. Simultaneously, they were engaged in extended deterrence to prevent the other from challenging their allies. The entire Cold War was a period of general deterrence, while the various crises were situations of immediate deterrence.

Classical Deterrence Theory

A great deal has been written about deterrence over the years. Much of the deterrence literature can be categorized as a single theory: **classical deterrence theory**. Classical deterrence theory is a large and diverse literature spanning more than five decades. I focus here on the primary arguments, not all the nuances of individual works. Because theoretical approaches and assumptions vary widely from one theorist to another, however, classical deterrence theory can also be divided into two subgroups: structural deterrence theory and decision-theoretic deterrence theory.

Structural Deterrence Theory

Structural deterrence theory, closely aligned with realism, argues that a balance of power brings peace; if two states are equal in power, each will be deterred since neither will be able to gain an advantage (e.g., Brodie 1959; Intriligator and Brito 1984, 1987; Kaufmann 1956; Mearsheimer 1990; Snyder 1961). Intriligator and Brito (1984, 1987) formalize the arguments of structural deterrence theory in a model of a missile war. The conclusions of their model, and of structural deterrence theory in general, are nicely displayed in a single figure of the weapons plane, shown in Figure 8.1. Their model considers two states, A and B, in a mutual deterrence relationship. Each state has a given number of nuclear missiles: The number of missiles held by state A is M_A, and B's quantity of missiles is M_B. Intriligator and Brito assume that one state will attack if it can use a

Figure 8.1 Weapons Plane

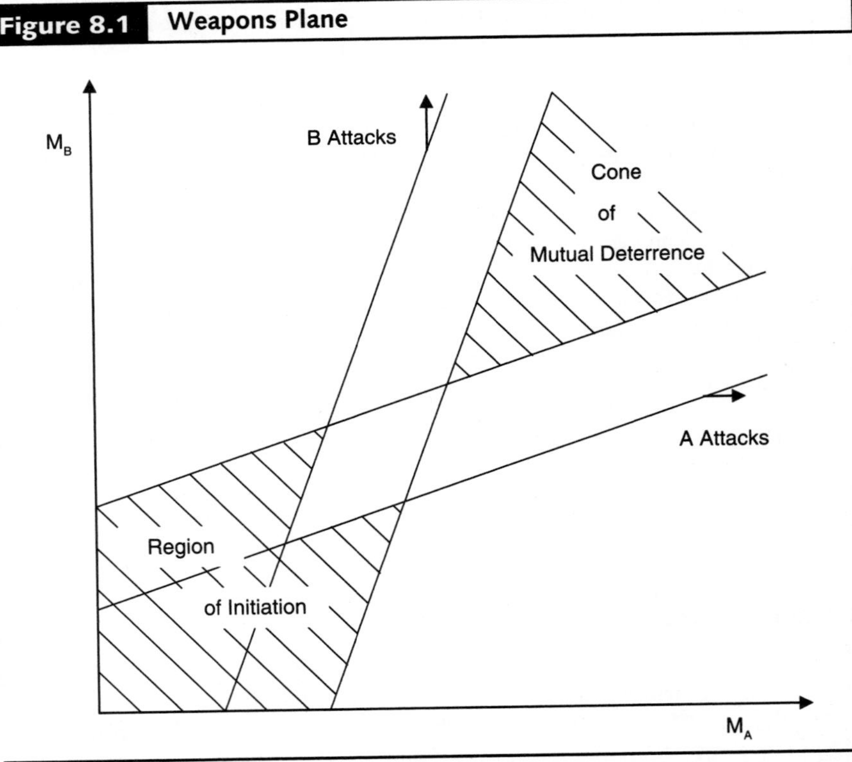

Source: Adapted from Michael D. Intriligator and Dagobert L. Brito, "Can Arms Races Lead to the Outbreak of War?" *Journal of Conflict Resolution* 28 (1984), 74.

first strike to eliminate the other state's ability to strike back. A state will be deterred, however, if it cannot eliminate the other state's retaliatory capability in a first strike. Thus, a second-strike capability is the key to deterrence.

As Figure 8.1 illustrates, if both states have only a few missiles (i.e., neither has a second-strike capability), they are in the region of initiation and deterrence is unstable—each state will seek to preempt the other. If only one state has a second-strike capability, it will be undeterred but the other state (i.e., the state without a second-strike capability) will be deterred. Finally, if both states have second-strike capabilities, then they will be in the cone of mutual deterrence and both will be deterred. Neither state would rationally attack since an attacker would bring about its own destruction. This reflects the idea of **Mutual Assured Destruction (MAD)**.

Structural deterrence theory argues that nuclear deterrence is inherently stable. While "in a conventional world, a country can sensibly attack if it believes that success is probable," with nuclear weapons "a nation will be deterred from attacking even if it believes that there is only a possibility that its adversary will retaliate" (Waltz 1988, 626). Thus, the key to deterrence is a second-strike capability, and once this is achieved (Wohlstetter 1959), deterrence is straightforward since the enormous costs associated with nuclear war make an attack irrational. Accidental war, then, represents the only real threat to deterrence.

There are three major predictions of structural deterrence theory. First, parity relationships, when coupled with high war costs, are especially peaceful. This is precisely the situation of mutual assured destruction. Second, asymmetric power relationships are associated with crises and war. In an asymmetric relationship, one side will be unable to deter the other, as illustrated in the A attacks and B attacks regions of Figure 8.1. Finally, as the absolute costs of war increase, the probability of war decreases, all else being equal. This simply pushes states farther and farther into the cone of mutual deterrence, making their relationship increasingly stable.

From these axioms flow several practical, policy-oriented conclusions. First, quantitative arms races, which serve to increase the cost of conflict, can help prevent wars. A quantitative arms race is when states are competing to build ever increasing numbers of weapons. How many missiles does it take to reach the lower-left corner of the cone of mutual deterrence? Perhaps it is one thousand for each side. But if each side had five thousand missiles (or ten thousand, or even fifty thousand), would that be better? Since nuclear weapons are so stabilizing, according to this logic, the more the better. Thus, classical deterrence theorists supported the tremendous growth of nuclear weapons during the Cold War, which peaked at 31,255 warheads for the United States in 1967 and 40,159 Soviet warheads in 1986 (Kristensen and Norris 2013).

By contrast, qualitative arms races, which threaten to provide one side or another with a first-strike advantage, increase the probability of preemptive war. A qualitative arms race is one in which the states are competing over the quality, rather than the quantity, of weapons. Examples of such qualitative developments include improvements in the accuracy of ballistic missiles and the introduction of MIRVs (multiple independent reentry vehicles) where one missile can carry multiple warheads instead of just one. The Minuteman III missile, first deployed by the United States in 1970, carried three warheads and became the world's first ICBM with MIRVs.

The third major policy prescription is that comprehensive and effective defense systems make conflict more likely. Classical deterrence theorists have argued against developments of missile defense systems each time the idea has been raised. They believe that defense systems are problematic because they undermine mutual assured destruction by potentially making the state possessing them immune to retaliation. Furthermore, the logic of classical deterrence theory suggests that the selective proliferation of nuclear weapons can help prevent war and promote peace. Since nuclear weapons worked so well to keep peace between the United States and Soviet Union, they can also work well to keep peace between India and Pakistan and other pairs of countries.

Finally, accidental war is the gravest threat to peace. Since war is the worst possible outcome, states would never purposefully choose to go to war. Thus, war could only happen as the result of some accident. Because of concerns about accidental war, classical deterrence theorists have steadfastly argued for increases in command and control safeguards for nuclear forces, such as requirements that any use of nuclear weapons must be explicitly authorized by the president, launching of nuclear missiles requires two people, and so on.

Classical deterrence theory, however, is beset with a number of logical and empirical problems. The first empirical anomaly is that a balance of power is not a good predictor of peace; neither is preponderance a good predictor of war. In fact, just the opposite has been found to be the case, particularly during transitions in the level of power between states, as we saw in chapter 5. In addition, although classical deterrence theorists claim that nuclear weapons are inherently stabilizing, there is evidence that nonnuclear opponents of nuclear powers do not appear cautious or constrained in their hostile activity (Organski and Kugler 1980). Furthermore, the possession of nuclear weapons does not appear to impede escalatory behavior by nonnuclear opponents (Geller 1990; Paul 1994). Rauchhaus (2009, 269) conducts a quantitative test of the nuclear peace hypothesis and finds that "states with nuclear weapons are more likely to engage in militarized disputes (crises), to use force, and to be involved in uses of force that result in fatalities. This is true for situations of nuclear symmetry as well as asymmetry."

The logic of classical deterrence theory would lead us to expect states to welcome nuclear proliferation as a way to stabilize relationships between key pairs of states. Nuclear states, however, seldom pursue proliferation policies; rather, they remain committed to nonproliferation. Attempts by Iran, North Korea, and others to acquire nuclear weapons have been met by fierce resistance by other states. Thus, both of structural deterrence theory's main empirical predictions—that parity produces peace and that nuclear weapons are inherently stabilizing—are contradicted by the historical record. Before evaluating classical deterrence theory's logic, we now turn to an examination of decision-theoretic deterrence theory.

Decision-Theoretic Deterrence Theory

Decision-theoretic deterrence theory utilizes expected utility and game theory to construct models of deterrence (e.g., Brams and Kilgour 1988; Powell 1990, 2003; Schelling 1960, 1966). In particular, decision-theoretic deterrence theorists have argued that deterrence relationships are best modeled by preferences drawn from the game of Chicken (Figure 8.2). As we saw in chapter 3, Chicken models situations that are similar to two drivers driving at each other head on trying to force the other to swerve. In Chicken, two players, State A and State B, simultaneously choose between cooperating and defecting. For each state, winning is the best outcome (utility = 4) and conflict is the worst (u = 1). The status quo provides the second best outcome, with losing being the second worst. The outcomes *A Wins* and *B Wins* are Nash equilibria (represented by an asterisk).

Chicken models quite well the structural deterrence theorists' idea that nuclear war is irrational. Conflict is the mutually worst outcome for the two players; therefore, neither would ever rationally choose to go to war. In Chicken, the two Nash equilibria are for *A Wins* and *B Wins*. Thus, it is in each player's interest to be the one who wins, and classical deterrence theorists have developed a variety of ideas about how to accomplish this goal, such as "burning one's bridges" (Schelling 1966) or "ripping off the steering wheel" (Kahn 1962). To make sure that they win rather than the opponent during crises, states should pursue "commitment tactics," such as making an irrevocable commitment to a hard-line strategy, forfeiting control over actions, and feigning irrationality.

Figure 8.2 **Game of Chicken**

		State B	
		Cooperate	Defect
State A	Cooperate	**Status Quo** (3, 3)	**B Wins*** (2, 4)
	Defect	**A Wins*** (4, 2)	**Conflict** (1, 1)

Chicken is a simultaneous move game; thus, each actor must choose its strategy without knowing what choice the other actor makes. Modeling deterrence as a simultaneous move game, however, seems contrary to the logic of deterrence. Although deterrence is based on a contingent threat, there is no opportunity to punish the other player's defection if the strategies are selected simultaneously. Thus, deterrence theory needs to use sequential moves.

A simple deterrence game with Chicken-like preferences, where the states make sequential choices, is shown in Figure 8.3. In this case, State A has the first move, followed by State B. In mutual deterrence, either state in a dyad might be the first to initiate conflict. While it is possible to model both sides' decisions about whether to initiate in the same game, here we only consider each state's decision separately for the sake of simplicity. Thus, for mutual deterrence, states in a dyad would be playing the role of State A and State B simultaneously.

Using backward induction, one can see that at node 2, State B will choose to cooperate (C), giving her u = 2, rather than to defect (D), which would give her the worst possible outcome. Knowing this, State A will defect (D) at node 1, giving him u = 4, since cooperation (C) would only give him u = 3. Thus, *A Wins* is the equilibrium. As with the original game of Chicken (Figure 8.2), *Status Quo* is never an equilibrium outcome when the choices are made sequentially (Figure 8.3).

Focusing on decision-theoretic deterrence theory reveals additional empirical problems with classical deterrence theory. While Schelling and others advise leaders to utilize commitment tactics such as burning one's bridges, in reality leaders of states tend to be risk averse: They avoid commitments

Figure 8.3 Simple Deterrence Game With Chicken-Like Preferences

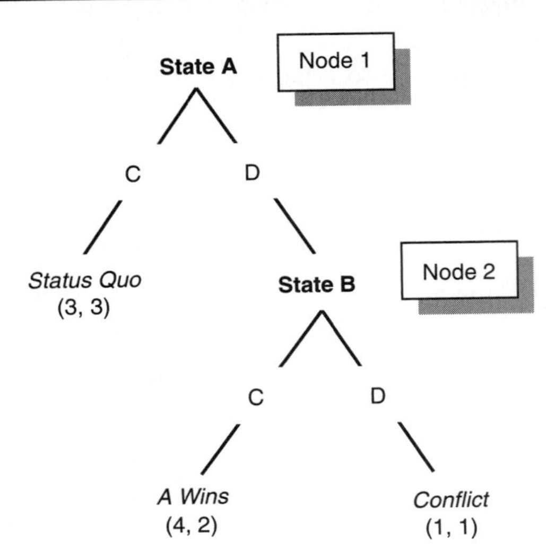

and almost always seek to maintain flexibility (Maxwell 1968; Snyder and Diesing 1977). As Jervis (1988, 80) observed, "Although we often model superpower relations as a game of Chicken, in fact the United States and the USSR have not behaved like reckless teenagers."

More importantly, decision-theoretic deterrence theory reveals the logical problems inherent in classical deterrence theory. Classical deterrence theorists have claimed that since nuclear warfare entails such high costs that it is the worst possible outcome for all sides, "nuclear weapons are a superb deterrent" (Mearsheimer 1990, 20). However, if conflict—nuclear or conventional—is indeed the worst possible outcome, deterrence *cannot ever* be successful. Any state that is attacked will always capitulate rather than bring about its own worst outcome, and knowing this, states will always attack.

Thus, classical deterrence theory is unable to explain successful deterrence (i.e., the survival of the status quo). Since the status quo is never an equilibrium outcome in Chicken, the logic of classical deterrence theory indicates that deterrence is quite unstable. But the history of the Cold War indicates otherwise. Freedman (1983) notes that the American attempt to make credible a nuclear guarantee to defend Western Europe worked far better in practice than an assessment of the theory might lead one to expect.

This gets to the heart of the logical problem with classical deterrence theory. Defender faces a quandary: A concession provides a more favorable outcome than the conflict that would follow defiance, but if Challenger knows that Defender will concede, Challenger will always attack and deterrence will always fail. While the logic of classical deterrence theory indicates that deterrence is quite unstable, classical deterrence theorists have nonetheless claimed that deterrence is quite stable. Thus, the logic of the theory contradicts what the theorists "conclude" from the theory. The fundamental problem for classical deterrence theory is reconciling the instability of the cooperative outcome in Chicken with the postulated robust stability of mutual deterrence. Zagare and Kilgour (2000) label this dilemma the **paradox of mutual deterrence.** Classical deterrence theory offers two primary solutions.

The first proposed solution is for Defender to make an irrevocable commitment to a hard-line strategy. Such irrevocable commitments are made by burning bridges to limit one's options by eliminating the ability to back down even though that would be the preferred alternative (e.g., Kahn 1962; Schelling 1966). So Defender makes an irrevocable commitment to defy and must communicate this commitment to Challenger.[1] Figure 8.4 shows the strategic form of the simple deterrence game shown in Figure 8.3.

[1] The movie *Dr. Strangelove,* in which the Soviet Union fails to communicate its development of a doomsday machine to the United States, highlights nicely the importance of communicating irrevocable commitments.

Figure 8.4 Strategic Form of Simple Deterrence Game

		State B	
		Cooperate	Defect
State A	Cooperate	**Status Quo** (3, 3)	**B Wins*** (3, 3)
	Defect	**A Wins*** (4, 2)	**Conflict** (1, 1)

Challenger is faced with the choice of remaining at the *Status Quo* by cooperating or defecting and starting a *Conflict*. Since choices of cooperate by Challenger and defy by Defender are mutual best responses, this strategy pair denotes a Nash equilibrium with *Status Quo* as the outcome. The problem, however, is that if Challenger does attack, then Defender's only rational response would be to concede, leading to *Defender Concedes*. Although the strategy pair (cooperate, defy) is a Nash equilibrium, it is not subgame perfect because it involves an irrational choice by Defender off the equilibrium path. The only subgame perfect equilibrium is the strategy pair (defect, concede), which results in *Defender Concedes*.

The second solution to this dilemma offered by classical deterrence theorists involves threats that leave something to chance (Schelling 1960). These threats allow Defender to circumvent the problem of irrational action by threatening to take action "that raises the risk that the situation will go out of control and escalate to a catastrophic nuclear exchange" (Powell 2003, 90). Thus, rather than relying on a threat to make an irrational choice for war, Defender can simply make a rational choice to raise the risk of war and leave the question of whether war starts or not to chance. Powell (1990, 2003) demonstrates that threats that leave something to chance not only lead to successful deterrence, but the resulting equilibrium is perfect as well.

The idea of threats that leave something to chance, however, rests upon the possibility of accidental war. There is a lot of speculation about this possibility (e.g., Blair 1993; Bracken 1983; Sagan 1993; Schelling 1966), with the outbreak of World War I often cited as the prime example of an inadvertent war. For example, Bracken (1983, 65) states that the idea of accidental nuclear war would sound unrealistic "were it not for the history of the outbreak of World War I." Trachtenberg (1991, 99), however, conducts an extensive examination of the coming of World War I and concludes that

> when one actually tests these propositions against the empirical evidence, which for the July Crisis is both abundant and accessible, one is struck by how weak most of the arguments turn out to be. The most

remarkable thing about all these claims that support the conclusion about events moving "out of control" in 1914 is how little basis in fact they actually have.[2]

The prime example of accidental war turns out to be not such a good example after all. Rather than support the idea, the outbreak of World War I actually undermines the idea of threats that leave something to chance.

Schelling suggests that in most crises, one side or another will be willing to run greater risks of mutual assured destruction to achieve its goals. Therefore, credibility is determined by whose interests are more greatly threatened by an ongoing crisis and by who is more willing to take risks to protect their interests. Hence, deterrence becomes a "competition in risk taking" as states employ brinksmanship in what the subtitle to Powell's 1990 book calls "the search for credibility." But a threat is said to be credible if it is believed (George and Smoke 1974; Jervis 1985; Schelling 1966). Given that a nuclear attack invites one's own destruction, the threat to choose to do so is not believable, and is thus not credible. Schelling (1960) argues that while this is true, the threat to increase the risk of inadvertent war can in fact be believable. But since historical evidence shows that World War I—the prime example of accidental war—arose as a result of conscious decisions, not chance (Trachtenberg 1991; Zagare 2011), these threats that leave something to chance seem to be not credible after all.

Classical deterrence theory and its associated ideas, such as mutual assured destruction and brinksmanship, represent the conventional wisdom about deterrence. However, as we have seen, classical deterrence theory is plagued by logical inconsistency and empirical inaccuracy, undermining the theory's ability to explain deterrence. Although conventional wisdom is unable to explain deterrence, there is another theory that one can turn to: perfect deterrence theory.

Perfect Deterrence Theory

Perfect deterrence theory, offered by Zagare and Kilgour (2000), is an alternative to classical deterrence theory developed to overcome the latter's logical deficiencies. In particular, perfect deterrence theory departs from classical deterrence theory in its view of **credibility**. Zagare and Kilgour (2000) argue that threats are believable, and thus credible, when they are rational to carry out (also see Betts 1987; Lebow 1981; Smoke 1987). Connecting credibility with rationality in this way is consistent with the treatment of credibility in game theory (e.g., Gibbons 1992; Rasmusen 1989; Selten 1975).

This connection between credibility and rationality can be seen by reexamining the game in Figure 8.3. If Defender prefers *Defender Concedes* to *Conflict*, and this is known to Challenger, Challenger has no reason

---

[2] Mulligan (2010) and Zagare (2009, 2011) reach similar conclusions.

whatsoever to believe that Defender will carry out her threat; thus Defender's threat is not credible. On the other hand, if Defender prefers *Conflict* to *Defender Concedes,* and this is known to Challenger, Challenger would believe Defender's threat: Defender's threat would be credible. The simple deterrence game with a credible threat by Defender is shown in Figure 8.5.

In this game, Defender will choose to defy at node 2 because a concession would result in her least preferred outcome. Knowing this, Challenger will cooperate at node 1 because the outcome that results (*Status Quo*) is preferred to the outcome that results from defection (*Conflict*). Thus, if Defender has a credible threat, *Status Quo* is the sole equilibrium outcome.

Accordingly, the solution to the paradox of mutual deterrence is the presence of credible threats (Zagare and Kilgour 2000). When both sides possess credible and capable threats, *Status Quo* emerges as the rational outcome. Furthermore, the resulting equilibrium is subgame perfect, and it does not rest upon untenable assumptions of the possibility for accidental war.

Zagare and Kilgour label their theory "perfect" because of their insistence on the use of perfect equilibria. This insistence stems from the observation that "perfectness rules out threats that are not credible" (Rasmusen 1989, 87). Powell (1990, 2003), however, has also employed perfect equilibria, but his work certainly falls within classical, rather than perfect, deterrence theory because he assumes that conflict is always the worst outcome. The real hallmark of perfect deterrence theory is the insistence that credibility varies, and that credibility is determined by a state's preference between conflict and backing down. I focus here on complete information conditions, where each actor's preferences are common knowledge. Zagare and Kilgour (2000) also examine situations of incomplete information, in which

Figure 8.5 Simple Deterrence Game With a Credible Threat

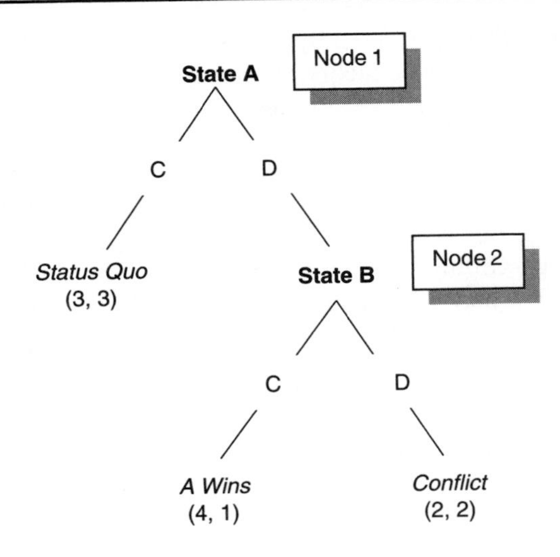

each actor forms beliefs about its opponent's preference between conflict and backing down. In that case, a state's credibility results from its opponent's belief about its type. Although the details are different, the implications are quite similar.

While classical deterrence theory is rooted in the basic assumption that the high costs of nuclear war make conflict the worst outcome for everyone, perfect deterrence theory is rooted in the assumption that different states have different preferences. While some may indeed prefer backing down to fighting, others prefer to fight and only these latter states have credible threats. Although there are other differences between the two theories (Zagare 2004), this one assumption makes a tremendous impact on the predictions and explanations offered.

Perfect deterrence theory also highlights the importance of capability. **Capability** is the ability to hurt. A state's threat is only capable if its opponent prefers *Status Quo* to *Conflict*. In each of the games we have examined to this point, we have assumed this to be true. Figure 8.6 shows the same simple deterrence game that we have been examining, but this time Defender does not have a capable threat.

In this game, State B's threat is once again credible; B will rationally choose to defy at node 2. At node 1, State A is again faced with a choice between the *Status Quo* following cooperation and *Conflict* following defection. Since B's threat is not capable, however, A prefers *Conflict* to the *Status Quo*. Thus, A will defect at node 1 and *Conflict* will result. Zagare and Kilgour (2000) demonstrate that capability is a necessary condition for successful deterrence.

Perfect deterrence theory highlights two keys to explaining deterrence: credibility and capability. In addition, the theory includes a variety of

Figure 8.6 **Simple Deterrence Game Without a Capable Threat**

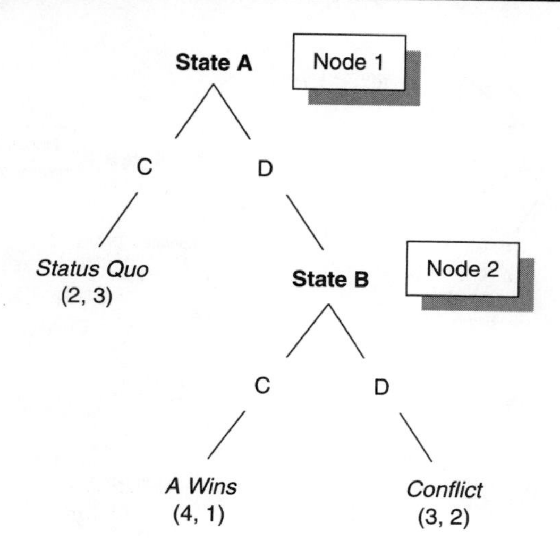

different games analyzing mutual deterrence, unilateral deterrence, and extended deterrence. We briefly consider each in turn.

Mutual Deterrence

To examine mutual deterrence, Zagare and Kilgour (2000) developed the Generalized Mutual Deterrence Game. In this game, shown in Figure 8.7, two undifferentiated states (A and B) begin by simultaneously choosing whether to cooperate with or defect from the status quo. The dashed line connecting State B's choices at nodes 2a and 2b indicates that nodes 2a and 2b are in the same information set. This means that State B has no knowledge of State A's choice at node 1 when making its first choice. If both actors choose to cooperate initially, the *Status Quo* remains unchanged. Similarly, if both states choose to defect initially, *Conflict* results. If, however, State A cooperates at node 1 and State B defects at node 2a, State A then has an opportunity to retaliate at node 3a. Likewise, if State A defects at node 1 and State B cooperates at node 2b, State B has an opportunity to retaliate at node 3b. At that point, the state can either cooperate, in which case the other state wins (resulting in *A Wins* or *B Wins*), or defect, resulting in *Conflict*.

In cases of mutual deterrence, each side has an incentive to challenge the status quo. Accordingly, each state would prefer to modify the status quo at minimal cost. The costs to State A when State B concedes (or vice versa) are minimal, but the (potential) gains are large. Hence, the benefits associated with *A Wins,* for State A, or *B Wins,* for State B, almost certainly outweigh the costs. As a result, both states are assumed to prefer winning to the status quo.

Figure 8.7 Generalized Mutual Deterrence Game

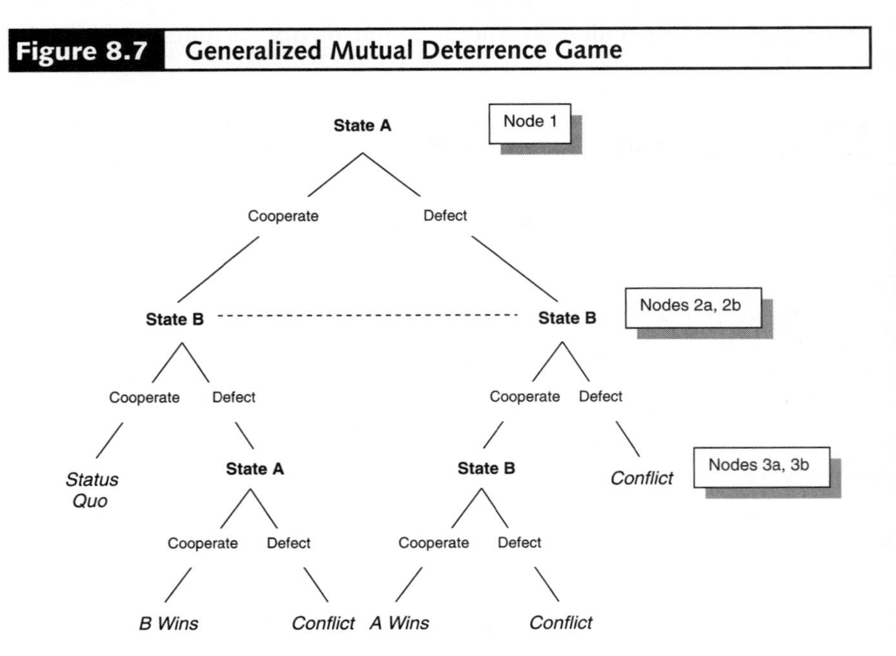

As discussed previously, perfect deterrence theory highlights the importance of capability and credibility. A state's threat is capable if the threatened party believes that it would be worse off if the threat were carried out than if it were not. Therefore, if State B has a capable threat, State A prefers *Status Quo* to *Conflict*, and if State A has a capable threat, State B prefers *Status Quo* to *Conflict*.

Credible threats are believable, and in order to be believable, they must be rational to carry out (Zagare 1990). Therefore, a state that prefers conflict to backing down (e.g., a State B that prefers *Conflict* to *A Wins*) has a credible threat, and is said to be of type Hard. On the other hand, a state that would rather back down than fight (e.g., a State A that prefers *B Wins* to *Conflict*) has an incredible threat, and is said to be of type Soft. With complete information, there are four distinct subgame perfect equilibria, one for each pair of State A and State B types. These equilibrium outcomes are shown in Figure 8.8.

This figure indicates that if one state has a credible threat (i.e., is Hard) and the other does not, the former wins. However, if neither state's threat is credible (i.e., both are Soft), either *A Wins* or *B Wins* is possible. But most significant here is that if both states are Hard, either *Status Quo* or *Conflict* is possible. That is to say, while the equilibrium resulting in the *Status Quo* is better for both sides, non–status quo outcomes are nonetheless always possible in mutual deterrence under rational play.

These complete information results rest on the assumption that player types are common knowledge; that is, that both players know with certainty whether or not the other player's threat is credible. Unfortunately, such certainty about other states is rarely, if ever, achieved in international relations. In other words, conditions of incomplete information are prevalent in the international arena. Furthermore, incomplete information can have a large impact on the outcomes that emerge from rational action.

For these reasons, Zagare and Kilgour (2000) extend their analysis to examine the Generalized Mutual Deterrence Game with incomplete information. Incomplete information is modeled by allowing each player to be uncertain of the other player's type (each player still knows his

Figure 8.8 **Subgame Perfect Equilibrium Outcomes for Mutual Deterrence Game With Complete Information**

		State B	
		Hard	Soft
State A	Hard	*Status Quo* or *Conflict*	*A Wins*
	Soft	*B Wins*	*A Wins* or *B Wins*

own type). State A estimates the probability that State B is Hard, p_B, and State B estimates the probability that State A is Hard, p_A (where $0 < p < 1$ in each case).

With conditions of incomplete information, the implications of the mutual deterrence game are quite similar. When both states have highly credible threats, deterrence success is once again likely, but certainly not guaranteed. And likewise, successful deterrence becomes notably less likely at lower levels of credibility (for one or both sides). Higher levels of credibility are therefore associated with raised probabilities of deterrence success, without the stability of mutual deterrence ever being fully assured.

Unilateral Deterrence

To examine situations of unilateral deterrence, Zagare and Kilgour (2000) developed the Unilateral Deterrence Game. Figure 8.9 shows the structure of this game, which has two players, Challenger and Defender. At node 1, Challenger can choose whether to cooperate or defect: If he

Figure 8.9 Unilateral Deterrence Game

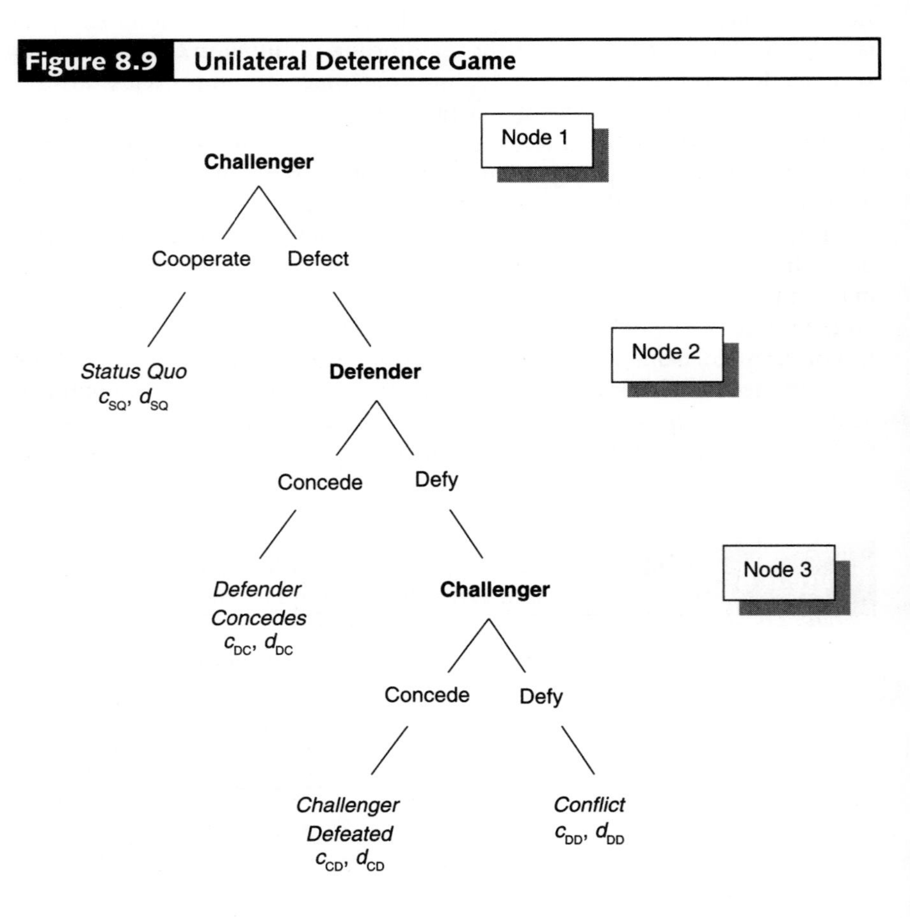

cooperates, the *Status Quo* remains unchanged; if he defects, Defender has an opportunity to respond. At node 2, Defender can choose whether to concede or defy. If she concedes, the outcome is *Defender Concedes,* but if she defies, the next choice is Challenger's. At node 3, Challenger can choose to concede, resulting in *Challenger Defeated,* or defy, resulting in *Conflict.*

Each state is assumed to have a capable threat, and therefore, each prefers *Status Quo* to *Conflict.* Further, Zagare and Kilgour allow threat credibility to vary by considering both Hard and Soft states. The equilibria for the Unilateral Deterrence Game with complete information are completely determined by player types. There are four distinct subgame perfect equilibria, one for each pair of Challenger and Defender types (for example, Hard-Hard or Hard-Soft), as shown in Figure 8.10.

Under complete information conditions, deterrence is stable in three of the four possible situations. If Defender has a credible threat, Challenger never defects, and *Status Quo* always prevails. But, contrary to claims by some (e.g., Lebow 1981), credibility is not a necessary condition for deterrence success. Even when Defender lacks a credible threat, *Status Quo* is still stable if Challenger's threat is incredible as well. Only in the case where only Challenger has a credible threat is another outcome (*Defender Concedes*) possible.

Once again, Zagare and Kilgour analyze the model to include the effects of incomplete information as well. Although the consideration of incomplete information adds a great deal of nuance to the model, the implications are quite similar regardless of the information conditions. With incomplete information, the stability of the status quo is assured as long as Defender's threat is highly credible. At lower levels of Defender credibility, however, successful deterrence is much less likely to be achieved.

Perfect deterrence theory reveals an important difference in the stability of mutual and unilateral deterrence. In mutual deterrence situations, even if both states have perfectly credible threats, stability (i.e., the survival of the status quo) cannot be ensured, whereas in unilateral deterrence situations, *Status Quo* is the *only* rational outcome if Defender's threat is sufficiently credible.

Figure 8.10 Subgame Perfect Equilibrium Outcomes for Unilateral Deterrence Game With Complete Information

		Defender	
		Hard	Soft
Challenger	Hard	*Status Quo*	*Defender Concedes*
	Soft	*Status Quo*	*Status Quo*

Thus, the stability of mutual deterrence is never certain in the way that it can be for unilateral deterrence, regardless of informational conditions.

Extended Deterrence

To situations of extended deterrence, Zagare and Kilgour (2000) have developed the Asymmetric Escalation Game. As in the simple deterrence game, Challenger begins play by choosing whether to cooperate (C) or defect (D). But, as Figure 8.11 shows, Defender's response options in the Asymmetric Escalation Game are more varied: At decision node 2, Defender can concede (C), it can respond in kind (D), or it can escalate (E). Depending on Defender's choice, Challenger can either escalate first (at node 3a) or counterescalate (at node 3b), or not. If Challenger escalates first, Defender has an opportunity, at node 4, to counterescalate. The outcomes associated with the various choices in the Asymmetric Escalation Game are summarized in Figure 8.11.

Figure 8.11 Asymmetric Escalation Game

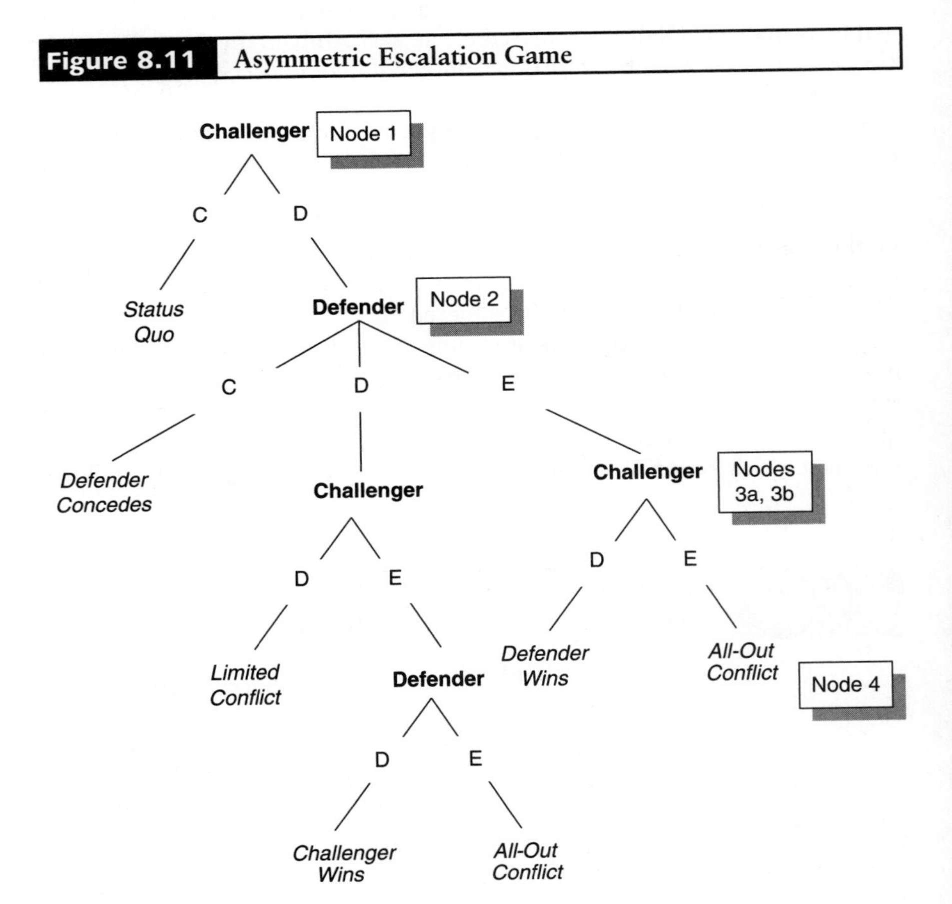

Recall that extended deterrence situations detail a state's attempt to deter an opponent from attacking a third state. This game is appropriate for these situations because of the three options available to Defender at node 2. Defection by Challenger at node 1 represents a challenge against a protégé of Defender's. If Defender concedes, she abandons the protégé to her fate against Challenger. Defender can stand beside her protégé by either responding in kind, standing up to Challenger on equal terms (generally equated with using conventional forces), or escalating against Challenger (including the possible use of nuclear weapons).

In this game, Defender has threats at two different levels because of the option to escalate. Defender's first-level threat deals with her threat to respond in kind at node 2. A Hard Defender at this level prefers *Limited Conflict* to *Defender Concedes,* while a Soft Defender has the opposite preference. Defender's second-level threat deals with her threat to counter-escalate at node 4 and is determined by her preference between *Challenger Wins* and *All-Out Conflict*. Defender can be Hard at both levels, Soft at both, or Hard at one level and Soft at the other.

Table 8.1 shows the subgame perfect equilibrium outcomes for the Asymmetric Escalation Game with complete information. Once again, credible threats make deterrence success more likely but are neither necessary nor sufficient for success.

In the 1950s, the Eisenhower administration pursued a policy of **massive retaliation**. The policy was enunciated by US Secretary of State John Foster

Table 8.1 Subgame Perfect Equilibrium Outcomes of the Asymmetric Escalation Game

CHALLENGER'S TYPE (2ND LEVEL)	DEFENDER'S TYPE (1ST LEVEL)	DEFENDER'S TYPE (2ND LEVEL)	SUBGAME PERFECT EQUILIBRIUM OUTCOME
Hard	Hard	Hard	*Status Quo*
Hard	Hard	Soft	*Defender Concedes*
Soft	Hard	Hard	*Status Quo*
Soft	Hard	Soft	*Status Quo*
Hard	Soft	Soft	*Defender Concedes*
Hard	Soft	Hard	*Defender Concedes*
Soft	Soft	Soft	*Status Quo*
Soft	Soft	Hard	*Status Quo*

Source: Adapted from Frank C. Zagare and D. Marc Kilgour, *Perfect Deterrence* (Cambridge: Cambridge University Press, 2000), 194.

Dulles and relied on strategic nuclear weapons rather than conventional forces to maintain stability. Zagare and Kilgour model massive retaliation by assuming that Defender is always Soft at the first level. Analyzing the game with incomplete information, they argue that Defender's strategic position is poor because "even perfect end-game credibility may be insufficient to deter a determined Challenger" (Zagare and Kilgour 2000, 215). Massive retaliation is problematic largely because it presents Defender with a stark choice between suicide and surrender.

In the 1960s, the Kennedy administration shifted to a policy of flexible response. **Flexible response** entailed strengthening conventional forces and reducing reliance on nuclear weapons. NATO formally adopted flexible response in 1967, and it has remained the cornerstone of American deterrence posture ever since. Zagare and Kilgour model flexible response by assuming that Defender is always Hard at the first level. Incomplete information analyses reveal that flexible response is much more stable than massive retaliation. It does not, however, guarantee deterrence success; much still depends on the credibility of Defender's second-level threat.

Empirical Support

Theories are useful to the extent that they help to understand and explain reality. Although to this point we have focused on the logic of deterrence theory, empirical testing is needed (Harvey 1998). Two empirical approaches have been used in the deterrence literature: case studies and quantitative analysis. The case study literature has focused on in-depth analyses of particular deterrence situations, but these studies have generally endeavored to criticize, rather than test, rational deterrence theory (e.g., George and Smoke 1974; Lebow 1981; Lebow and Stein 1990).

Furthermore, an unfortunate divide exists between formal theories and the quantitative analysis of deterrence. There are several reasons for this disconnect: (1) while deterrence theory has typically focused on general deterrence, quantitative studies have focused almost exclusively on immediate deterrence; (2) quantitative studies have generally not tested rational deterrence theory, but rather have tested independently developed hypotheses; and (3) when studies have harkened back to rational deterrence theory, they have typically examined classical deterrence theory, a formal framework characterized by logical inconsistency and empirical inaccuracy.

An example of this continuing divide is Danilovic's (2002) quantitative study of extended deterrence. Danilovic argues that a state's inherent credibility—determined in large part by its regional interests—is a far more important predictor of deterrence outcomes than attempts to shore up credibility through the use of commitment strategies recommended by Schelling (1960, 1966) and other classical deterrence theorists. However, she does not explicitly tie her quantitative analysis to any theory of deterrence. Thus, it is left to the reader to bridge this divide.

Fortunately, through her focus on credibility, Danilovic (2002) directly addresses a key distinction between perfect and classical deterrence theory. As discussed above, the reason that classical deterrence theorists focus on commitment tactics as ways to shore up credibility is that they assume that all threats are inherently incredible. Zagare and Kilgour (2000), on the other hand, argue that states' threats are only credible to the extent that the state actually prefers conflict to backing down. Accordingly, Danilovic's analysis can be seen as an attempt to determine what factors (relative power, regional interests, and democracy) lead states to prefer conflict over backing down.

Since rational deterrence theory is primarily expressed through game-theoretic models, empirical tests of the theory need to be able to test predictions that result from such models. There are two primary avenues available for such tests: evaluation of equilibrium predictions and evaluation of relationship predictions (Morton 1999). Evaluation of equilibrium predictions entails comparison of the outcome predicted in equilibrium at each observation with the outcome actually observed, while testing relationship predictions entails testing relationships between variables that are implied by the theory. Once one relates Danilovic's argument about inherent credibility to Zagare and Kilgour's arguments about credibility, it becomes clear that her analysis is at least indirectly an evaluation of perfect deterrence theory's relationship predictions.

Selection of deterrence cases has been a subject of much disagreement, primarily between quantitative and case study researchers. Much of an important 1989–1990 debate in the journal *World Politics* dealt with the proper identification of immediate deterrence cases (Achen and Snidal 1989; George and Smoke 1989; Huth and Russett 1990; Jervis 1989; Lebow and Stein 1989, 1990). Unfortunately, however, the existing deterrence literature provides little guidance on the selection of general deterrence cases. This is particularly notable because of the importance of general deterrence, as discussed above.

Morgan (1983) sees the maintenance of armed forces as indicative of general deterrence behavior. One can safely assume that every state wishes to deter attacks against itself—this is the basic rationale for the maintenance of armed forces. This is essentially identical to the assumption within the alliance portfolio literature that every state has a defense pact with itself, as discussed in chapter 6. Therefore, the difficult part of general deterrence case selection is not determining who makes deterrent threats (everyone does), but rather what states the threats are directed against. General deterrent threats are directed against any state that might consider an attack, but it can be difficult to identify exactly which states might do so. The key to general deterrence case selection then is to identify which states may consider attacking which other states. I argue that the states that might consider an attack on a particular state are those that have the opportunity to fight that state. Fortunately, Quackenbush (2006a) has developed the concept of politically active dyads (as discussed in chapter 4), which capture opportunity as a necessary condition for international conflict. Therefore, politically active dyads can be used as a case selection mechanism for studies of general deterrence.

Another concern that has often driven debates between those employing case-study (Lebow and Stein 1989, 1990) and quantitative (Huth and Russett 1990) methodologies is the determination of deterrence success and failure. Game-theoretic models of deterrence, however, do not make predictions regarding the success or failure of deterrence, per se; rather, the particular outcome of an interaction is predicted.[3] Furthermore, classification of certain cases as successful deterrence runs into a variety of conceptual and selection bias issues (Danilovic 2001).

Quackenbush (2010a, 2011b) tests the equilibrium outcome predictions of perfect deterrence theory. He uses binary and multinomial logit methods to examine the prediction of militarized interstate disputes and of particular game outcomes, and the results indicate that the predictions of perfect deterrence theory are generally supported by the empirical record.

Most attempts to test deterrence theory have utilized large-N, quantitative methods. Such analysis is useful because of the generalizable nature of conclusions derived from it. Case studies, however, would also be useful for future research. Since quantitative analyses demonstrate that perfect deterrence theory is generally supported by the historical record, there is evidence that the theory can usefully be applied to particular cases. This can be done through the application of the insights of perfect deterrence theory to detailed case studies of particular historical episodes in order to understand these intrinsically interesting events more completely. For example, one study applies perfect deterrence theory to an analysis of the war in Kosovo to answer questions—such as why NATO's threat of bombing was unable to deter Serbia, and later why Serbia escalated ethnic cleansing once the bombing started—to which others have struggled for answers (Quackenbush and Zagare 2006). Another study explains the July 1914 crisis leading to World War I through an analytic narrative based on perfect deterrence theory (Zagare 2011). The analysis demonstrates that although general war was not sought by any of the actors, the war was no accident.

Conclusion

In this chapter, we examined deterrence in the international system. There are two main theories of deterrence: classical deterrence theory and perfect deterrence theory. While both are rational-choice theories, they differ in several respects, particularly regarding their treatment of credibility. Classical deterrence theory assumes that conflict is the worst possible outcome for both sides, meaning that no retaliatory threats are credible. However,

[3] For example, in the Unilateral Deterrence Game (Figure 8.9), the only rational outcome is *Status Quo* if Challenger prefers *Status Quo* to *Defender Concedes*. Although Challenger's decision to not challenge the status quo in this case is not really "successful deterrence," it is predicted by perfect deterrence theory.

this assumption leads directly to the paradox of mutual deterrence, wherein classical theory is unable to explain deterrence success.

The paradox of mutual deterrence is solved by perfect deterrence theory, which argues that the credibility of a state's threat depends upon its preference between backing down and conflict. Therefore, we need to move away from the assumption that conflict is the worst possible outcome. By doing so, Zagare and Kilgour (2000) are able to develop a logically consistent theory of general deterrence that usefully explains the dynamics of deterrence in mutual, unilateral, and extended deterrence situations.

Although the distinction between classical and perfect deterrence theories is stark, this distinction—or at least the challenge of perfect deterrence theory—is often ignored. The contrast between the two theories is not simply a matter of differences over assumptions and abstract theoretical issues. The policy recommendations of perfect deterrence theory and classical deterrence theory are diametrically opposed on many issues. For example, perfect deterrence theory supports minimum deterrence and significant arms reductions but opposes overkill capability and proliferation, while classical deterrence has the opposite recommendation in each case (Zagare 2004). Further, classical deterrence theory argues that national missile defense undermines the stability of deterrence (Powell 2003), whereas an application of perfect deterrence theory demonstrates that national missile defense can enhance deterrence stability (Quackenbush 2006c). However, policy discussions in academia and government are generally based on classical deterrence theory. Given its strong empirical support, coupled with logical and empirical limitations of classical deterrence theory, perfect deterrence theory provides a much better basis for analyzing various aspects of national security policy.

Clearly, classical deterrence theory constitutes the conventional wisdom regarding deterrence. Nonetheless, classical deterrence theory is badly flawed. The (unnecessarily restrictive) assumption that conflict is always the worst possible outcome needs to be discarded. It has not proven useful for developing logically consistent and empirically accurate theory. By contrast, perfect deterrence theory provides a logically consistent alternative to understand the dynamics of deterrence. Therefore, perfect deterrence theory provides the most appropriate basis for further theoretical development, empirical testing, and application to policy.

Key Concepts

Capability

Classical deterrence theory

Credibility

Decision-theoretic deterrence theory

Deterrence

Direct deterrence

Extended deterrence

Flexible response

General deterrence

Immediate deterrence

Massive retaliation

Mutual assured destruction (MAD)

Mutual deterrence

Paradox of mutual deterrence

Perfect deterrence theory

Structural deterrence theory

Unilateral deterrence

9 Escalation of Disputes to War

Serbian nationalist Gavrilo Princip assassinated Archduke Franz Ferdinand of Austria-Hungary and his wife, Czech Countess Sophie Chotek, during their visit to Sarajevo on 28 June 1914, as depicted in this artist's rendition. The incident precipitated World War I, one of the most costly examples of the escalation of disputes to war.
Source: Associated Press

In the preceding chapters, we examined several primary causes of international conflict, including contiguity, power, and alliances. In some cases we have distinguished between militarized interstate disputes and war, but for the most part we have focused on various causes of international conflict in general, rather than disputes or war in particular. Our focus in this chapter shifts to the escalation of disputes to war. While militarized interstate disputes are important in their own right, war involves much greater costs and has much greater consequences. Therefore, it is important to understand why wars in particular occur, not only conflict in general. Bremer's (1995) process of war model, which has provided the framework for this book, highlights the connection between militarized disputes and war. Remember that war is a dispute that involves at least one thousand battle deaths. What accounts for the escalation of intensity that turns a dispute into war?

We begin by looking at patterns of escalation, focusing on the frequency of wars compared to disputes in general. We then move on to examine different explanations of why (some) disputes escalate to war. We first consider explanations based on contextual factors, focusing on steps to war theory. We then turn our attention to explanations focused on strategic interaction between states. We examine attempts to characterize the bargaining strategies used by states within crises as well as models of escalation before looking at how game theory can help us understand the escalation of disputes to war.

Patterns of Escalation

As we saw in chapter 2, there are many examples of crises and wars in history. One such crisis is the July Crisis of 1914, which led to World War I. The crisis was sparked by the assassination of Archduke Franz Ferdinand, heir to the Austrian throne, by a Serbian nationalist on 28 June 1914. Germany pledged its support to Austria-Hungary on 5 July, and after weeks of debate, Austria-Hungary gave Serbia a forty-eight-hour ultimatum to respond to its demands on 23 July. Serbia rejected the ultimatum on 25 July, and Austria-Hungary declared war on Serbia on 28 July, one month after the crisis began. Russia mobilized its forces on 30 July, followed by German and French mobilizations on 1 August. Germany began its invasion in the west on 2 August, and Britain declared war on Germany two days later, on 4 August (Gilbert 1994; Williamson and Van Wyk 2003). Thus, what began as a dispute between Serbia and Austria-Hungary escalated to a massive war involving all of the great powers and the death of more than nine million people.

Of course not every crisis escalates to war. For example, during a routine reconnaissance flight on 14 October 1962, the United States discovered that the Soviets were installing nuclear missiles in Cuba. The ensuing Cuban Missile Crisis was probably the closest the world has come to nuclear war. President John F. Kennedy thought there was a distinct chance of nuclear war. The United States considered using air strikes or an invasion to take the missiles out directly or simply launching diplomatic protests in the United Nations and elsewhere. It finally decided on the implementation of a blockade to prevent further Soviet missiles, personnel, and equipment from arriving in Cuba. The blockade (called a *naval quarantine*, since a blockade is technically an act of war) was instituted on 24 October, and the crisis ended on 28 October as Soviet Premier Nikita Khrushchev announced that the Soviet Union would withdraw the missiles from Cuba. Thus, war was successfully avoided as US Secretary of State Dean Rusk claimed that the Soviets had blinked (Allison and Zelikow 1999).

A third type of example that is useful to consider in this context is a militarized interstate dispute that involves armed combat but does not escalate to war because it does not reach the required threshold of casualties. One such case was Operation Just Cause, the invasion of Panama by the United States in December 1989. The United States invaded on 20 December to remove Manuel Noriega from power in Panama for a variety of human rights abuses and drug trafficking. Noriega was captured on 3 January 1990, and the operation quickly came to an end (Donnelly, Roth, and Baker 1991; McConnell 1991). Although the invasion involved combat between United States military forces and the Panamanian Defense Force, casualties (23 American and 314 Panamanian military battle deaths) were not large enough to meet the threshold of a war.

We can look at the militarized interstate disputes data to get an overall impression of escalation to war. One of the variables in the militarized interstate disputes (MID) data that we highlighted in chapter 2 was the hostility level. The **hostility level** codes which type of action—no militarized action, threat of force, display of force, use of force, or war—was used in the dispute. This is coded separately for each state in a dispute. Thus, one state might use force while the other state only threatens to use force, or possibly takes no militarized action at all.

Table 9.1 shows the distribution of militarized interstate disputes by hostility level. The column labeled "DISPUTES" shows the highest level of hostility reached by any state in the dispute. Note that there are zero disputes in the no militarized action row, because at least one state must take a militarized action for it to be a dispute. There are 2,332 militarized disputes from 1816 to 2001. Of these, 103 (4.4 percent) involve threats of force and 569 (24.4 percent) involve displays of force as the highest level of hostility. The most common hostility level is use of force, appearing in 1,553 cases

Table 9.1 **Militarized Interstate Dispute Hostility Levels, 1816–2001**

HOSTILITY LEVEL	DESCRIPTION	DISPUTES	DISPUTE DYADS
1	No militarized action	0	0
2	Threat to use force	103	131
3	Display of force	569	806
4	Use of force	1,553	2,087
5	War	107	541
	Total	*2,332*	*3,565*

Source: Compiled by author.

(66.6 percent). Only 107 disputes, however, escalated to war, which is 4.59 percent of all disputes.

The final column in the table shows dispute dyads, which identify each pair of states engaged on opposite sides of a dispute. For example, a bilateral dispute has only one dispute dyad. However, a dispute with two countries (A and B) on one side and one country (C) on the other has two dispute dyads: A versus C and B versus C. Finally, a dispute with three countries on one side (A, B, and C) and four on the other (D, E, F, and G) has twelve different dispute dyads: A versus D, A versus E, A versus F, A versus G, B versus D, B versus E, B versus F, B versus G, C versus D, C versus E, C versus F, and C versus G. There are 3,565 MID dyads from 1816 to 2001. Of these, 131 (3.7 percent) involve threats of force and 806 (22.6 percent) involve displays of force as the highest level of hostility. Once again, use of force is the most common hostility level, appearing in 2,087 cases (58.5 percent). Only 541 dispute dyads, however, escalated to become war dyads. Note that although this percentage (15.18 percent) is still relatively low, the percentage of dispute dyads that escalate to war is higher (column 4) than the percentage of disputes that escalate to war (column 3). This indicates that multilateral disputes represent a higher percentage of wars than disputes more generally.

Clearly, wars are just a small subset of all disputes. So why do some disputes escalate to war, while most do not? There have been a number of studies seeking to explain the escalation. Siverson and Miller (1995) classify these explanations into two basic types. The first relies on contextual factors, and the second focuses on interaction between states. We turn our attention to each.

Contextual Factors Leading to Escalation

The process of war, developed by Bremer (1995) and discussed in chapter 1, highlights the underlying context within a dyad. These political, economic, social, and geographic conditions not only affect the outbreak of interna-

Box 9.1

Case in Point: Kosovo War

One conflict that escalated to war was the Kosovo War, pitting the United States and allies against Yugoslavia. The origins of the conflict over Kosovo are both complex and ancient, involving both ethnic rivalries and great power political machinations that date back centuries. In 1987, Serbian Yugoslavian president Slobodan Milosevic rose to power and skillfully used the wave of Serbian nationalism sparked by unrest among the majority of ethnic Albanians in Kosovo for personal political advantage. Soon thereafter, Milosevic terminated Kosovo's autonomy (1989) and abolished its parliament (1990). Milosevic's repressive policies made unrest among the ethnic Albanians worse. The Kosovo Liberation Army (KLA) was formed by ethnic Albanians to seek independence for Kosovo in 1991 and became increasingly violent in the mid-1990s.

An intrastate war between the Yugoslav government and the KLA began on 28 February 1998. The United States and its NATO allies tried to resolve the conflict by deterring Serbia from continuing the conflict. Despite negotiating the Rambouillet Accord to bring an end to the conflict, Yugoslavia refused to sign. Instead, it launched Operation Horseshoe, a campaign of ethnic cleansing to remove ethnic Albanians from Kosovo, leaving only ethnic Serbs there.

The interstate war began on 24 March 1999 as the United States and NATO began a bombing campaign against Yugoslavia. Dashing international hopes that the air campaign would be short, Yugoslavia appeared unmoved and even escalated its ethnic cleansing. Up to 10,000 Kosovar Albanians were killed and close to 850,000 were displaced. The war entered a new phase after a NATO summit on 23 April, at which it was decided to markedly increase the breadth and intensity of air strikes. In addition, the United States and NATO began planning for a ground invasion. The war finally ended on 10 June, as Yugoslavia agreed to accept American and NATO demands. The resulting settlement in large part instituted the provisions of the Rambouillet Accords, including the withdrawal of Yugoslav forces from Kosovo and the deployment of an international security force (KFOR) to Kosovo to monitor adherence to the settlement.

tional conflict in the first place, they can also make the escalation of disputes to war more or less likely. Accordingly, many studies have focused on **contextual factors** that affect the likelihood of escalation. While I generally use the term *dispute*, a number of studies have used the International Crisis Behavior data (introduced in chapter 2) to examine the escalation of crises to war (Brecher, James, and Wilkenfeld 2000).

Given the common focus on the influence of power on the outbreak of international conflict, it is not surprising that it has been a common focus in studies of escalation as well. As with the outbreak of conflict in the first place, debates about the impact of relative capability on the escalation of disputes to war have centered on arguments about whether a balance of power or a preponderance of power is more pacifying. There have been a number of studies about the effect of relative capabilities leading to very mixed results (e.g., James 1988; Moul 1988; Siverson and Tennefoss 1984), but arguments that preponderance makes escalation less likely are better supported.

A related focus of some studies is polarity. Is escalation to war more likely in bipolar or multipolar systems? Stoll (1984) finds some evidence that polarity mattered for escalation in the nineteenth century, although the relationship disappears in the twentieth century. Bueno de Mesquita and Lalman (1988) provide evidence that offers an important reason why: Systemic theories (such as those focusing on polarity) do not explain well, while dyadic theories explain much better.

Although alliances would seem a natural contextual factor to examine, there has not been much research on their impact on escalation. Several studies, however, have shown that alliances make escalation more likely. Smith (1995, 1996) argues that states are less willing to initiate a challenge if they believe that the target is going to be joined by one or more alliance partners in a potential war. Therefore, states are much less likely to challenge other states that have outside alliances, unless they judge those alliances to be unreliable. Accordingly, if a country with an alliance is in a dispute, that alliance is not able to deter the challenger from escalating, since they have already decided to challenge the target despite the alliance. Senese and Vasquez (2008) develop a separate argument for why alliances increase the likelihood of escalation, which we examine later in the chapter.

Given the strong effect of geographic contiguity that we examined in chapter 4, it makes sense that disputes between contiguous foes would be much more likely to escalate. Indeed, Senese (1996) finds that disputes between contiguous states are more likely to escalate to war than those between noncontiguous states. Senese (2005), however, finds that the effect of contiguity on escalation largely disappears once territorial issues are properly accounted for. Territory makes escalation more likely, while contiguity's effect is confined to the initial onset of the dispute.

The effect of arms races on the escalation of disputes to war has been subject to a great deal of debate. In a series of analyses, Wallace (1979, 1981, 1982) finds that arms races make escalation more likely, while Diehl (1983, 1985) finds no relationship between arms races and the escalation of disputes to war. Sample (1997) continues the debate, and finds that once a five-year time lag is introduced, arms races make escalation of disputes to war more likely. As these analyses indicate, findings regarding the impact of arms races on the escalation of disputes to war have been very mixed.

In chapter 7, we found that domestic politics play an important role in the outbreak of international conflict. The strength of the democratic peace would lead one to expect that jointly democratic dyads would be unlikely to escalate. Surprisingly, Senese (1997) finds that joint democracy has no effect on the escalation of disputes to war. In a subsequent study, Senese (1999) finds that joint maturity—the stability of the two regimes, regardless of their type—makes escalation in a dyad less likely, but democracy continues to have almost no effect on escalation. Like contiguity, democracy appears to not affect the escalation of disputes to war, despite being very important at the dispute onset stage.

Other scholars have examined the impact of nuclear weapons on escalation. Quantitative studies of extended immediate deterrence have found no relationship between nuclear weapons and escalation (Huth 1988; Huth and Russett 1984; Signorino and Tarar 2006). Geller (1990) finds that disputes between nuclear weapons states are more likely to escalate short of war, while nuclear weapons appear to have no effect on nonnuclear states. In contrast, Asal and Beardsley (2007) find that nuclear weapons make crises less likely to escalate to war. Overall, while findings are somewhat mixed, nuclear weapons do not appear to affect decisions to escalate.

A final contextual factor to consider here is the issue area. Gochman and Leng (1983) found that disputes involving vital issues—threats to territorial integrity or political independence—are significantly more likely to escalate to war than other disputes. Goertz and Diehl (1992), focusing on territory, have similar findings. Ethnic issues have also been found to make escalation to war more likely (Carment 1993; Carment and James 1995). Furthermore, Brecher (1993) finds that the more issues in contention within a dyad, the more likely they are to escalate to war. Of course, these strong findings for the importance of issues are exactly what we would expect, since countries will be much more motivated to fight when the stakes are very important.

Several of these contextual factors—especially territory, alliances, and arms races—are the focus of steps to war theory (Senese and Vasquez 2008; Vasquez 1993), which we examine in more detail in a later section of the chapter. It is important, however, to first consider the importance of selection effects and their impact on studies of the escalation of disputes to war.

Dealing With Selection Effects

The empirical analysis of the escalation of disputes to war is complicated by the presence of selection effects. By **selection effects**, we mean the situation in which "the cases we observe . . . are the result of choices made by people who would have chosen differently if circumstances had been different" (Fearon 2002, 24). As we discussed in the previous section, alliances appear to make escalation more likely since unreliable alliances are targeted in the first place. This is a selection effect, as these cases were selected as cases of disputes by the initiator because of the unreliability of the target's alliance. Examining the escalation of disputes to war is always subject to selection effects, as particular cases become disputes because of the prior choices of states and leaders to initiate. Reed (2000) argues that selection effects are a key reason for mixed findings on escalation.

Reed (2000) develops a way to deal with selection effects by analyzing a unified statistical model of conflict onset and escalation. Table 9.2 shows Reed's (2000) main empirical results. The first column shows the results from a model considering only the onset of militarized interstate disputes.

Table 9.2 Reed's Unified Model of Onset and Escalation

VARIABLE	DISPUTE ONSET	ESCALATION TO WAR	ONSET AND ESCALATION
Onset α	-0.486 (0.033)**		-0.484 (0.032)**
Power parity	0.353 (0.083)**		0.356 (0.090)**
Joint democracy	-0.611 (0.066)**		-0.611 (0.066)**
Joint satisfaction	-0.166 (0.065)**		-0.165 (0.066)**
Alliance	0.040 (0.052)		0.042 (0.054)
Development	-0.010 (0.005)*		-0.010 (0.005)*
Interdependence	-1.472 (3.420)		-1.432 (4.368)
Escalation α		-0.543 (0.056)**	0.648 (0.096)**
Power parity		-0.086 (0.218)	-0.333 (0.189)*
Joint democracy		-1.279 (0.440)**	-0.305 (0.342)
Joint satisfaction		-0.582 (0.316)**	-0.051 (0.303)
Alliance		-0.864 (0.166)**	-0.637 (0.153)**
Development		0.057 (0.012)**	0.048 (0.009)**
Interdependence		-34.504 (28.944)	-3.887 (14.829)
ρ Selection effect			-0.772 (0.053)**
Log-likelihood	-2,810.693	-436.185	-3,194.134
N	20,990	947	20,990

Notes: $*p < 0.05$, $**p < 0.01$. Standard errors are in parentheses.

Source: Adapted from William Reed, "A Unified Statistical Model of Conflict Onset and Escalation." *American Journal of Political Science* 44 (2000), 88.

Joint democracy, joint satisfaction, and economic development all have significantly negative effects, indicating that they make disputes less likely to occur. Power parity has a positive effect, indicating that states approximately equal in power are more likely to get involved in disputes. Finally, alliance and economic interdependence are found to have no significant effect on the likelihood of conflict.

The second column of Table 9.2 shows the results from a model considering only escalation of disputes to war. Once again joint democracy and joint satisfaction have significantly negative effects, indicating that they make disputes less likely to escalate to war. The finding for economic interdependence is also consistent between the two models, with no significant effect. The other three variables, however, are found to have different effects on escalation than they had on dispute onset. Whereas power parity was found to make disputes more likely, it appears to have no effect on the likelihood of escalation. Alliances appear to have no effect on dispute onset, but they make escalation less likely. Finally, the effect of economic development completely switches, increasing the likelihood of escalation.

For the model in the third column, the onset of the dispute and the escalation of the dispute to war are estimated simultaneously. This model is central to Reed's (2000) analysis and is important because the second model does not account for selection effects in its estimation of escalation. There are major changes in the estimated effects of the variables from model 2 to model 3. Whereas power parity was found to have an insignificant effect when considering escalation in isolation, it is found to be significantly negative in the unified model. Joint democracy and joint satisfaction are still negative, but no longer significant. Thus, their effect is realized at the onset stage, making dyads less likely to get into a dispute in the first place, but they have no effect on the likelihood of escalation once a dispute begins.

Note that there are no changes from the estimated effects on onset from model 1 to model 3. Selection effects are not an issue when looking at the onset of conflict because states do not choose to be in the overall pool of international system members (except in some much broader sense of societal development). As Reed (2000, 92) argues, however, "It is essential for researchers interested in escalation to consider first how states become involved in disputes."

The Steps to War

A prominent explanation that focuses on the effect of dyadic context on escalation of disputes to war is **steps to war theory**, developed by Vasquez and Senese (Senese and Vasquez 2008; Vasquez 1993). The idea of steps to war is that war results from a series of steps that countries may take in their relations with one another. Wars do not just happen automatically or in only one way. Rather, there are a variety of steps on the road to war, some of which make war more or less likely.

The steps to war explanation is based on the rise of security issues and the ways that leaders deal with them. Leaders believe their state's security is threatened and start to take measures to protect themselves. Vasquez (1993, 1995) argues that territorial issues are the ones that pose the greatest threat. Realism argues that states should deal with security issues by pursuing power politics. Two key ways to pursue power politics include the formation of alliances and military buildups.

Steps to war theory, however, argues that these realist prescriptions are precisely the steps to war. "Power politics behavior increases the probability of war because it leads each state to feel more threatened and more hostile toward the other side and to take actions against its opponent" (Senese and Vasquez 2008, 14). Thus, territorial issues are expected to be the most conflict prone, but if they are dealt with through negotiation and other nonpower politics means, they do not have to result in war. On the other hand, the formation of alliances, engagement in arms races, and repeated militarized disputes represent the steps to war that make territorial issues extremely war prone (Senese and Vasquez 2008; Vasquez 1993).

Senese and Vasquez (2003, 2008) begin with an in-depth study of the interplay between territory and contiguity. They control for selection effects and find that territorial issues make both the onset of disputes and the escalation of disputes to war much more likely. Unfortunately, they do not control for the onset stage in their later analyses. In each of their analyses, they examine not only the escalation of the current militarized dispute, but the escalation of any dispute to war within a five-year window of time.

Senese and Vasquez (2008) conduct a number of analyses to test steps to war theory. We begin looking at their results by focusing on the entire time period of available data. Table 9.3 shows two logit models examining the escalation of the current or any MID within five years to war for the entire time period of 1816 to 2001. The difference between model 1 and model 2 is the inclusion of the arms race variable. Territorial disputes have a positive and highly significant effect in both models as they expected.

Steps to war theory argues that alliances are very important for escalation, although they operate in different ways. Accordingly, Senese and Vasquez (2008) include four different variables dealing with alliances. They find that when two states are allied with each other and are only in the same alliances, escalation to war is significantly less likely. If one or both of the states in the dyad have outside alliances, however, then escalation of the dispute to a war within five years becomes significantly more likely. Furthermore, this danger of outside alliances holds even when the two states are allied to each other. Their findings for outside alliances, however, are not robust; once the arms race variable is included in model 2, none of the three outside alliance variables remains significant.

The steps to war explanation also highlights the number of prior militarized interstate disputes as an important variable that makes escalation more likely. Senese and Vasquez (2008) include two variables to examine this, both of which are consistently significant in models 1 and 2. The positive and significant result for the number of prior MIDs variable indicates that the greater the number of times that the two states have fought previously, the more likely that the current dispute will escalate to war (within five years). The negative and significant result for the squared term, however, shows that as the number of prior disputes increases, the size of the

Table 9.3 Escalation of the Current or Any MID Within Five Years to War, 1816–2001

VARIABLE	MODEL 1	MODEL 2
Territorial MID	1.36*** (0.111)	1.08*** (0.137)
Regime MID	-0.07 (0.231)	0.01 (0.288)
Other MID	0.41 (0.324)	1.05*** (0.428)
Allied and only in same alliances	-1.38*** (0.340)	-2.83*** (0.743)
One side has outside alliance	0.29** (0.148)	0.250 (0.181)
Both sides have outside alliances	0.29* (0.163)	-0.03 (0.204)
Allied to each other and outside alliance	0.36* (0.194)	0.06 (0.248)
# of prior MIDs	0.03** (0.016)	0.08*** (0.020)
# of prior $MIDs^2$	-0.001** (0.000)	-0.002** (0.001)
Arms race		1.61*** (0.226)
Constant	-2.49*** (0.157)	-2.65*** (0.195)
Chi-square (df)	199.55*** (9)	180.32*** (10)
Pseudo R^2	0.0757	0.1023
N	2,918	1,983

Notes: $*p < 0.10$, $**p < 0.05$, $***p < 0.01$. Standard errors are in parentheses. The reference category is a policy MID and no alliances.

Source: Adapted from Paul D. Senese and John A. Vasquez, *The Steps to War: An Empirical Study* (Princeton, NJ: Princeton University Press, 2008), 192.

effect becomes smaller. Thus, the relationship between number of prior disputes and escalation is curvilinear, in the form of an inverted U shape; the probability of escalation goes up initially as the number of prior disputes increases, but then it starts going back down with further increases.

The final variable that Senese and Vasquez (2008) include deals with arms races, and only appears in model 2. The effect of arms races appears to be strong, greatly increasing the likelihood of escalation to war. Interestingly, the inclusion of the arms race variable causes the outside alliance variables to become insignificant. A possible reason is that there are many missing values for their arms race variable, so the sample size is cut by about one-third in the number of cases.

Senese and Vasquez (2008) also conducted separate analyses of the pre–Cold War (1816–1945), Cold War (1946–1989), and post–Cold War (1990–2001) time frames in order to determine if their findings hold across time. Rather than continue looking at tables of regression results, we examine their findings for the different time periods by focusing on the predicted probabilities that stem from their logit results. Table 9.4 shows the probability of escalation of the current or any militarized dispute within five years to war in the period from 1816 to 1946. The three columns compare the predicted probabilities of escalation for territorial, policy, and regime disputes. Regardless of alliances, prior conflict, or arms race, territorial MIDs are always the most likely to escalate to war. Furthermore, the presence of outside alliances for both sides approximately triples the likelihood of escalation, while a longer history of prior disputes and arms races further increase the probability. Escalation of a dispute involving a territorial issue in a dyad with outside alliances, a long history of conflict (fifteen MIDs), and an arms race is very likely, with a probability of 0.921. In contrast, escalation of a dispute involving a regime issue in a dyad without outside alliances is unlikely, with a probability of 0.069. Thus, the period between the Napoleonic Wars and World War II fits steps to war theory very well.

The corresponding probabilities for escalation of disputes to war in the Cold War period from 1946 to 1989 are shown in Table 9.5. Once again, territorial disputes have the highest probability of escalation across

Table 9.4 Probabilities for Escalation of Disputes to War, 1816–1945

CONDITIONS	TERRITORIAL MID	POLICY MID	REGIME MID
No alliances	0.165 (0.120–0.221)	0.092 (0.063–0.127)	0.069 (0.025–0.140)
Both sides with outside ally	0.486 (0.384–0.583)	0.326 (0.248–0.407)	0.253 (0.115–0.442)
Both sides with outside ally and sixth MID	0.572 (0.474–0.667)	0.406 (0.324–0.489)	0.320 (0.158–0.525)
Both sides with outside ally and fifteenth MID	0.692 (0.588–0.790)	0.536 (0.426–0.640)	0.436 (0.234–0.663)
Both sides with outside ally, fifteenth MID, and arms race	0.921 (0.864–0.959)	0.857 (0.766–0.922)	0.788 (0.598–0.918)

Source: Paul D. Senese and John A. Vasquez, *The Steps to War: An Empirical Study* (Princeton, NJ: Princeton University Press, 2008), 197.

Table 9.5 Probabilities for Escalation of Disputes to War, 1946–1989

CONDITIONS	TERRITORIAL MID	POLICY MID	REGIME MID
No alliances	0.178 (0.102–0.281)	0.034 (0.017–0.061)	0.112 (0.052–0.196)
Both sides with outside ally	0.045 (0.023–0.075)	0.008 (0.004–0.013)	0.027 (0.012–0.049)
Both sides with outside ally and sixth MID	0.069 (0.040–0.110)	0.012 (0.006–0.021)	0.042 (0.020–0.073)
Both sides with outside ally and fifteenth MID	0.121 (0.071–0.189)	0.022 (0.012–0.040)	0.075 (0.036–0.132)
Both sides with outside ally, fifteenth MID, and arms race	0.139 (0.047–0.285)	0.027 (0.007–0.066)	0.088 (0.023–0.208)

Source: Paul D. Senese and John A. Vasquez, *The Steps to War: An Empirical Study* (Princeton, NJ: Princeton University Press, 2008), 204.

the board, although unlike the earlier period regime disputes are found to have a higher probability of escalation than policy disputes. The results otherwise are much different. If both countries in the dyad have outside alliances, then the probability goes down significantly (from 0.178 to 0.045 for a territorial dispute), contradicting Senese and Vasquez's (2008) expectations. Furthermore, while the addition of a history of conflict and arms races increases the likelihood of escalation, the predicted probabilities remain lower than for a dyad with no alliances. Clearly, the effect of alliances on escalation during the Cold War period is contrary to the expectations of steps to war theory.

Examination of the post–Cold War period allows an initial look at which of the previous two time periods is more broadly applicable. Table 9.6 shows the probabilities for escalation of the current or any MID within five years to war in the post–Cold War period from 1990 to 2001. Once again, the probabilities for escalation of territorial disputes are higher than for policy disputes (regime disputes are excluded because they never escalated to war during this limited sample). In addition, when both states have an outside alliance, escalation becomes more likely. If both states, however, have outside alliances and are also allies themselves, then the probability of escalation is even higher, which is a surprising finding. Furthermore, if the states have an average history of prior conflict (mean number of prior MIDs), then the likelihood of escalation decreases. Thus, Senese and Vasquez's (2008) findings for the post–Cold War period fit steps to war theory better than during the Cold War, but not nearly as well as the earliest period.

Table 9.6 Probabilities for Escalation of Disputes to War, 1990–2001

CONDITIONS	TERRITORIAL MID	POLICY MID
Allied and only in same alliances	0.071 (0.008–0.207)	0.011 (0.001–0.040)
Both sides with outside ally	0.119 (0.052–0.220)	0.018 (0.003–0.048)
Both sides with outside ally and mean MID	0.067 (0.024–0.138)	0.009 (0.002–0.029)
Allied but also outside alliance	0.313 (0.180–0.473)	0.057 (0.010–0.142)
Allied but also outside alliance and mean MID	0.196 (0.083–0.358)	0.031 (0.006–0.090)

Source: Paul D. Senese and John A. Vasquez, *The Steps to War: An Empirical Study* (Princeton, NJ: Princeton University Press, 2008), 211.

Overall, the increased likelihood of territorial disputes to escalate to war is the one finding of Senese and Vasquez (2008) that unambiguously supports steps to war theory. On the other hand, the findings regarding alliances are quite mixed, despite the importance that the theory attaches to them. While the different effects of alliances during the Cold War might be related to the two primary alliance blocs led by the United States and the Soviet Union during the period, this difference is not explained by steps to war theory.

Strategy, Interaction, and Escalation

While the studies in the previous sections focus on particular characteristics of the states and the underlying context of their relationship to try to explain escalation, other studies do so by focusing on the strategic interaction between states. As Siverson and Miller (1995, 105) note, these studies center "on the interaction processes characterizing the relations between and among states after a conflict has begun." Thus, the focus is on determining whether the behavior of states within a crisis or dispute drives the likelihood of escalation, regardless of what context brought that dispute about in the first place.

Leng (1993, 2000a) develops a typology of **influence strategies** by which states approach crisis bargaining. He identifies five different strategies and examines their use in a sample of forty crises (with seventy-four separate decisions) from 1816 to 1980 (Leng 1993). The most aggressive type of strategy is bullying. A state using this strategy meets any response by the opponent other than compliance with a more severe threat or punishment. This is the most frequently used strategy in Leng's analysis, being used twenty-six times. Although six of those resulted in victory, seventeen resulted in war. Thus, Leng (1993) finds that bullying strategies are the most likely to escalate.

The second type of strategy is reciprocating, which involves responding to the other side in kind. Thus, if the opponent is willing to make cooperative moves in the pursuit of compromise, then the actor responds with cooperation, while a state employing a reciprocating strategy is just as willing to respond to coercion with coercion. This strategy in Leng's analysis occurred twenty-five times, eleven of which resulted in compromise and five in victory. Thus, Leng (1993) finds that reciprocation is the most successful strategy, which echoes a common theme in previous research (such as Axelrod 1984).

The third type of strategy is appeasing, which is the opposite of bullying. A state employing an appeasing strategy tries to use positive inducements to satisfy the opponent, assuming "that the opponent has finite demands that can be satisfied at an acceptable cost to the appeaser" (Leng 1993, 142). After Britain and France failed to stop Hitler in the Munich Crisis in 1938 (in which France and Britain gave Germany the Sudetenland from Czechoslovakia), appeasement has had a bad name. This is tied for the least common strategy, being used by only two states in Leng's (1993) sample, both times leading to defeat.

The fourth strategy is trial-and-error. Trial-and-error strategies have been used nineteen times with very mixed results. More than half of the states using trial-and-error had negative outcomes (six defeats and five wars), while others had much more positive outcomes (one victory and six compromises). The final strategy Leng (1993, 143) identifies is stonewalling, which entails "essentially ignoring all inducements from the other party, without initiating any influence attempts." This is also an uncommon strategy, being used only twice resulting in war both times.

Models of Escalation

Leng (1993, 2000b) develops **models of escalation**, which identify four main patterns of behavior that repeat in crises throughout history. He classifies these patterns according to two primary criteria. The first criterion is escalation, which considers both the level and the rate of escalation. The second is reciprocity, which accounts for both "the distance between the behavior patterns of the two sides" and "the degree of congruity in the direction in which each is moving at a particular time in the crisis" (Leng 1993, 71).

The first model of escalation is the fight. These are crises in which there is both high escalation and high reciprocity. Eight of the fifty crises (in an expanded data set) that Leng examines fit into this category. Of these eight crises, six ended in war. One of the others ended in a diplomatic victory, while the final crisis ended in a compromise. An example of a fight is the crisis leading up to the Six-Day War, shown in Figure 9.1. Through the first week of May 1967, both sides maintained a low level of hostility (around 20).

Figure 9.1 Example of a Fight: Six-Day War Crisis, 1967

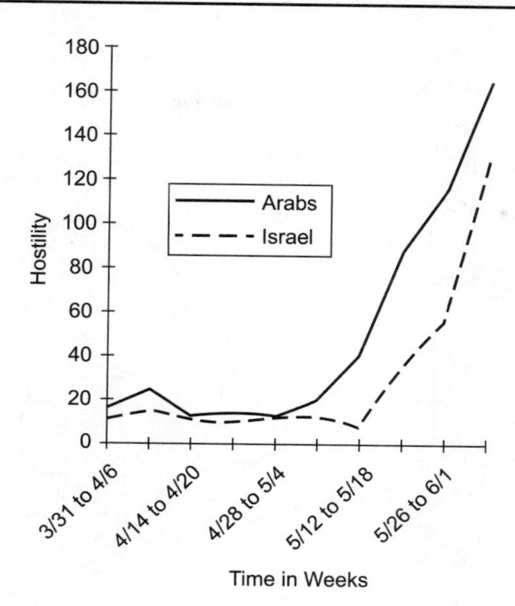

Source: Russell J. Leng, "Escalation: Crisis Behavior and War." In *What Do We Know About War?*, ed. John Vasquez (Lanham, MD: Rowman and Littlefield, 2000), 242. Reproduced with permission of ROWMAN & LITTLEFIELD PUB in the format Book via Copyright Clearance Center.

In the second week of May, Arab hostility began to rise dramatically, followed closely by Israel. The two lines move together, indicating reciprocity, and shoot to a high hostility score (120+), indicating a high level of escalation.

The second model of escalation is resistance, which is characterized by high escalation but low reciprocity. Leng identifies thirteen of these, of which seven ended in war, five ended with one side yielding to the demands of the other, and the final one ended in stalemate. The Cuban Missile Crisis provides an example of resistance, as shown in Figure 9.2. The initial part of the figure follows the same pattern as shown in Figure 9.1, and the United States also escalates to a high hostility score of over 120. The reciprocity is low, however, because the Soviet hostility score levels off after barely passing 60.

The third pattern of crisis behavior that Leng identifies is the standoff. These crises are characterized by high reciprocity in the interactions of the two sides, but relatively low escalation. These are crises that appear to be effectively managed. Eight of the twelve cases in Leng's data ended in compromises, two with diplomatic victory, one with war, and one ended in a stalemate. The Alert Crisis between the United States and Soviet Union

Figure 9.2 **Example of Resistance: Cuban Missile Crisis, 1962**

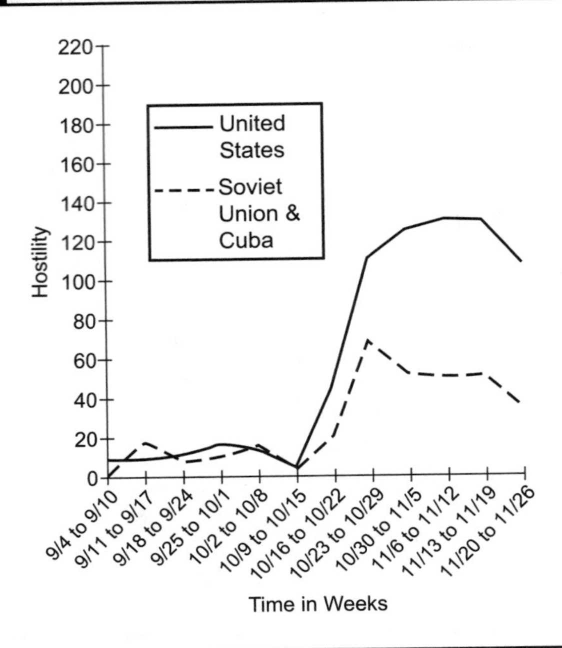

Source: Russell J. Leng, "Escalation: Crisis Behavior and War." In *What Do We Know About War?*, ed. John Vasquez (Lanham, MD: Rowman and Littlefield, 2000), 243. Reproduced with permission of ROWMAN & LITTLEFIELD PUB in the format Book via Copyright Clearance Center.

in 1973 provides an example of a standoff, and is shown in Figure 9.3. This crisis concerned the Yom Kippur War between Israel (ally of the United States) and Arab states (Soviet allies), as the two superpowers were concerned that the other's client states (particularly Egypt and Israel) were violating the terms of the cease-fire (Rabinovich 2004). There is a high degree of reciprocity, as both lines move (or do not move) together, but very low escalation, as both sides maintain hostility scores around 20 or below throughout the month-long conflict.

The final model of escalation is the put-down. In these crises, the composite scores for both escalation and reciprocity are relatively low. One side employs coercive tactics, but the other responds with a more accommodative mix of behavior. Four of the seven cases in this category ended in diplomatic victory, where a more powerful state was able to bully its adversary into submission. The other three put-downs, however, ended in war. An example of a put-down is the Anschluss Crisis in 1938, as shown in Figure 9.4. This crisis concerned the absorption of Austria into Germany, which Hitler demanded but Austrian leaders had some initial hesitation about. There is low escalation (hostility less than 50) and low reciprocity, as the lines for Germany and Austria do not move together.

Figure 9.3 **Example of a Standoff: Alert Crisis, 1973**

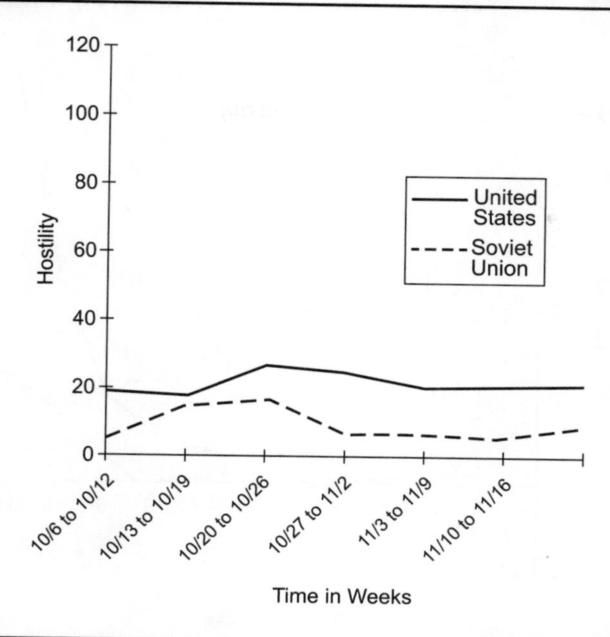

Source: Russell J. Leng, "Escalation: Crisis Behavior and War." In *What Do We Know About War?*, ed. John Vasquez (Lanham, MD: Rowman and Littlefield, 2000), 244. Reproduced with permission of ROWMAN & LITTLEFIELD PUB in the format Book via Copyright Clearance Center.

Leng (1983; 2000a) also examines patterns of coercive bargaining within recurrent crises. In these studies, he examines the behavior of states from one crisis to the next. Do states change their approach from one crisis to the next, or do they essentially repeat the same behaviors? How can we explain these strategy changes or lack thereof? There are competing ideas regarding the answers to these questions.

The first idea deals with conventions or norms of behavior. In particular, many scholars have argued that properly managed crises can avoid wars. Therefore, some scholars have argued that states would develop conventions to help them properly manage crises (e.g., Bell 1971). For example, during the Cold War, the United States and Soviet Union developed a variety of conventions for resolving their many crises, most famously including the hotline connecting Washington and Moscow enabling direct communication between the leaders.

The second basic idea of how states adapt from one foreign policy decision to the next is experiential learning, which involves "a change in beliefs resulting from the observation and interpretation of experience" (Leng 2000a, 5). In other words, this approach to learning is that if what was done before worked, then do not change, but if it did not work, then change.

Figure 9.4 Example of a Put-Down: Anschluss Crisis, 1938

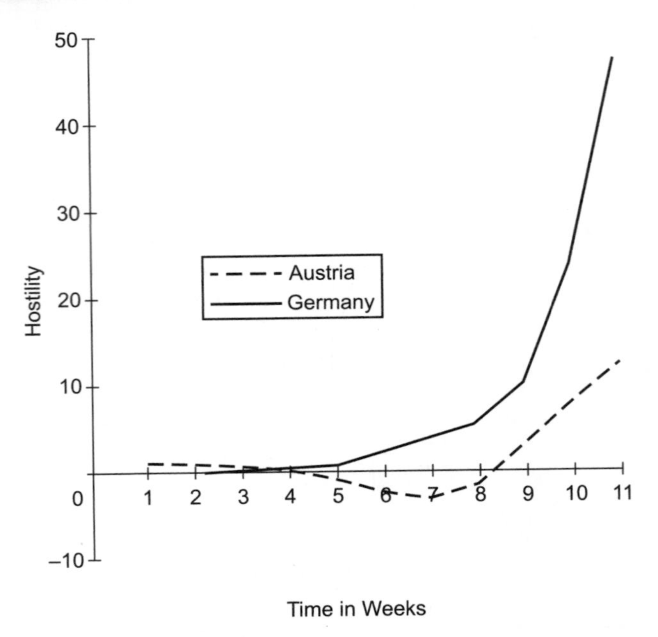

Source: Russell J. Leng, "Escalation: Crisis Behavior and War." In *What Do We Know About War?*, ed. John Vasquez (Lanham, MD: Rowman and Littlefield, 2000), 245. Reproduced with permission of ROWMAN & LITTLEFIELD PUB in the format Book via Copyright Clearance Center.

While this approach has been used by several scholars (e.g., Jervis 1976; Levy 1994), it has difficulties. In particular, it fails to specify how one can know if an approach worked or, if a change is needed, in what direction the leaders will change.

Leng (1983; 2000a) introduces an alternative explanation that he calls the **realpolitik experiential learning** model. Leng considers four basic types of crisis outcomes: diplomatic victory, diplomatic defeat, compromise, and war. Based on those four outcomes, he argues that there are six basic scenarios for how states can learn from one crisis to the next. First, if the previous crisis resulted in a diplomatic victory, then the state will employ the same level of coercion as before (it worked, so don't fix it). If the previous crisis was a diplomatic defeat, however, then a change is needed, and Leng argues that that change will lead to the use of more coercion.

The third and fourth scenarios are based on a compromise outcome in the previous crisis. If it was a compromise without a significant retreat, then the leaders can be satisfied with the outcome and therefore are expected to employ the same level of coercion the next time. On the other hand, if the compromise was only achieved by making a significant concession, it is a less satisfactory outcome. Thus, Leng expects that the state will use a more coercive strategy the next time. The final two scenarios deal with cases in which the previous crisis resulted in a war. If the war was unwanted, then the state is expected to use a more coercive strategy. Finally, if the previous war resulted from a premeditated attack, then Leng expects the state's approach to be less coercive the next time.

Thus, Leng (1983, 2000a) expects states to use either the same or a greater level of coercion from one crisis to the next in almost all situations. Leng (1983) tests these predictions using twenty-four different crises and finds strong empirical support. Leng (2000a) conducts a more focused examination of series of crises between three pairs of states: the United States and Soviet Union, Egypt and Israel, and India and Pakistan. He again finds strong support for the realpolitik experiential learning model: Fourteen of eighteen cases he examined agreed with the model's predictions. The Soviet-American dyad is the one that does not seem to conform as well to the realpolitik experiential learning model, which Leng argues is due to the exceptionally strong aversion to war that the two sides had during the Cold War.

Game Theory and Strategic Interaction

Game theoretic models have also been used extensively to theorize about the escalation of disputes to war. Although Leng (1993; 2000a) highlights the importance of strategy, he does not formally model the strategic interaction between states. Nonetheless, other scholars have more explicitly examined the role of strategic interaction in the escalation of disputes to war. An important advantage of using game theoretic models to examine escalation is that they provide a way to explicitly theorize about selection effects, because selection effects usually arise because of strategic interaction (Fearon 2002). Techniques have been developed to integrate statistical models into game theoretic models of strategic interaction, allowing the theory and empirical analysis to be combined into the same logical structure (Signorino 1999; Signorino and Tarar 2006; Smith 1998).

Snyder and Diesing (1977) undertake an early game-theoretic analysis of crisis decision making. They argue that crises can be represented by 2 × 2 games, in which there are two states making a simultaneous choice between two strategies. There are a number of possible 2 × 2 games, but Snyder and Diesing focus on only a few important ones. Each game presents the players with the choice between cooperating and defecting (or standing firm), and the difference between them is determined by the states' preferences between the four resulting outcomes. The first game they examine is Prisoner's Dilemma, which we examined in chapter 3. These games are marked by the preference to stand firm over backing down, and mutual defection is the sole equilibrium. Snyder and Diesing identify several crises in this mold, such as the Berlin Crisis between the Soviet Union and United States from 1958 to 1962.

The second game they examine is Chicken, in which each state prefers to back down rather than stand firm if the other defects. The two equilibria in Chicken involve a victory for one side; the Munich Crisis in 1938, where Germany was awarded the Sudetenland by Britain and France, is an example. A closely related game that Snyder and Diesing (1977) examine is Called Bluff, where one side has the same preferences as Chicken while the other has Prisoner's Dilemma preferences. In these crises, the side that prefers to fight rather than concede will emerge the winner, which is their explanation for why the United States was able to emerge from the Cuban Missile Crisis victorious.

The fourth game that Snyder and Diesing examine is Bully. This game represents situations in which one side is much stronger than the other. The stronger side (the bully) prefers mutual defection to mutual cooperation, while the weaker side has Chicken-like preferences. Victory for the bully is the only equilibrium, which Snyder and Diesing argue characterizes the British victory over France in the Fashoda crisis of 1898, in which France challenged British colonial control of Sudan. Finally, Deadlock is like Prisoner's Dilemma except that both sides prefer mutual defection to mutual cooperation (i.e., both are bullies). In this case, mutual defection is the obvious and sole equilibrium, so war is expected. Snyder and Diesing argue that the crisis between the United States and Japan in 1940 stemming from the American oil embargo and demands that Japan withdraw from China (which led to World War II in the Pacific) is a case of Deadlock.

Snyder and Diesing (1977, 129) also examine three alliance games in which alliance partners interact to "coordinate strategy against or toward a third party." The first of these is Leader, which is like Chicken except that both sides prefer to cooperate while the other stands firm over mutual cooperation. The second alliance game, Hero, is just like Leader except that mutual defection is preferred by both sides over mutual cooperation. Finally, Protector is the alliance version of Bully, where one side is much stronger than the other.

While Snyder and Diesing's (1977) approach was advanced at the time, their reliance on 2 × 2 games is needlessly limiting. Even if we are willing to assume that each state has only two options to choose from, the assumption that every crisis involves only simultaneous choices is problematic. A key trend that is revealed time and again in the history of international conflict is that crises take the form of moves and countermoves between states. Thus, they are better represented as sequential games rather than simultaneous ones.

More recent game-theoretic models of escalation take into account the sequential nature of crisis bargaining. For example, we examined the difference between general deterrence and immediate deterrence in chapter 8. Recall that immediate deterrence deals with attempts to prevent war once a challenge to the status quo has already been initiated, while general deterrence deals with attempts to deter challenges from occurring in the first place. Thus, general deterrence deals with the outbreak of international conflict, whereas immediate deterrence relates to the escalation of disputes and crises to war. The unilateral deterrence game and other games developed by

Zagare and Kilgour (2000) that we examined in chapter 8 explicitly model both the onset of a conflict and escalation to war.

The international interaction game developed by Bueno de Mesquita and Lalman (1992), discussed in chapter 3, also examines the onset of a conflict and escalation to war within the same game-theoretic model. The game, shown in Figure 3.6 on page 62, begins with decisions by each side about whether or not to make a demand. If both states make initial demands, this enters the crisis subgame, in which each side must choose whether or not to use force to pursue their demands.

Another important limitation of the analyses by Snyder and Diesing (1977) and others is that they assume complete information, where both sides fully know each other's preferences. More recent models of escalation focus on incomplete information, in which one or both sides have uncertainty about the opponent's preferences. Morrow (1989) develops an early game-theoretic model of crisis escalation in which both sides have private information about their military capabilities. Another limited information model of crisis bargaining is developed by Fearon (1994), who demonstrates the importance of audience costs, as we discussed in chapter 7.

To illustrate how incomplete information works in a model of crisis bargaining, consider the model developed by Bueno de Mesquita, Morrow, and Zorick (1997) and shown in Figure 9.5. The game represents crisis decision making between two states, i and j. State i begins the game, and must choose whether or not to make a demand. If i makes a demand, then state j must decide whether to accept the demand or not. If j rejects the demand, then a crisis begins and both sides must choose to initiate negotiation or hostilities. These choices are simultaneous; although i is shown making her decision first and then j, the dashed line connecting j's final two decision nodes shows that j does not know i's choice when making his own (they are in the same information set).

In their analysis, Bueno de Mesquita, Morrow, and Zorick (1997) focus on the role that capabilities and perceptions play in the likelihood of escalation. Their central variable is K, which "represents the observable military advantage of one or the other side in the event that the antagonists use force to resolve their dispute" (Bueno de Mesquita, Morrow, and Zorick 1997, 18). As we discussed previously, balance of power and power transition theorists disagree about the impact of power advantage on the likelihood of escalation. Bueno de Mesquita, Morrow, and Zorick's model shows that larger military advantages initially decrease the likelihood of violence and then begin to increase the probability. Furthermore, they conduct a quantitative analysis of militarized disputes in Europe from 1816 to 1970 and find the nonmonotonic relationship between power and the likelihood of war that their theory leads them to expect.[1] Nonetheless, Molinari (2000) provides important corrections to their theory that call into question the relationship between power and escalation.

---

[1] A monotonic relationship goes only one way, either always increasing or always decreasing. A nonmonotonic relationship changes direction at one or more points, sometimes increasing and sometimes decreasing.

Figure 9.5 Crisis Escalation Game

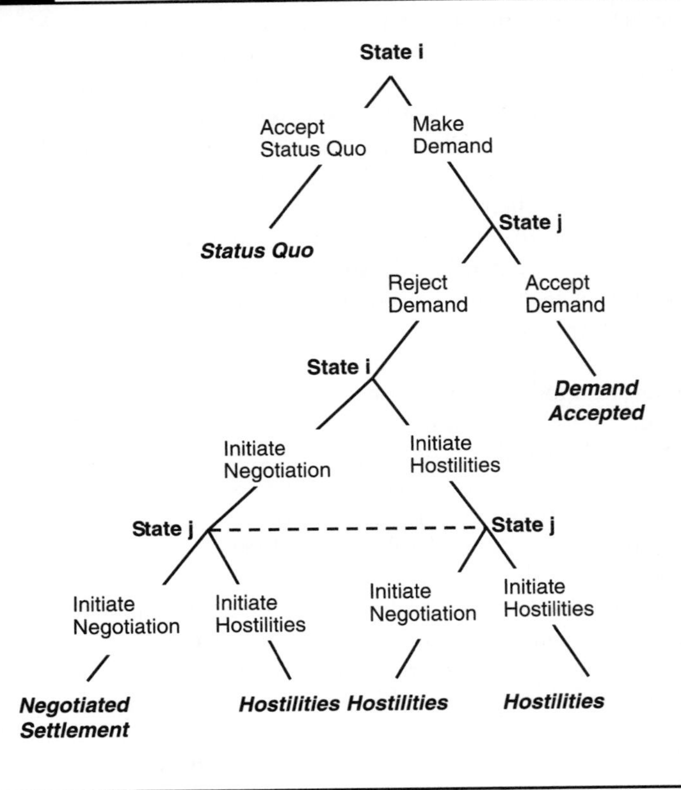

Source: Bruce Bueno de Mesquita, James D. Morrow, and Ethan R. Zorick, "Capabilities, Perception, and Escalation." *American Political Science Review* 91 (1997), 17.

Conclusion

Examining the escalation of disputes to war is important because it can help us understand why some disputes become wars but many others do not. Patterns of escalation reveal that, although the most common hostility level in disputes is the use of force, wars make up only a small subset of disputes. There have been many studies attempting to explain why (some) disputes escalate to war.

Many of these studies focus on the effect that contextual factors such as power, alliances, contiguity, arms races, democracy, and other issues have on the escalation of disputes to war. Examining escalation is complicated by the presence of selection effects, which Reed (2000) dealt with by developing an integrated statistical model of dispute onset and escalation. One of the most prominent contextual approaches to explaining escalation is steps to war theory (Senese and Vasquez 2008; Vasquez 1993), which focuses on the impact of territory, alliances, arms races, and recurrent conflict on the likelihood of escalation.

Other studies focus on the importance of strategy and patterns of interaction between states for explaining escalation. Leng (1983, 1993, 2000a) highlights different strategies that states employ in crises and finds that reciprocity is the most effective. He also examines how states change their strategies from one crisis to the next. Game-theoretic models of escalation also fit into this category. These are particularly important because they enable scholars to explicitly examine the strategic interaction between states and models such as those developed by Bueno de Mesquita and Lalman (1992) and Zagare and Kilgour (2000) and enable escalation to be explained within a broader theory of international conflict.

Key Concepts

Contextual factors

Hostility level

Influence strategies

Models of escalation

Realpolitik experiential learning

Selection effects

Steps to war theory

Part III

The Conduct and Aftermath of War

10 Military Doctrine and Strategy

Supreme Allied Commander General Dwight Eisenhower discusses D-Day plans with key leaders with the aid of a giant map at his London Headquarters on 1 February 1944. D-Day was one of the largest and most complex strategic operations in military history.

Source: Popperfoto/Getty Images

To this point, we have examined many different causes of international conflict as well as the escalation of disputes to war. The process of war, however, does not end when war begins and so we now turn to the evolution of war itself. Looking at the conduct of war necessarily involves a closer look at military matters than we have engaged in to this point. In this chapter, we focus on military doctrine and strategy, which cover the various ways that countries try to use their military forces in warfare to achieve political objectives. The topics examined in this chapter set the stage for our exploration of the evolution of war, including war's duration and outcome in chapter 11, and its termination in chapter 12.

D-Day, the invasion of France by the United States and Britain on 6 June 1944 during World War II, is quite famous and has been depicted in movies such as *Saving Private Ryan* and been the subject of numerous books, games, and more. People often have a general familiarity with the basics of what happened but have little understanding of why it happened. Why did the Allies choose to invade France? And, given that they chose to invade France, why did they invade in Normandy in particular? The basic answer is that the western Allies needed to invade France to (1) open up a second front of battle to relieve some of the pressure on their Soviet ally on the eastern front, (2) enable them to advance into Germany, and (3) position themselves on the continent to prevent Soviet domination of Europe after the war (Weinberg 1994; Willmott 1984). They chose to invade Normandy because it had good beaches, was near a port, was located on the northern coast of France, and was less heavily defended than Pas de Calais, the other attractive (and more direct) alternative. These considerations are the realm of strategy.

The chapter begins with an examination of the study of warfare, focusing on reasons why it has been understudied in political science. We then discuss military strategy—including strategic theory, basic types of strategy, and reasons for states' choices of strategy—before turning our attention to the basic kinds of military power and the principles of war. The chapter concludes with an examination of models of combat, focusing on combat effectiveness and force employment.

The Study of Warfare

Warfare is the action of carrying on or engaging in war. At its heart, warfare is about the preparation for, and conduct of, organized violence. Thus, studying warfare requires a focus on armies and other military forces, as well as military campaigns, battles, and engagements. Understanding warfare is a vital element of understanding international conflict more generally. Although the causes of war have been studied extensively, warfare has been understudied in political science (Biddle 2007; Gartner 1998; Stoll 1995). This lack of attention to war once it starts is the result of several factors, among which are perceptions of disciplinary boundaries and expectations that different causal factors are at play.

Concerns about causal factors stem from the belief that the factors that account for the evolution of war are different from the factors that account for other phenomena more central to the study of international relations, which generally focuses on causes such as power, alliances, democracy, and contiguity, as we have seen in preceding chapters. While these factors can have important effects on the evolution of war as well, phenomena such as morale, training, terrain, and weather are important factors with potentially large effects on war outcomes and duration.

A related concern that some people have about the study of warfare stems from boundaries between academic disciplines. The study of warfare rests uncomfortably within the field of political science and international relations. As Freedman (2002, 334) observes, the political science "purist might be appalled by the arbitrary mixture of politics, sociology, economics, psychology, and history that regularly influences decisions in crisis and combat." The very nature of warfare, however, is political. As Clausewitz (1976, 87) states, "War is not merely an act of policy but a true political instrument, a continuation of political intercourse, carried on with other means." Therefore, warfare very much falls within the realm of political science.

Finally, some have avoided studying warfare itself because of moral objections to it. According to this line of thinking, looking at the causes of war is okay and necessary to enable its prevention and the achievement of peace. But studying warfare itself is thought to be tantamount to warmongering. As Morillo (2006, 1) states, "There exists deep suspicion that to write about war is somehow to approve of it, even to glorify it."

But war is very important and merits study. Wars—particularly large ones—have far reaching consequences that have shaped and reshaped the world in which we live. Furthermore, understanding warfare is important because expectations about war outcomes and costs affect its outbreak. Fortunately, led by the pioneering work of Stam (1996) and the bargaining model of war (e.g., Filson and Werner 2002; Wagner 2000), warfare and the evolution of war have received much greater attention within political science in the past two decades.

Military Strategy

Military strategy is the application of military power—which includes land power, naval power, and air power—to achieve political objectives. Strategy provides the bridge between military means and political goals. It is the art of using military force against an intelligent foe toward the attainment of policy objectives. The logic of strategy is universal, applying to wars of varying sizes, styles, and time periods. War is an act of force to compel the adversary to do one's will, and strategy is a practical endeavour focused on winning wars. Strategy serves as the essential link between political objectives and military force, but it is complex and difficult to master (Gray 2012; Kane and Lonsdale 2012; Sloan 2012). A key reason for this complexity is that there are no golden rules to follow that magically achieve victory. In addition, interaction with the enemy makes it difficult to achieve objectives, since they are also attempting to accomplish their own objectives.

Warfare is often broken up into three levels. The **levels of warfare** are much like the general concept of levels of analysis that we examined in chapter 1 (see Figure 10.1). The lowest level is the tactical level of warfare, the level at which battles and engagements are planned and fought. Next is the operational level, which is the level at which campaigns and major operations are planned, sequenced, and directed. Finally, there is the strategic level, which deals with the broad picture of how to achieve the state's policy interests. We can distinguish between military strategy and grand strategy. Military strategy is the application of military resources to help achieve grand strategic objectives, while grand strategy is the application of national resources to achieve national or alliance policy objectives. The strategic level is the usual realm of international relations.

To illustrate the levels of warfare, we can return to examine D-Day. As we discussed at the outset of the chapter, the decision to invade and where to do so was the realm of strategy. But the massive planning and coordination that was necessary to make the invasion happen—moving more than 150,000 troops across the English Channel in a single day, supported by over 7,000 naval vessels and over 11,000 aircraft (Murray and Millett 2000), supplying them once there, and coordinating their actions to secure key objectives and break out from the Normandy beachhead—took place at the operational level of war. Finally, the tactical level covers the procedures and techniques that individual soldiers and units used to cross the beach under fire, take out enemy positions in bunkers and hedgerows, and otherwise operate in close proximity to enemy forces.

Figure 10.1 Levels of Warfare

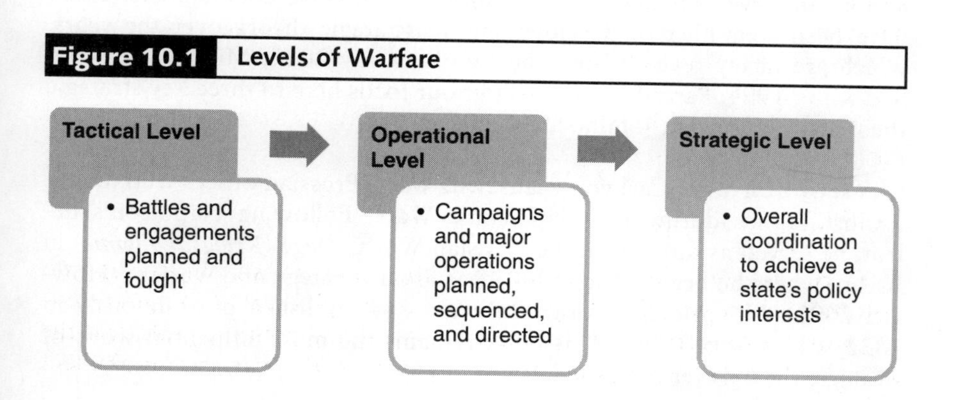

Another thing that makes strategy complex is the presence of disharmony among the levels of warfare. The tactical level must serve the operational level, but tactical outcomes do not necessarily determine operational and strategic outcomes. At the same time, the operational level must serve the strategic level. Operational victories, however, do not always lead to strategic victory, just as strategic victory can be achieved despite operational losses. For example, Germany did an excellent job at the tactical and operational levels in World Wars I and II but was unable to translate these successes—such as the conquest of France in 1940—into strategic victory (Kane and Lonsdale 2012). More recently, the United States has had difficulty translating tactical and operational successes in its counterinsurgency campaign in Afghanistan into strategic victory. The strategic level coordinates a plan for the use of military force (which occurs at the tactical and operational levels) to achieve the policy objectives set forth.

Ultimately, the complexity of strategy stems from the fact that it is multidimensional. A number of people have identified different dimensions of strategy. Gray (1999) developed a list of seventeen different dimensions of strategy in his book *Modern Strategy* and argues that mastering strategy requires understanding these multiple dimensions. Gray identifies five dimensions related to people and politics: people, society, culture, politics, and ethics. The next six dimensions relate to war preparation: economics and logistics, organization, military administration, information and intelligence, strategic theory and doctrine, and technology. The final group of Gray's (1999) dimensions of strategy relates to war proper: military operations, command, geography, friction, the adversary, and time.

Strategic Theory

Sorting through these various dimensions in order to make sense of strategy and warfare can be overwhelming to an experienced general, let alone a student simply trying to understand international conflict. This is why **strategic theory**, which seeks to understand the complexities of strategy, is important. Strategic theory provides ways that we can sift through the many dimensions of strategy to focus on what the most important elements are, and it can serve as a guide for education of battlefield commanders. There have been a number of developments in strategic theory over the years, which are nicely reviewed in other works (Gray 2012; Murray, Berstein, and Knox 1994; Paret 1986). We limit our focus here to three key strategic theorists: Clausewitz, Jomini, and Sun Tzu.

Clausewitz on War. Carl von Clausewitz was a Prussian officer who fought against France during the Napoleonic Wars. Following Napoleon's defeat, he served as director of the Prussian War College (*Kriegsakademie*) in Berlin, where he began writing his ideas about strategy and warfare (Howard 2002). His primary work, *On War,* was published posthumously in 1832 (Clausewitz 1976). This book remains the most influential work of strategic theory ever published.

Smith (2005) argues that Clausewitz views war in four different contexts: It is killing and dying; a clash of armies; an instrument of policy; and a social activity. Ultimately, war is an extension of politics, and therefore must serve policy. Policy may refer to any objective for which war is waged, whether that is religious issues, territorial disputes, resources, or some other aim.

Clausewitz introduced a number of concepts that have become central to the study of strategy. We review several of the most important here, beginning with his trinity of passion, probability, and rationality. Passion is most often associated with the people of a country, whose animosities move states to fight. Probability is associated with the military because soldiers must constantly deal with uncertainty and friction. Finally, rationality is associated with the government, which determines the aims of war and the means for waging it. Clausewitz argues that the interaction among these three tendencies determines the character of a war. Furthermore, the intensity of and relationship among these three tendencies change according to the circumstances of the war (Mahnken 2007).

Clausewitz argued that the key to victory in war is attacking the enemy's **center of gravity**, "the point against which all our energies should be directed" (Clausewitz 1976, 595–6). The term translated as "center of gravity" in *On War* is *Schwerpunkt* in the original German, and there is some debate over the proper interpretation of the concept (Echevarria 2003). Nonetheless, Clausewitz (1976) identifies two centers of gravity as the most important:

> [T]he acts we consider most important for the defeat of the enemy are the following:
>
> 1. Destruction of his army, if it is at all significant
> 2. Seizure of his capital if it is not only the center of administration but also that of social, professional, and political activity. (596)

In addition to the enemy's army and capital city, other possible centers of gravity include the enemy's allies, leadership, and public opinion. For example, in the Vietnam War, North Vietnam had no possibility of capturing the American capital city but recognized that the key center of gravity to focus on was American public opinion, which they were successful in turning against the war as American casualties continued to mount. The United States withdrew from the war in 1973 in part due to pressure from its unsupporting public.

Clausewitz (1976) also distinguished between limited and unlimited wars. Others have employed the same basic reasoning to distinguish between wars with a coercive objective and those with a brute force objective (Sullivan 2007, 2012). In wars for limited aims, soldiers and statesmen must translate battlefield success into political leverage over the adversary. As a result, they must continually reassess how far to go militarily and what to demand politically. Such wars end through formal or tacit negotiation and agreement. Wars for unlimited aims are fought to overthrow the adversary's regime or achieve unconditional surrender. For example, the 1991 Gulf War was a limited war, as the United States and allies sought simply to force Iraq into leaving Kuwait. In contrast, the 2003 Iraq War was more unlimited, as the United States and allies sought to remove Saddam Hussein from power.

A final key contribution of Clausewitz for us to consider here is his concept of friction. **Friction** makes any activity in war difficult, preventing armies from achieving their maximum operational efficiency. As Dupuy (1987, 7) notes, "Combat activities are always slower, less productive, and less efficient than anticipated." Sources of friction include difficulties in the physical environment, ignorance of important information, difficulties created by enemy activity, and other factors. Friction can be limited by good and ample equipment, high morale, rigorous training, imaginative planning, historical education, combat experience, and sensitivity to potential problems.

Jomini's Scientific Principles. The second primary strategic theorist was Antoine-Henri Jomini, who was the strategic theorist of choice during much of the nineteenth and early twentieth centuries. Jomini was from Switzerland, although he later served as a staff officer in the French army under Napoleon and as a general in the Russian army. Unlike Clausewitz, who is known for his one main work, Jomini produced a number of books and other works until his death in 1869 (Shy 1986).

Jomini sought to uncover the scientific principles underlying war. He studied military history, particularly the successes of Napoleon and Fredrick the Great, and developed a number of principles that he argued should lead to victory if followed. In particular, Jomini stressed the advantages of interior lines of communication. Their significance comes from the fact that they most readily permit the commander to mass his forces against the enemy's weaknesses at the decisive point.

Jomini also examined irregular warfare and the effects of terrain. For irregular warfare, he concludes that the decisive use of force is not always appropriate, and that appropriate treatment of the local population may contribute significantly to achieving one's objectives. These conclusions significantly foreshadow the conclusions and recommendations of modern counterinsurgency theory (e.g., Kilcullen 2010; Nagl 2005). Jomini's strategic ideas, however, have fallen out of favor over time because the repeated message in his work is too prescriptive and gives too much credence to set principles. In particular, the influence of Jomini declined steeply after World War I, as his principles—which had been the basis of thinking about war in all of the major European armies at the time—were unable to guide military commanders in how to deal with the modern weapons, total war, and attritional warfare of the war (Alger 1982; Shy 1986).

Sun Tzu and the Indirect Path. The final strategic theorist for us to consider here is the ancient Chinese writer Sun Tzu. His primary work, *Art of War* (Sun Tzu 1963), is a very short work that totals less than forty pages in

English. Sun Tzu argued that the ideal victory requires no fighting, since more is gained through victory without battle or violence. This is quite in contrast to Clausewitz, who stressed that victory in war was not possible without battle or violence.

Sun Tzu advocated an indirect approach to warfare, rather than the more direct approach of Clausewitz and Jomini. In particular, he argued that success in war comes from shattering the adversary's will to fight rather than destroying his army. This is done by attacking the enemy's strategy. He also stressed the importance of intelligence, arguing that the acquisition of knowledge—about the enemy, the environment, and oneself—is the key to success in strategy. Thus, the outcome of a war is knowable in advance if the leader makes a complete and accurate assessment of the situation. Because it is important to prevent the enemy from gaining this information, deception has a central role in war. Given the constant use of deception, Clausewitz is far more pessimistic about intelligence than is Sun Tzu (Jordan et al. 2008).

An example of the usefulness of deception is provided by Operation Fortitude, an elaborate deception plan to coincide with Operation Overlord, the Allies' invasion of France in 1944. The fictional First US Army Group was created in the southeast of England, supposedly to land in Pas-de-Calais after German reserves were committed to Normandy. The plan included an elaborate radio deception scheme, involving the apparent movement of units from their true locations to southeastern England, the display of dummy landing craft and associated simulated wireless traffic and signing of roads and special areas, bombing of the Pas-de-Calais beach area and railway network immediately before D-Day, increased activity around Dover to suggest embarkation preparations, and night lighting to simulate activity at night where dummy landing craft were situated. The plan succeeded in deceiving the German leadership, who kept strong forces in Pas-de-Calais for weeks after D-Day, ready to meet the invasion there that was never coming (Weinberg 1994).

Of course much has changed in the centuries since Clausewitz, Jomini, and Sun Tzu wrote. Modern technology has produced developments in firepower, mobility, communications, and other abilities that these writers could not possibly have foreseen. In addition, wars involving nonstate actors have become increasingly common. These developments, however, have not rendered classical strategic theory obsolete because the fundamental nature of war has remained the same (Jordan et al. 2008; Kane and Lonsdale 2012).

Types of Strategy

As suggested by the previous discussion, the subjects of strategy, operational art, and tactics are quite broad and have the potential to be overwhelming. A complete analysis of them is beyond the scope of this book and is unnecessary for many political science applications. Stam (1996), however, breaks strategy down into three basic types—maneuver, attrition, and punishment—and his typology has been commonly used in recent political science applications. These strategy types are summarized in Figure 10.2.

Maneuver Strategy. A **maneuver strategy** focuses on the movement of forces to achieve an advantageous position against the enemy (Stam 1996). Mearsheimer (1983, 36) calls this strategy blitzkrieg, and explains the basic idea as follows:

> The blitzkrieg is predicated upon the assumption that the opponent's army is a large and complex machine that is geared to fighting along a well-established defensive line. In the machine's rear lies a vulnerable network, which comprises numerous lines of communication, along which supplies as well as information move, and key nodal points at which the various lines intersect. Destruction of this central nervous system is tantamount to destruction of the army.

Thus, maneuver warfare is focused on breaking through the enemy's lines and penetrating deep into the enemy's rear areas. This enables targeting of the enemy's logistical and communications systems. In this way, a maneuver strategy focuses on striking to achieve disproportionate effect.

The German campaign in France in May through June 1940 provides an excellent example of a maneuver strategy. Strategy was crucial to the campaign in France, and France and Britain's strategy was deeply flawed. Germany's strategy, on the other hand, was excellent and took advantage of the Allies' weaknesses. The Allies anticipated that Germany would essentially repeat its approach at the beginning of World War I, sweeping through Belgium on a broad arc and advancing toward Paris. To meet this, the Allies' plan was to advance into Belgium on the left and defend the Maginot Line along the border with Germany on the right. The Germans, however, had different ideas. Germany attacked on 10 May 1940, but rather than a right hook, the German main thrust was through the Ardennes forest in the center (see Map 10.1).

Although the Allies expected it would take them about a week to reach the Meuse River, the Germans reached it in less than three days. They secured crossings of the Meuse between Dinant and Sedan by 16 May, and began their drive toward the coast. The leading German forces reached the English Channel on 21 May, trapping the British Expeditionary Force and French forces away from the rest of France. Germany then focused on eliminating the pocket of British and French forces around Dunkirk before turning to the south; the British evacuation of more than 150,000 soldiers at Dunkirk was crucial for their further conduct of the war. On 4 June, the Germans attacked along the Somme River and broke through the French lines by 12 June leading to rapid advance across France until the French surrender on 22 June.

Important developments in the past century have strengthened the capacity for maneuver. Most importantly, the internal combustion engine made

Map 10.1 French Campaign, 1940

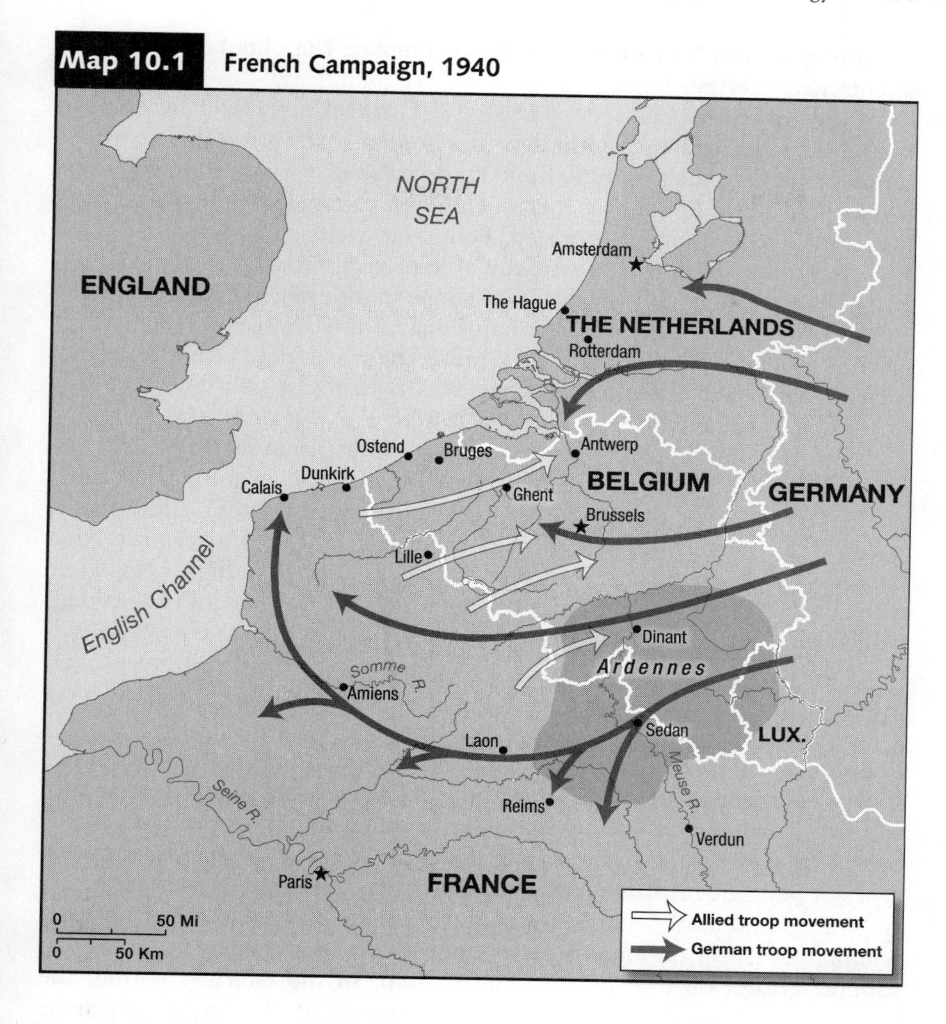

tanks and other armored vehicles and trucks possible, which are integral to the mobility of modern armies. The development of aircraft and helicopters made air mobility possible, allowing armies to move at speeds and distances that are not possible when confined to land. Finally, communications developments such as wireless radios and digital communications equipment have also greatly increased, enabling military forces to take advantage of these increases in mobility yet still maintain contact with each other despite being separated by large distances.

Attrition Strategy. The second type of strategy is an **attrition strategy**, which is focused on defeating the enemy by killing personnel and destroying material (Mearsheimer 1983; Stam 1996). By wearing down the enemy army, either through a decisive battle or a series of battles over time, a country employing an attrition strategy seeks to eliminate the opponent's capacity to continue fighting.

In the American Civil War (1861–1865), the Union won by employing an attrition strategy. Although the Confederates were able to win a series of battles, such as Bull Run, Fredericksburg, Chancellorsville, and Chickamauga, the Union was much better able to replace the losses (in both personnel and equipment) suffered. Following victories at Gettysburg and Vicksburg, the Union was eventually able to wear down the Confederacy. After Union General Ulysses S. Grant's nine month siege of Petersburg and General William Tecumseh Sherman's march to the sea in 1864–1865, General Robert E. Lee surrendered, ending the war, since further resistance for the exhausted Confederate Army was both logistically and tactically impossible (Keegan 2009; McPherson 1988).

There is a fine line between maneuver and attrition at times. Successful maneuver can lead to inflicting large enemy losses, while unsuccessful maneuver can force a country to adopt an attrition strategy. Additionally, attrition can open up opportunities for maneuver. The key, however, is what the focus of an army is on: Attrition strategies emphasize firepower, while maneuver strategies emphasize mobility (Antal 1992). Generally speaking, maneuver warfare is more effective than attrition warfare. However, it is more difficult in terms of command and control, training, equipment, and other factors.

Punishment Strategy. The final type of strategy that Stam (1996) identifies is a **punishment strategy**. Whereas both maneuver and attrition strategies focus on defeating the enemy's army (and other military forces), a punishment strategy focuses on increasing the enemy's costs. There are two basic types of punishment strategy: strategic bombing and guerilla warfare.

Strategic bombing entails attempts to use air power to achieve a war-winning effect on the opponent, as opposed to simply using air power to support efforts on the surface (land or sea). In the interwar period, the Italian theorist Giulio Douhet wrote *Command of the Air*, in which he envisioned masses of bombers shattering the enemy's morale by striking at their logistic, manufacturing, and population centers. Although details may change, particularly with regard to intentional bombing of civilians, modern strategic air power theorists continue to envision being able to win wars by increasing the enemy's costs beyond the point they are willing to bear (Jordan et al. 2008; Pape 1996; Sloan 2012).

Guerilla warfare, or insurgency, is a type of punishment strategy sometimes adopted by weaker belligerents that is intended to weaken the resolve of their political adversaries and lead to the withdrawal of occupying or government forces. Guerilla tactics include hit-and-run raids, ambushes, and remote attacks using mortars, rocket launchers, and improvised explosive devices. Mao Zedong was a primary theorist of insurgency, arguing that insurgents could overcome by being patient, knowing when and where to attack, and cultivating internal and external support.

Figure 10.2 Types of Military Strategy

Maneuver Strategy	Attrition Strategy	Punishment Strategy
• Focus on movement of forces to achieve an advantageous position	• Focus on defeating the enemy through killing personnel and destroying material	• Focus on increasing enemy costs through strategic bombing or guerilla warfare

Why Do Different States Use Different Strategies?

A great deal of attention has been paid to understanding different types of strategies and the effects that they have on the evolution of war, which we will examine in more detail in the next chapter. Others have examined the factors that lead states to use different strategies. Most of these studies have been qualitative, examining different key cases. The period between World War I and World War II—the interwar period—has been a particularly common focus because of its similarities with the post–Cold War period. As Murray and Millett (1998b, 2) state, "Military institutions had to come to grips with enormous technological and tactical innovation during a period of minimal funding and low resource support. Some succeeded, creating a huge impact on the opening moves in World War II. Others were less successful and some institutional innovation resulted in dismal military failure."

Posen examines the sources of **military doctrine**, which he defines as "the subcomponent of grand strategy that deals explicitly with military means. Two questions are important: *What* means shall be employed? and *How* shall they be employed?" (1984, 13, emphasis in the original). Other political scientists have followed Posen in distinguishing between two basic types of doctrine: offensive and defensive.[1]

Posen (1984) bases his expectations for the sources of doctrine on balance of power and organizational theory. He argues that the need for standard scenarios encourages military organizations to prefer offensive doctrines because they help you get your standard scenario, but also deny the enemy his standard scenario. This is important because military organizations prefer to fight their own war and prevent the adversary from doing so. Furthermore, he argues that expansionist powers will prefer offensive doctrines, while status-quo states will generally prefer defensive doctrines

[1] Posen also discusses a doctrine of deterrence, although this is ignored in subsequent works such as those by Stam (1996).

if geography or technology makes such doctrines attractive. Posen (1984) concludes that Germany's expansionist aims during the interwar period led it to focus on an offensive doctrine. In contrast, Britain and France were status-quo powers and selected defensive doctrines. The geography of the two states was much different, however. France, with a direct border with Germany, constructed the Maginot Line to bolster its defense. Britain relied largely on its air force and navy for defense, since it was protected from land invasion by the English Channel.

Murray and Millett (1998a) examine military innovation in the interwar period. Contributors to the volume examine innovation with regard to armored warfare, amphibious operations, strategic bombing, close air support, aircraft carriers, submarines, and electronics. Common themes that emerge are that the two factors most important for successful innovation were "the presence of specific military problems the solution of which offered significant advantages" (Murray 1998, 311) and a military culture that placed value and emphasis on innovation. The two factors that most contributed to failures to innovate were the misuse of history and excessive rigidity within the military leadership and culture (Murray 1998). Similarly, Corum (1992) focuses on the development of German doctrine during the interwar period and discusses how the leadership of Hans von Seeckt, chief of staff of the army, was crucial to successful German innovation.

In contrast to these qualitative studies that focus on particular countries in particular time frames, Reiter and Meek (1999) conduct a quantitative empirical test of the determinants of military strategy during the twentieth century. They find that only 12 percent of countries employed maneuver strategies. Thus, a maneuver strategy appears to be an uncommon choice, yet it is an important one as maneuver strategies can make victory in war far more likely, as we will examine in the next chapter.

They find three factors have been the most important for the choice of maneuver strategies. The first of these is democracy. Democracies are more than three times more likely to choose maneuver strategies than are nondemocracies. In addition, advanced industrialized democracies are even more likely to select maneuver.

The strength of a country's economy also matters greatly. The more industrialized a state is, as measured by its steel production, the more likely it is to adopt a maneuver strategy. A state with an average level of steel production (and average in each of the other variables as well) has a predicted probability of maneuver of 0.090. In contrast, a state with steel production one standard deviation above the mean has a predicted probability of 0.391, whereas a state two standard deviations above the mean has a predicted probability of a maneuver strategy of 0.785.

The direct experience of a country with a maneuver strategy is also significant. This is based on the wars that the state has fought previously. Victories when using maneuver against attrition or losses when using attrition against maneuver are lessons that favor maneuver. Opposite outcomes (victory with attrition or loss with maneuver) are lessons against maneuver, while other outcomes provide neutral lessons (Reiter and Meek 1999). Countries with lessons against maneuver are very unlikely to adopt a maneuver strategy, while countries with lessons favoring maneuver are likely to adopt maneuver strategies.

Reiter and Meek (1999) find that other potential sources of maneuver strategies such as a military regime, terrain, militarized dispute participation, or vicarious experience have no significant impact on a state's choice of strategy. Overall, their findings suggest that ideas of balance of power and organizational interests can only partially explain states' choices of military strategy.

Principles of War

Military practitioners and theorists have tried to understand warfare through a variety of different methods, one of which is the development of principles of war "to guide the effective application of military power" (Jordan et al. 2008, 8). Modern US Army doctrine identifies nine principles of war: objective, offensive, mass, economy of force, maneuver, unity of command, security, surprise, and simplicity (Department of the Army 2008). Although their labels are different, other countries, such as the United Kingdom and the former Soviet Union, have developed very similar lists. These principles have evolved over time and are the key area in which Jomini continues to influence strategic studies (Alger 1982). We discuss each of the principles of war in turn.

The first principle of war is *objective,* which asserts that military operations should be directed "toward a defined, decisive, and obtainable objective" (Department of the Army 2008, A-1). This principle comes directly from Clausewitz's (1976) focus on the political nature of warfare. The specific objectives that military operations are directed toward should contribute to the broader political and military objectives. As we discussed previously, the tactical level must serve the operational level and the operational level must serve the strategic level.

The principle of the *offensive* reflects the basic idea that military operations should be proactive rather than reactive. Success in warfare is more likely if a focus is placed on aggressive operations designed to seize and hold the initiative. There is an important tradeoff between attack and defense in warfare. Victory in land warfare requires offensive operations. Even where an army conducts a prolonged defensive campaign in order to wear down the enemy, a return to offensive operations is needed to exploit this attrition. At the strategic level, the growing scale of war has generated economic and political costs that often place a premium on ending wars as quickly as possible. At the tactical and operational levels, a force that does not attack is likely to cede the initiative to the enemy. However, it is more difficult to attack than to defend. The frictions associated with issues such as supply, command and control, coordination, and so on tend to be magnified in attack because attacking generally requires that forces maneuver. In addition, defending forces can use the terrain to multiply the effectiveness of their forces (Jordan et al. 2008).

The next two principles are closely related. *Mass* indicates that military power should be concentrated at the decisive time and place. Success is more likely if a preponderance of force can be gathered for the most significant battles and not dispersed at less important points. *Economy of force* means that forces allocated to secondary objectives or positions should be minimized in order to facilitate the concentration of forces at the decisive point.

The next principle of war is *maneuver,* which is the basic idea that friendly forces should use movement to place the enemy at a disadvantage. This might involve trying to move onto the enemy's flanks as a way of avoiding a frontal assault, or using combinations of fire and movement to suppress the enemy and advance on a position. This is particularly important because a flank or rear attack is more likely to succeed than a frontal attack. Maneuver is not only important when attacking, but in defense as well.

Also important is *unity of command,* which indicates that the forces allocated to a specific operation or purpose should be allocated a single overall commander so that the activities of different elements are properly coordinated. In various wars in the past, rivalries between the army and the navy have undermined cooperation between branches, and operations between allied nations have lacked effective coordination. These problems are largely solved by having a single overall commander.

Security specifies that friendly forces should be protected from enemy action that might disrupt or harm them. This principle encompasses a wide range of activities, such as providing land forces with adequate antiaircraft protection, employing information security techniques, establishing secure communications, and patrolling aggressively to inhibit enemy reconnaissance efforts.

Surprise indicates that military forces should be used in a manner, time, or place that is unexpected by the enemy. Surprise substantially enhances combat power, and can be important at the tactical, operational, and strategic levels of war.

The final principle of war is *simplicity,* which indicates that unnecessary complexity should be avoided in preparing, planning, or conducting operations. Those factors that create general friction in warfare, such as human involvement, fear, chance, and uncertainty, mean that the more complex a military operation is, the more likely it is that something will go wrong.

These principles of war were developed through experience and study of military history, rather than scientific analysis. They harken back to Jomini's quest for scientific principles, but they are not meant to be followed blindly (Alger 1982). Rather, they are rules of thumb that must be adapted to particular situations. Nonetheless, they nicely summarize key elements of strategic theory for both battlefield commanders and students of warfare.

Modeling Combat

Combat is the heart of warfare, as opposing military forces of all sizes engage each other in battle. Some scholars, including many in fields other than political science, have created models of combat to predict outcomes of battles, measure combat power, and make policy recommendations regarding military forces (Stoll 1995). Modeling combat entails a focus on the tactical and operational levels of warfare, rather than the strategic level (Luttwak 1980).

Some attempts to model combat are very simplistic. For example, an often used rule of thumb is the 3:1 rule, which argues that in order to prevail in battle the attacker must have a three to one numerical superiority (Dupuy 1989; Epstein 1989; Mearsheimer 1989). However, since "combat is too complex to be described in a single, simple aphorism" (Dupuy 1987, 7), other models of combat are much more complicated. We focus on two: Dupuy's (1987) measure of combat effectiveness, and Biddle's (2004) argument about modern-system force employment.

Combat Effectiveness

Dupuy (1979, 1987) develops a statistical model, which he calls the **Quantified Judgment Model (QJM)**, to examine battle outcomes. The model accounts for a variety of important variables detailing both operational factors characterizing the opposing forces and environmental factors accounting for the natural environment in which the battle took place. Environmental factors include terrain, weather, and season. Operational factors include posture and fortifications, mobility, vulnerability, air superiority, surprise, fatigue, and weapons sophistication.

There are additional factors such as leadership, training, experience, morale, and manpower quality that theory suggests are important but are difficult to measure. To get at these intangibles, which he terms **combat effectiveness**, Dupuy (1987) compares the actual result of each battle with the predicted result based on the quantified judgment model. This results in a combat effectiveness value (CEV), which is the ratio of the actual battle result to the predicted result. If the result ratio for one side is larger than its predicted result, this means that that side performed better than was predicted, and its CEV for that battle is greater than 1.0. If the actual result is less than the predicted result, then the CEV is less than 1.0.

Dupuy (1987) looks at the combat effectiveness of a variety of armies (and units within armies) in different wars in history. Table 10.1 shows the combat effectiveness values of 24 American, British, and German divisions during World War II. These units appeared in between 3 and 17 engagements. German units consistently had the highest combat effectiveness values, meaning that they performed much better than would be expected based on operational and environmental factors alone. In contrast, British divisions displayed the lowest combat effectiveness of the three armies, echoing claims elsewhere that the British Army underperformed throughout World War II (Millett and Murray 1988).

For another example of combat effectiveness, we turn to the Six-Day War (1967), discussed in Box 10.1, and the Yom Kippur War (1973), which were part of a series of Arab-Israeli wars that exemplify the advantage of

Table 10.1 CEVs of Selected Divisions in World War II

DIVISIONS	NUMBER OF ENGAGEMENTS	AVERAGE CEV
United States		
1st Armored	3	0.86
3rd Infantry	4	0.86
4th Armored	8	0.73
34th Infantry	5	0.81
45th Infantry	11	0.72
85th Infantry	6	0.79
88th Infantry	4	1.14
Average		*0.84*
British		
1st Infantry	8	0.82
5th Infantry	3	0.61
7th Armored	3	0.83
46th Infantry	6	0.96
56th Infantry	9	0.60
Average		*0.76*
German		
Herman Goering Panzer	5	1.49
Panzer Lehr	4	1.02
3rd Panzer Grenadier	17	1.17
4th Parachute	5	0.93
11th Panzer	4	1.31
15th Panzer Grenadier	11	1.12
16th Panzer	7	1.07
29th Panzer	3	0.82
65th Infantry	6	0.98
94th Infantry	8	1.38
361st Infantry	3	0.95
362nd Infantry	3	0.98
Average		*1.10*

Source: Trevor N. Dupuy, *Understanding War: History and Theory of Combat* (New York: Paragon House, 1987), 115.

combat effectiveness. In the Yom Kippur War, the Arab states attacked Israel, successfully achieving surprise. Syria attacked in the Golan Heights, and a key part of the attack was by the Syrian 7th Infantry Division (later joined by the 3rd Armored Division) against the Israeli 7th Armored Brigade. In the ensuing battle on 6–9 October 1973, known as the Valley of Tears, the Israeli forces emerged victorious despite being heavily outnumbered (Asher and Hammel 1987; Rabinovich 2004). At one point on the second day of the battle, approximately forty Israeli tanks were confronted with around five hundred Syrian tanks. How were Israeli forces able to consistently defeat their Arab opponents in battle despite being consistently outnumbered?

Box 10.1

Case in Point: Six-Day War

The Six-Day War was the third Arab-Israeli war, and grew out of rising tensions in the Middle East in the spring of 1967. Although at a distinct disadvantage based on numbers, the Israeli military triumphed due to its high combat effectiveness. On 16 May, Egypt demanded that the United Nations Emergency Force, which had been deployed in the Sinai Peninsula and Gaza Strip to discourage hostilities between Egypt and Israel, withdraw. The withdrawal was completed by 19 May. Three days later, on 22 May, Egypt blocked the Strait of Tiran to Israeli shipping for the first time since the 1956 war. This action was an act of war under international law.

On 30 May, King Hussein of Jordan and Egyptian president Gamal Abdel Nasser signed a mutual defense pact in which Egypt gained joint command of the Jordanian army. In addition, Iraq joined a military alliance with Syria, Egypt, and Jordan. Egyptian troops poured into the Sinai Peninsula. Jordanian forces were largely spread out across West Bank Palestinian villages instead of being concentrated in more strategically important locales. Confident of victory, the Jordanians resolved to cut off western Jerusalem by attacking Israeli positions in the north and south of the city at the start of the fighting. But Syria failed to coordinate with Egypt despite their defense pact. Like the Jordanians, the Syrians adopted ambitious war plans, opting for an offensive operation as opposed to a more limited plan to fend off Israeli attacks on the Golan Heights. Morocco, Libya, Saudi Arabia, and Tunisia all sent troops to the Sinai.

In 1967, Israel was badly outnumbered by its Arab neighbors. It had a total army strength of 264,000 personnel (including 214,000 reserve troops), as well as 300 combat aircraft and 800 tanks. For the Arab states, Egypt had 240,000 personnel, while Syria, Jordan, Lebanon, and Iraq added a further 307,000. In addition, the Arab states had 957 combat aircraft and 2,504 tanks.

Nonetheless, the Israeli cabinet opted for war in a 12 to 2 vote, and on 5 June 1967, Israel invaded Egypt. The invasion began with a massive air campaign against Egyptian air bases, which succeeded in destroying much of the Egyptian Air Force in the first few hours of the war. Israeli Army columns, led by three armored divisions and three separate brigades (one armored and two infantry) charged across the Sinai Peninsula. Israel succeeded in reaching the Suez Canal on 8 June.

(Continued)

(Continued)

Map 10.2 Six-Day War, 1967

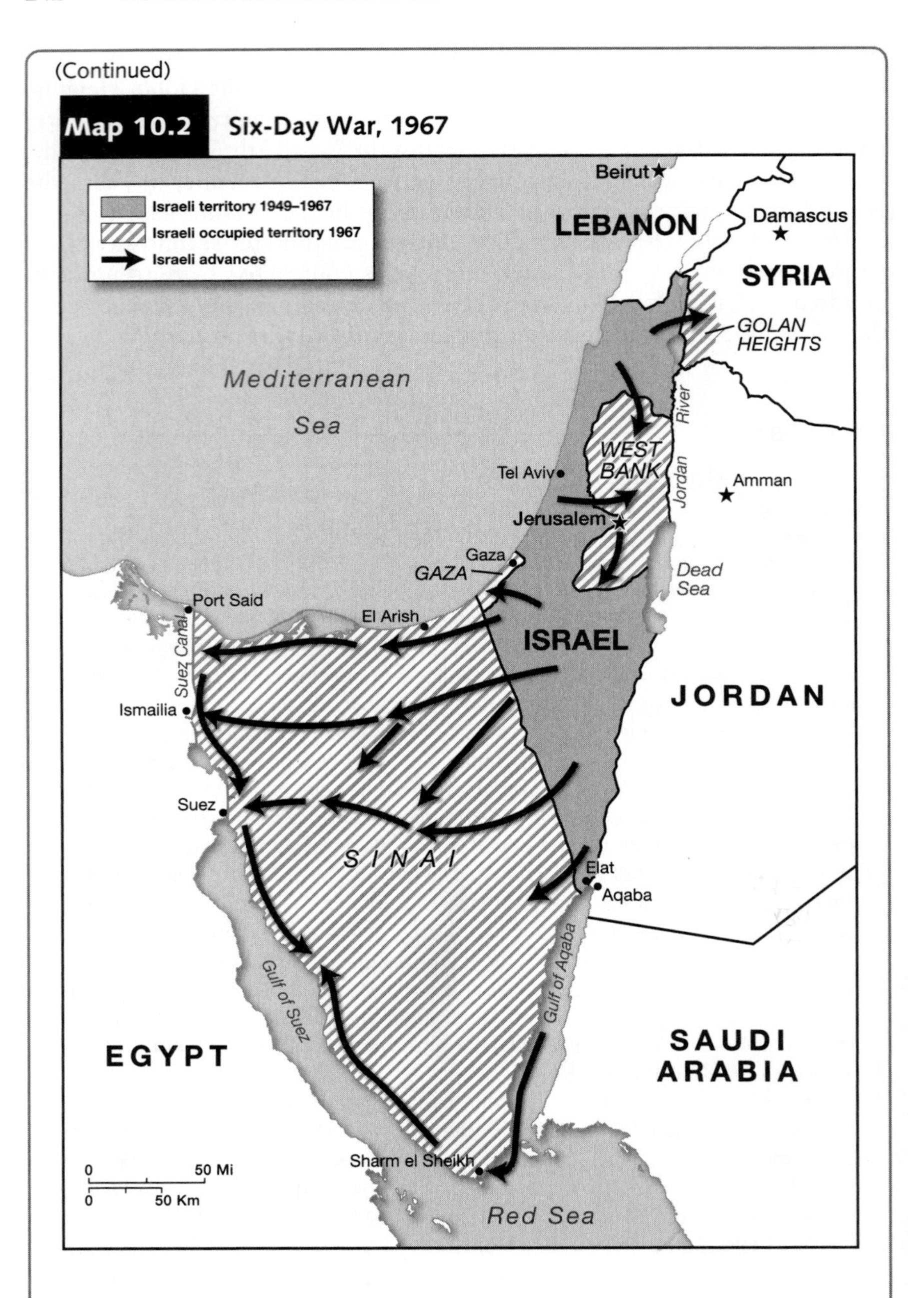

Israel's attack on Jordan also began on 5 June. Israel launched a pincer movement into the West Bank, with one armored brigade attacking from the north and another attacking from the south. They quickly overwhelmed the Jordanian forces, overrunning the entire West Bank by 7 June. Two days later, on 9 June, Israel

(Continued)
invaded Syria. Israeli forces quickly breached Syrian defenses in the Golan Heights and found themselves on the road to Damascus by 10 June. The following day, a cease-fire was signed.

The result was a decisive Israeli victory. In terms of casualties and losses, Israel lost 779 killed, 2,563 wounded, and 15 prisoners taken, with forty-six aircraft lost. In contrast, the Arab states suffered 21,000 killed, 45,000 wounded, and 6,000 prisoners taken, with over four hundred aircraft destroyed. In terms of territory, Israel occupied the Sinai peninsula, the Gaza Strip, the West Bank, and the Golan Heights—a growth of Israel's territory about 300 percent. Models of combat can help us understand why Israel was able to win the Six-Day War despite being outnumbered so greatly.

Table 10.2 Israeli CEVs Versus Arabs in 1967 and 1973 Wars

	JORDAN	EGYPT	SYRIA	IRAQ
1967	1.54	1.75	2.44	—
1973	1.88	1.98	2.54	3.43

Source: Trevor N. Dupuy, *Understanding War: History and Theory of Combat* (New York: Paragon House, 1987), 121.

Dupuy's concept of combat effectiveness provides a way that we can gain some understanding of why Israeli forces performed so well in the Six-Day War and others. The average combat effectiveness values of Israeli forces versus Egyptian, Jordanian, Syrian, and Iraqi forces during the 1967 and 1973 Arab-Israeli wars are shown in Table 10.2. These CEVs demonstrate that Israeli units consistently outperformed each of the Arab countries in both wars, although their performance in the Yom Kippur War was even better than in the Six-Day War. In addition, Israeli performance against Syrian and Iraqi forces was much better than their performance against Egypt and Jordan.

The combat effectiveness value is useful because it provides a way to quantify differences in quality between different units and armies. It has important limitations as well. In particular, it does not specify why some armies are more effective than others. Although Dupuy (1987) indicates that training and leadership are key factors influencing combat effectiveness, the keys to effective training and leadership are left unspecified.

Force Employment

One approach to modeling combat that is more specific about what makes some armies more effective than others is developed by Biddle (2004, ix), who focuses on the importance of **force employment**, which refers to "the doctrine and tactics by which forces are actually used in combat." Biddle contrasts his expectations based on force employment with arguments by others that battle outcomes are driven by the numerical balance between the opposing sides or by either dyadic or systemic technology. Biddle's (2004) analysis is focused on the period of modern warfare in the twentieth century and beyond. Over the course of the nineteenth century, firepower increased greatly, culminating with the emergence of modern warfare in World War I (Gray 2012). All sides were forced to seek methods to deal with the greatly increased lethality of weapons. The particular problem that most plagued military commanders was determining how one can move and attack in battle in the face of this greatly increased firepower.

Biddle (2004) argues that the solution is modern-system force employment. Modern-system force employment is complex and difficult to implement and entails a variety of steps at the tactical and operational levels. Keys to implementation at the tactical level include cover and concealment, dispersion, suppression, independent small-unit maneuver, and combined arms. Cover (e.g., protection from enemy fire, as provided by hiding behind a wall) and concealment (protection from enemy observation) are important because they deny enemy forces targets. Dispersion makes this cover usable by breaking up larger formations. Additionally, dispersion reduces vulnerability by putting fewer targets in the blast radius of any given shell or in the beaten zone of any given machine gun. Suppression reduces attackers' exposure by forcing opponents to take cover. Also important is independent small-unit maneuver, which allows subunits to find their way forward by sprinting between terrain features. Finally, the use of combined arms reduces net vulnerability by teaming together weapon types with contrasting strengths and weaknesses.

Keys at the operational level are depth, reserves, and differential concentration. Depth is important on both offense and defense because it provides successive waves of attackers (on offense) or defensive lines (on defense) with which to overcome enemy forces. Having reserves withheld behind the front lines is also crucial because this enables forces to be shifted to vital sectors of the front. Differential concentration entails concentrating forces disproportionately against certain parts of the lines to secure a breakthrough, for example.

Technological Change and Modern-System Force Employment

Biddle (2004) argues that modern-system force employment damps the effects of technological change by insulating its users from the full lethality of their opponents' weapons. Militaries that fail to implement the modern system have been fully exposed to the firepower of modern weapons, with increasingly severe consequences as those weapons' reach and lethality have expanded.

Modern-system offensive tactics are extremely complex and demand high levels of training and skill to be implemented properly. Part of this complexity arises from the speed limitations that the modern system imposes because modern-system tactics create a sharp tradeoff between speed and losses. Attackers can reduce exposure, and thus their losses, only by slowing their net rate of advance when in the presence of the enemy. The modern system is a complex doctrine that demands high levels of proficiency for proper implementation for both attackers and defenders.

Of course technology has changed greatly over the course of the past century, but have these technological changes made modern-system force employment less important? Technological change since 1918 has had three main effects: increased firepower, increased mobility, and improved information technology.

With regard to increased firepower, many argue that long-range, precision-guided weapons have revolutionized warfare, demanding radically new doctrines and tactics. Sharp lethality growth, however, is nothing new; it gave rise to the modern system in the first place. This continuing trend has progressively increased the modern system's importance. While weapons lethality has been increasing against all targets in absolute terms, it has grown much faster against massed targets in the open than dispersed targets in cover. The net result has been to afford modern-system militaries an increasing edge in relative vulnerability as weapons have grown more lethal (Biddle 2004).

There is no question that increases in mobility have created a situation in which today's forces are far more mobile than 1918's. The foot soldier of 1918 could manage perhaps 6 km/hr in an assault or 25 km/day in an unopposed forced march. By 1940, the German Panzer III tank could make 40 km/hr on roads and travel up to 175 km without refueling. By 1991, the American M1A1 Abrams tank could reach 67 km/hr on roads with an unrefueled range of 465 km. Many have seen these mobility increases as revolutionary, particularly the tank and helicopter. Modern-system defensive operations, however, are designed to compel attackers to slow down in order to survive. This offensive speed reduction is at the heart of modern-system defense and is relatively insensitive to the nominal speed of the attacker's platforms. Thus, increased mobility has important effects on exploitation, but not breakthrough. Greater mobility, like heavier firepower, increases the premium on modern-system force employment (Biddle 2004).

Finally, the past century has seen greatly improved information technology. Has new information technology made the modern system obsolete? Certainly there have been tremendous improvements in communications and related technologies. Better information, like greater firepower and mobility, however, renders the modern system more important, not less (Biddle 2004).

Force Employment and Battle Outcomes

Biddle's (2004) primary theoretical predictions regarding the importance of force employment for explaining battle outcomes are summarized in Figure 10.3. A modern-system attacker is characterized by low exposure, moderate closure rate, and narrow frontage. In contrast, a non-modern-system attacker is exposed, with a high closure rate and broad frontage. A modern-system defender defends in depth, with a large reserve withhold and low exposure. A non-modern-system defense is shallow, with a small reserve withhold and high exposure.

In the upper left cell of the figure, we have a modern-system attacker versus a modern-system defender. Biddle's (2004) prediction in this case is a contained offensive. Attacker and defender casualties are expected to be moderate, although they will increase as the amount of territorial gain increases. Only a moderate territorial gain is predicted, limited by attacker preponderance. Overall, the duration of the campaign is potentially long, with little sensitivity to technology and moderate sensitivity to preponderance.

If a non-modern-system attacker confronts a modern-system defender, the prediction is again a contained offensive, as shown in the upper right cell of the figure. In this case, Biddle (2004) predicts the attacker's casualties will be very high while the defender's casualties will be very low. In addition, a low territorial gain is predicted, and only a short campaign duration. Attacker casualties increase rapidly with advancing technology. There will be little sensitivity to preponderance, however, as a prohibitive attacker numerical advantage is required to prevail.

In contrast, if the attacker employs the modern system while the defender does not, a breakthrough is predicted. In this case, we would expect low attacker casualties, and exploitation prospects will improve with advancing technology. If exploitation succeeds, then very high defender casualties, very large territorial gains, and a very short campaign are predicted. If exploitation fails, then high defender casualties, with territorial gains proportional to preponderance, and moderate campaign duration are predicted. Additionally, there is little sensitivity to preponderance; even an attacker inferior in numbers can prevail.

Figure 10.3 **Theoretical Predictions for Force Employment and Battle Outcomes**

		ATTACKER	
		Modern System	Non-Modern System
DEFENDER	Modern system	Contained offensive	Contained offensive
	Non-modern system	Breakthrough	Conditionally contained offensive

Source: Adapted from Stephen Biddle, *Military Power: Explaining Victory and Defeat in Modern Battle* (Princeton, NJ: Princeton University Press, 2004), 74.

The final scenario, with a non-modern-system attacker facing a non-modern-system defender, is shown in the lower right cell of Figure 10.3. Biddle's (2004) prediction in this case is a conditionally contained offensive. Casualties, territorial gain, and campaign duration are all sensitive to both technology and preponderance. If there is a marked numerical or technological imbalance, then attrition, territorial, and duration outcomes favor the superior side. Otherwise, we would expect high defender and very high attacker casualties along with low territorial gain and potentially long duration. Casualties will increase with territorial gain and duration, while duration increases with territorial gain. A breakthrough is possible if the attacker is very superior in terms of material (numbers or technology), but exploitation of a breakthrough is unlikely.

Operation Michael. Biddle (2004) tests his theory regarding force employment in a variety of ways, starting with three case studies of major battles in the past century. His first case study examines Operation Michael, the German offensive against the British sector on the western front of World War I from 21 March through 9 April 1918. This is a useful case to examine because it should be an easy case for preponderance and technology theories to predict but should be a difficult case for force employment theory to predict.

A focus on weapons technology would predict a British defensive victory. Systemic technology greatly favored defenders. World War I is usually seen as the defining example of a technologically determined defense dominance given its domination by infantry and artillery. Dyadic technology was about even, with German and British weapons systems having approximately the same age and performance characteristics.

The numerical balance was also favorable for the British. The theaterwide force to force ratio was 1.17:1. By contrast, the theaterwide balance in 1915 was 1.56:1; in 1916, 1.33:1; and in 1917, 1.29:1. The Germans in 1918 faced one of the war's most attacker-averse theater numerical balances. The local balance of forces was about 2:1, which is again a low balance. Attackers had much larger advantages in most other World War I offensives. For example, at the Battle of Verdun in 1916, the Germans had a 4.25:1 advantage. Thus, a British defensive victory is predicted.

Force employment tells a different story, however. The Germans implemented the modern system very thoroughly at the tactical level. Indeed, the infiltration tactics that Germany developed during World War I (Lupfer 1981) formed the foundation of modern-system force employment. They used dispersed formations, made extensive use of terrain, used combined arms techniques, and allowed subunits to maneuver independently. Germany implemented the modern system somewhat at the operational level. The fifty-mile assault frontage was among the widest of the war (most previous major offensives had between five- and fifteen-mile fronts) and significantly exceeded the minimum for effective logistical support.

The British failed to implement modern-system methods at either the tactical or operational level. They made no systematic effort to disperse forces, most British positions were exposed to continuous observation, and the British defensive position—less than one half the depth of typical late-war German defenses—was quite shallow. The British also employed one of the most forward-oriented defenses of the war, with only a small reserve withhold. Since this battle pitted a modern-system attacker facing a non-modern-system defender, a German offensive breakthrough is predicted (Biddle 2004).

In the actual battle, Germany broke through the British lines, as predicted by force employment theory, but contrary to expectations based on numbers and technology. This result was literally unprecedented in almost four years of continuous warfare on the western front. The breakthrough, however, failed to produce a decisive exploitation.

Operation Goodwood. Biddle's (2004) second case study is about Operation Goodwood, the British offensive attacking German forces around Caen, 18–20 July 1944. The British were trying to secure a breakthrough, enabling the Allies to finally break out of Normandy. Again, this should be an easy case for preponderance and technology theories to predict but a difficult case for force employment theory to predict.

Weapons technology favored the British. Systemic technology favored attackers, and the attacking British forces had 1,277 tanks and 4,500 aircraft. The British aircraft per supported soldier is probably an all-time high for a major ground attack. British attackers deployed about twice as many tanks per capita as either attackers or defenders at the Battle of El Alamein, over three times more than at Kursk, and over ten times more than the Germans in France or Russia in 1940 and 1941. The British also enjoyed a slight edge in the dyadic technological balance across the entire range of weapons types. While the German Panther and Tiger tanks were superior, they comprised a minority of the German tanks. Most of the German tanks were less capable than their British counterparts. In addition, airpower was a major advantage in the British favor. Thus, a successful British attack is predicted.

There was also a major numerical advantage for the British at both the theater and local level. The theaterwide force-to-force ratio (FFR) was 3.8:1. At the point of attack, the British had a 22.7:1 advantage in troops and an astounding 82.7:1 edge in tanks. Along the entire assault frontage (including reserves), it was 3:1 in troops and 5.9:1 in tanks, which is still a significant advantage to the British and one of the most favorable attacker imbalances of the war. If numerical imbalance determines success, then it should certainly produce a British victory here. Thus, a successful British attack is predicted.

Once again, the expectation based on force employment is different. The British failed to implement the modern system at the tactical level. British armor simply rolled forward in dense, exposed waves. British infantry advanced in the open and tended to bunch up when taken under fire. Neither made much effort to use terrain to cover their advance. Additionally, cooperation between infantry and armor was poor. At the operational level, they had a too narrow assault frontage for a breakthrough attempt. The critical sector assigned to VIII Corps' three armored divisions—the units assigned to break through the German defense—was only 2 km wide. This allowed German positions to engage targets across the entire penetration corridor, forced the British to echelon the three armored divisions one behind the other, and created serious congestion, making movement and combined arms coordination difficult.

In contrast, the Germans implemented the modern system thoroughly. German defenses at Goodwood were disposed in depth with a very large fraction of troop strength withheld in mobile reserve. The depth of the German defenses was 16 km, among the deepest defenses of World War II. Their prepared positions were also extremely well covered and concealed, and their reserve movements made aggressive use of exposure-reduction techniques to reduce their vulnerability to Allied air attack. Given a non-modern-system attacker against a modern-system defender, German defensive success is predicted.

Goodwood produced a contained offensive that petered out with a net ground gain of only 8 to 10 km, providing yet more strong support for the importance of modern-system force employment. Despite an overwhelming advantage in numbers and a strong edge in technology, the British were unable to break through. In contrast, the American attempt to break out of Normandy with Operation Cobra on 25 July 1944 was wildly successful. The key difference in the two operations was that the Americans made excellent use of the modern system.

Operation Desert Storm. The final case study Biddle (2004) examines is Operation Desert Storm, the US/Allied campaign to liberate Kuwait in January and February 1991. The coalition attackers enjoyed superior numbers, technology, and force employment. Thus all three theories would predict Iraqi defeat. Weapons technology strongly favored the United States and its allies. The weapons used by the United States were introduced in 1973 on average, while Iraqi weapons were introduced on average in 1961. This twelve-year lead is the largest of the sixteen wars for which data are available (Biddle 2004). The numerical balance also favored the United States. The theater-wide FFR was about 2:1 in combat maneuver strength and about 3:1 in total personnel. In addition, Iraqi forces were spread very thinly.

In terms of force employment, Iraq implemented the modern system moderately well at the operational level but very poorly at the tactical level. Iraqi forces were disposed in moderate depth and much of Iraq's total combat power was held in mobile reserve. Iraqi defenses, however, provided neither the concealment, cover, independent small-unit maneuver, nor combined arms integration demanded by modern-system tactics. The Americans implemented the modern system very thoroughly at both the tactical and operational levels.

In the end, a massive breakthrough was achieved, supporting all three theories. But we can look more closely at the campaign to try to distinguish between the explanations. While Coalition ground force technology varied widely, losses did not. For example, the two US Marine divisions were equipped with M60A1 tanks rather than the much more advanced M1A1 tanks of the Army, yet the Marines suffered few casualties. Furthermore, many lightly armored vehicles, such as M2 Bradleys, engaged in extensive close combat yet suffered very few losses. Coalition forces also won decisively regardless of local force ratios; US units were able to engage and destroy Iraqi units four times their size. The Iraqis employed their forces in ways that left them exposed to the full lethality of US weapons, while US forces were employed in ways that limited the Iraqis' ability to exploit their own firepower. As we examined previously, technology magnifies the effect of force employment.

Testing Force Employment Theory With Simulation. Biddle (2004) also employs experiments to test force employment theory based on computer simulation of the Battle of 73 Easting. This was a decisive tank battle on 26 February 1991, featuring a hasty attack by the 2nd Armored Cavalry Regiment (ACR) against the Iraqi 18th Mechanized Brigade and 37th Armored Brigade of the Tawakalna Division. It is the best documented battle of the Gulf War. The simulations were conducted using the Janus system developed by the Lawrence Livermore National Laboratory, which has resolution down to the individual weapons system. The simulation allows us to conduct experiments to address important questions. Biddle focuses on two: What if the Iraqis had fully implemented the modern system? What if the United States had not had such advanced technology?

As we saw in the case study about Desert Storm, Iraq employed the modern system very poorly at the tactical level. What were the key problems? Biddle (2004) focuses on two key variables. The first is poor Iraqi defensive preparations, with armored vehicles perched on the surface behind exposed sand berms. The second is the failure of Iraqi covering forces to provide warning. Although the Iraqis used observation posts forward of their main defensive line, they provided no warning whatsoever of 2 ACR's approach. What if these errors had been corrected?

The United States also had a clear technological edge on Iraq. Biddle (2004) focuses on two key areas. The first is the M1A1 Abrams tank's thermal sights, which gave a significant targeting advantage to the United States. The second is air technology, which gave a significant advantage to the United States in additional means to destroy Iraqi equipment. What if the United States did not have these advantages in technology?

Experimental results were generated by running each scenario 10 times. The results from the seven different scenarios are shown in Table 10.3. In the base scenario, with no changes from the actual battle, the average number of armored fighting vehicle (AFV) losses for the two sides are 2 US and about 86 Iraqi vehicles, for a 42.90 loss exchange ratio (LER). This is quite

Table 10.3 Simulation Experiment Results

SCENARIO	KEY FEATURES	US AFV LOSSES	IRAQI AFV LOSSES	LER*
Base		2.00	85.80	42.90
A	Modern-system Iraqi tactics: both errors corrected	48.30	31.20	0.65
B	Partial Iraqi improvement: positions properly prepared	5.30	57.20	10.79
C	Partial Iraqi improvement: tactical warning provided	1.80	86.10	47.83
D	No US thermal sights; Iraqi warning error corrected	39.10	38.30	0.98
E	No US thermal sights; neither Iraqi error corrected	15.90	59.80	3.76
F	No US thermal sights or air; neither Iraqi error corrected	40.00	38.00	0.95

* The LER is the ratio of Iraqi to US losses.

Source: Adapted from Stephen Biddle, *Military Power: Explaining Victory and Defeat in Modern Battle* (Princeton, NJ: Princeton University Press, 2004), 184.

close to the actual losses of 1 US Bradley and 85 Iraqi tanks. If both Iraqi errors in force employment are corrected (scenario A), casualties increase markedly for the United States—by about twenty-four times—while casualties are more than halved for Iraq. If the only change is that the Iraqi positions are properly prepared (scenario B), Iraq fares better than the base case, but there is still a decisive edge for the United States. If tactical warning is provided for Iraq, but nothing else is changed (scenario C), there is no real change from the base case. Thus, correcting both errors changes things greatly, but correcting only one does not. The modern system needs to be implemented fully.

The next three scenarios focus on the impact of technology. If there are no US thermal sights and the Iraqi warning error is corrected (scenario D), Iraq does much better. If there are no US thermal sights but neither Iraqi error is corrected (scenario E), then Iraq does better, but not that much. Finally, if there are no US thermal sights or air and neither Iraqi error is corrected, then Iraq does much better than in the base case. Technology, however, does not have as large of an impact as force employment.

Assessing Force Employment Theory

Biddle (2004) performs a series of basic statistical tests using data on battles and wars in the twentieth century. Grauer and Horowitz (2012) conduct a more direct quantitative test of force employment theory through an analysis of decisive operations in interstate wars from 1917 to 2003. They find that states fully implementing the modern system win about 80 percent of the time, while states not implementing the modern system have won only 20 percent of battles. In sum, the importance of modern-system force employment is strongly supported by empirical evidence gained through case studies, experiments, and statistical tests. Furthermore, preponderance and technology theories are not well supported.

If the modern system is so effective, then why does force employment vary? The modern system poses difficult political and organizational problems that prevent many states from adopting it. Defense in depth, for example, requires states to yield territory early in hope of regaining it later by counterattack. This is systematically unpopular with residents of border areas and is often unattractive more broadly for states whose border conflicts are matters of wide nationalist concern. The modern system also requires extensive independent decision making by junior officers and senior enlisted personnel. For social and political reasons, many states are unwilling to tolerate such autonomy for so many individuals (Biddle 2004).

One limitation of these models of combat is that they do not explain how battle outcomes translate into war outcomes. Biddle (2004) shows that force employment is crucial to understanding battle outcomes. The winners in two of his case studies (the Germans in Operations Michael and Greenwood), however, went on to lose the war within a year of the battle taking place. Battle outcomes are certainly important, but war outcomes are more important. Unfortunately, no one has been able to develop a general theory of how battle outcomes translate into war outcomes. Given the great complexity of strategy that that we have reviewed in this chapter, perhaps this is to be expected.

Conclusion

Examining military strategy and doctrine entails a closer look at military matters than we have had in the previous chapters. Although warfare has been understudied in political science, it is inherently political in nature, and therefore is an important part of political science and international relations.

Military strategy is the application of military power—which includes land power, naval power, and air power—to achieve political objectives. Strategy is complex and difficult to master, although strategic theory can help us to understand its complexities. The work of Clausewitz, Jomini, and Sun Tzu has proven to be particularly useful in this regard. Stam's (1996) typology of maneuver, attrition, and punishment strategies has also proven to be quite useful for political science applications.

Models of combat build on strategic theory to predict outcomes of battles, measure combat power, and make policy recommendations regarding military forces. Dupuy (1987) measures combat effectiveness to show how superior leadership, training, experience, morale, and manpower quality allow armies to perform much better in battle than would be expected based on operational and environmental factors alone. Biddle (2004) argues that modern-system force employment is the key to victory in battle, and his theory is well supported by empirical evidence gained through case studies, experiments, and statistical tests.

These preliminaries are important to set up our look at the evolution of war in the next two chapters, including the duration and outcome of war in chapter 11 and the termination of war in chapter 12.

Key Concepts

Attrition strategy

Center of gravity

Combat effectiveness

Force employment

Friction

Levels of warfare

Maneuver strategy

Military doctrine

Military strategy

Punishment strategy

Quantified Judgment Model (QJM)

Strategic theory

Warfare

11 Evolution of War

Victorious Israeli infantry advance in Egypt in M-3 armored personnel carriers during the Six-Day War. Part of the study of international conflict is to understand war outcomes, including how smaller countries such as Israel are sometimes able to defeat much larger countries and coalitions.

Source: Getty Images

What happens once a war starts? This is an important question that has received much less attention in political science than the causes of war and is the focus of this chapter. The evolution of war covers everything that happens during a war, including plans for military operations, wartime decision making, and warfare itself. We examined explanations of battle outcomes in connection with our exploration of military doctrine and strategy in chapter 10. Much research on the evolution of war has focused on explaining war expansion or summary statistics such as war duration, war outcome, and war costs.

A useful foundation for our examination of the evolution of war is provided by Stoll (1995), who developed a framework to highlight relationships between key variables dealing with war's evolution. His framework is shown in Figure 11.1. The upper left of the figure highlights state decisions to enter and leave the war, which affects the power balance between the two sides. Change in participant capabilities, reflecting how the opposing states' power levels shift during the course of the war, is at the upper right of the figure and also affects the power balance. The central element in Stoll's

Figure 11.1 **A Framework for Studying the Evolution of War**

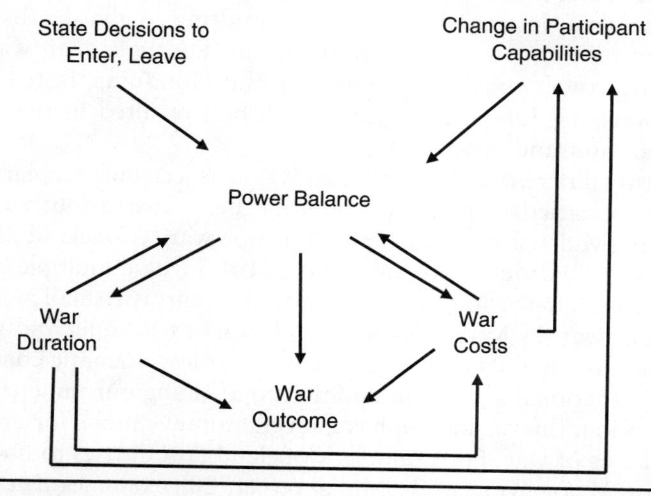

Source: Richard J. Stoll, "The Evolution of War." In Stuart A. Bremer and Thomas R. Cusack, eds., *The Process of War: Advancing the Scientific Study of War* (Amsterdam: Gordon and Breach, 1995), 133.

conception of the evolution of war is the power balance, which drives war duration, outcome, and costs. War duration and costs affect both the outcome and changes in participant capabilities.

Stoll's placement of power at the center of his framework reflects the conventional wisdom that power is central to every aspect of the evolution of war. This is a somewhat limited view, however, because there are a variety of nonmaterial factors—such as strategy, regime type, and terrain—which strongly affect the evolution of war. Therefore, while Figure 11.1 highlights power as the central element, it is only one of the factors that we will examine in this chapter.

Stoll's framework provides a basic road map for our examination of the evolution of war in this chapter. We will first look at war expansion and then at explanations of the duration of war. In the final main section of the chapter, we look at explanations of war outcomes, in particular contrasting traditional power-based views with Stam's analysis focused on a broad range of factors.

War Expansion

War expansion is the process through which wars grow to involve additional states that were not initially belligerents in the war. Before examining primary explanations of war expansion, it is important to address an important question: Are all wars comparable? Certainly, some wars are very large, while others are much smaller. For example, World War II is the largest war, involving over twenty-five countries, with fighting across the world (mostly in Europe, Asia, Africa, and throughout the world's oceans), lasting six years, and resulting in the deaths of more than sixty million people. In contrast, the Football War was fought between just two countries, El Salvador and Honduras, lasted only five days between 14 July and 18 July 1969, and resulted in the deaths of about two thousand people.

The vastly different scale of different wars has led some scholars to argue that these are strictly separate phenomena. Big wars, variously called systemic wars (Midlarsky 1990a, 1990b), major wars (Copeland 2000), and complex wars (Vasquez and Valeriano 2010), involve multiple states and typically have far-reaching consequences. In contrast, small wars, called nonsystemic wars by Midlarsky and dyadic wars by Vasquez and Valeriano, usually involve only two countries and have far less dramatic consequences for the international system. In addition to affecting our understanding of war expansion, this distinction has important implications for understanding the causes of war. For example, Copeland (2000) tries to uncover *The Origins of Major War*, while Midlarsky (1988) examines *The Onset of World War*. These works ignore small wars altogether, explicitly focusing exclusively on the causes of big wars.

Inherent in these studies is a belief that wars are "born big." Big wars are not simply small wars that grew larger; they are a completely different phenomenon with different causes. But does this make sense? If wars are born big, then nothing about the course of a war leads to its expansion. Rather, war expansion is predestined by the conditions that led to its start. This is a strong assumption that most scholars are unwilling to make (Stoll 1995). In addition, Bueno de Mesquita (1990) argues that focusing solely on big wars leads to selection bias and undermines efforts to gain a general understanding of international conflict.

How then can war expansion be explained? Two of the most common ways have focused on the factors that provide states with the opportunity and willingness to join the war or have focused on explaining potential joiners' cost-benefit calculations about whether or not to join. We examine each approach in turn.

Opportunity, Willingness, and War Diffusion

Siverson and Starr (1991) tried to predict the expansion of war by focusing on opportunity and willingness. As we discussed in chapter 4, opportunity is the possibility for militarized conflict between two states, while willingness refers to the desire by two states to engage in militarized conflict. In order for states in a dyad to become involved in an international conflict, they must have both the opportunity and willingness to fight. But opportunity and willingness also apply to getting involved in a war after it has already started.

To identify opportunity, Siverson and Starr (1991) look at **warring border states**. If one or more of a state's neighbors are involved in the war, then the chance that the country joins the war greatly increases. They also examine the type of border in question, whether it is a direct land border, across less than 200 miles of open water, or a colonial border. They argue that direct land borders are the most influential, while colonial borders are the least influential.

To identify willingness, they look at **warring alliance partners**. Countries that have at least one ally fighting in an ongoing war are much more likely to join the war than states without an ally involved. Siverson and Starr (1991) also examine the type of alliance in which an ally is at war, that is, whether it is a mutual defense pact, neutrality/nonaggression agreement, or entente. They argue that mutual defense pacts are the most serious commitments and therefore have the largest influence, while ententes are the agreements with the least influence.

A simple cross tabulation of their empirical results is shown in Table 11.1. States that have neither a warring border nation nor a warring alliance partner have only become involved in ongoing wars 8 times out of 2,328 cases, which is a 0.3 percent rate. In contrast, states that have either a warring border nation or a warring alliance partner (or both) have become involved in the war 86 out of 1,421 times, a 6.1 percent rate. Thus, states that have either the opportunity or willingness to join an ongoing war (as Siverson and Starr measure them) are nearly 17 times more likely to join the war.

Table 11.1 Treatments and War Involvement, 1816–1965

WAR INVOLVEMENT	WARRING BORDER NATION OR WARRING ALLIANCE PARTNER		
	NO	YES	TOTAL
No	2,320	1,335	3,655
Yes	8	86	94
Total	2,328	1,421	3,749

Source: Adapted from Randolph M. Siverson and Harvey Starr, *The Diffusion of War: A Study of Opportunity and Willingness* (Ann Arbor: University of Michigan Press, 1991), 55.

Additionally, states can have multiple treatments, multiple influences that may drag them into a war. Thus, a state might have one warring border nation and one warring alliance partner, for a total of two treatments, or four warring alliance partners for a total of four treatments, and so on. Siverson and Starr find that the more treatments that a state is exposed to, the more likely it is to join the war. States with only 1 treatment have joined the war only 1.2 percent of the time (9 of 712), states with 4 treatments have joined 14.9 percent of wars (10 of 67), and states with more than 6 treatments have joined the ongoing war half the time (15 of 30).

One limitation of Siverson and Starr's theory is that it treats war as a contagious disease that other states might "catch." However, wars do not just expand by accident or chance. Rather, they expand because states join in. For example, the Korean War began as a bilateral war when North Korea invaded South Korea on 25 June 1950. The war expanded because the United States chose to intervene on 27 June following condemnation of the North Korean attack by the UN Security Council. It expanded further because additional countries (such as Britain, Canada, Turkey, and the Netherlands) chose to contribute forces alongside South Korea, and China intervened on behalf of North Korea. The leaders of each joining state made conscious decisions to join.

Also, many times wars do not expand (or expand further) because no one chooses to join. For example, Germany and Italy sent forces to fight alongside Francisco Franco's Nationalists during the Spanish Civil War (1936–1939). Like Hitler and Mussolini, Franco was a fascist leader; therefore, it would seem natural that Spain would repay the favor and join the Axis powers in World War II. Franco, however, felt that German demands for Spanish bases were too great, and Spain ultimately decided to stay out of the war (Goda 1998; Weinberg 1994).

Joining War as a Rational Choice

Altfeld and Bueno de Mesquita (1979) take a completely different approach to explaining war expansion. They used formal expected utility models that predicted third-party decisions to join with the stronger or weaker side in an ongoing war to remain neutral. Their theory based on expected utility calculations has done well in explaining war expansion.

They focus on a situation in which a state, B, decides whether or not to join an ongoing war between two other states, A and C. State B must decide not only whether to join the war, but also which side to join if they choose to do so. To explain B's decision, Altfeld and Bueno de Mesquita utilize expected utility theory, which we introduced in chapter 3.

In making its decision, State B must consider its utility and costs for the different courses of action as well as the probability that its joining will make a difference. This leads to several important terms in Altfeld and Bueno de Mesquita's model. For the utilities, U_{ba} is the amount B values a victory by A, while U_{bc} is the amount B values a victory by C. For the costs, K_{ba} is the cost B expects to endure in helping A against C, while K_{bc} is the cost B expects to endure in helping C against A.[1]

Developing the expected utility logic based on these costs and benefits leads to three equations that are central to Altfeld and Bueno de Mesquita's analysis. First, if

$$\left[\frac{b}{a+b+c}\right](U_{ba} - U_{bc}) > K_{ba} - K_{bc},$$

then State B should be expected to join an ongoing dispute on the side of A. On the other hand, if

$$\left[\frac{b}{a+b+c}\right](U_{ba} - U_{bc}) < K_{ba} - K_{bc},$$

then State B should be expected to join an ongoing dispute on the side of C. Finally, if

$$\left[\frac{b}{a+b+c}\right](U_{ba} - U_{bc}) = K_{ba} - K_{bc},$$

then State B is expected to remain neutral.

The logic of these equations is easy to understand. They indicate that the decision to help one side or the other is driven by three key factors. The first factor is the strength of State B's motivation about whether A or C wins the war. This is reflected in the difference in the U terms, which appears in each equation. If $U_{ba} - U_{bc}$ is positive and large, then B strongly prefers that A wins; if it is negative and large, then B strongly prefers that C wins; if it is zero, then B is indifferent between a victory by either side.

The second factor is B's wherewithal to exert influence over the outcome. Altfeld and Bueno de Mesquita assume that this is driven by State B's power relative to States A and C, given by the ratio $b/(a + b + c)$. In this equation, a is State A's power, b is State B's power, and c is State C's power. The

[1]This discussion of Altfeld and Bueno de Mesquita (1979) is based on the discussion in Bueno de Mesquita (2006, 286–90, 539–45).

stronger B is relative to the others, the larger this ratio will be. This reflects the idea that the more power B has, the more she will be able to influence the outcome of the war.

The third factor driving State B's decision is the prospective costs of fighting in the war. The difference in the K terms, K_{ba} - K_{bc}, reflects how costly it is for B to fight alongside State A relative to fighting with State C. If the difference is positive and large, then supporting A is much more costly; if it is negative and large, then supporting C is much more costly; if it is zero, then the costs of supporting either side are equal.

Altfeld and Bueno de Mesquita (1979) conduct tests to examine whether their theory is empirically supported. Doing so requires one to calculate how much each prospective participant in a war values victory by one side or the other and how valuable the military contribution of each participant will be. They estimate the utilities for victory based on alliance portfolios, as discussed in chapter 6. The military contribution is estimated using the Correlates of War (COW) composite capabilities score discussed in chapter 5.

Their key results are summarized in Table 11.2. They find that when the third party joined the weaker side, their predictions were 80 percent correct. When the third party joined the stronger side, their model's predictions were 63 percent correct. And when the third party remained neutral, their predictions were 96 percent correct.

Yamamoto (1990) analyzed whether war expansion is better explained by deliberate, rational choice or by chance. Siverson and Starr's (1991) approach assumes the process is governed by chance, while Altfeld and Bueno de Mesquita's (1979) assumes it is driven by rational choice. Yamamoto (1990) compares and tests the implications of the two approaches and finds better support in the historical record for the rational model.

Gartner and Siverson (1996) examine an important implication of this rational model of war expansion. Although some wars expand to include many states, most do not. This should actually be expected because states are more likely to initiate wars against opponents who they expect will not receive support from others. Thus, the prevalence of bilateral wars in

Table 11.2 Predicted and Actual Third-Party War Choices

	PREDICTED CHOICE		
ACTUAL CHOICE	JOIN WEAKER SIDE	STAY NEUTRAL	JOIN STRONGER SIDE
Join weaker side	16	4	0
Stay neutral	1	104	3
Join stronger side	1	5	10

Source: Adapted from Michael Altfeld and Bruce Bueno de Mesquita, "Choosing Sides in War." *International Studies Quarterly* 23 (1979), 106.

history is the result of a selection effect, as initiators typically select wars to fight when they know it will remain bilateral.

War Duration

The next major factor regarding the evolution of war that has received a lot of attention from scholars is war duration. Some wars last for many years. Most dramatically, the Hundred Years War lasted from 1337 to 1453 and the Thirty Years War lasted from 1618 to 1648. More recently, the Vietnam War lasted ten years from 1965 through 1975, and the Iran-Iraq War lasted eight years from 1980 through 1988. Other wars last a very short time, only a matter of days for some—such as the Six-Day War in June 1967 or the Football War in July 1969—while others last for just a matter of weeks—such as the Seven Weeks War in June through August 1866 or the Iraq War in March and April 2003. Clearly, there is a great deal of variation in the duration of different wars, just as wars involve a wide range of different numbers of states. Additionally, **war duration** is important because it has important effects on war costs, leader popularity, regime stability, and war outcomes. This is the first summary characteristic of the evolution of war highlighted by Stoll (1995) and illustrated in Figure 11.1.

Several studies have attempted to predict the duration of wars. Early analyses of war duration were simplistic and highly limited in their ability to explain and predict duration (e.g., Horvath 1968; Morrison and Schmitt-lein 1980; Vuchinich and Teachman 1993). The study of war duration was advanced greatly by Bennett and Stam (1996), who developed a general model of war duration accounting for a variety of different variables. They argue that military strategy and terrain are the most important determinants of war duration. Maneuver strategies tend to create shorter wars, while punishment strategies tend to make wars last longer. Similarly, wars tend to be shorter when terrain is open but longer in rougher terrain. There also exists an important interaction between strategy and terrain as different strategies work best on different terrains.

Bennett and Stam also account for several other variables. Relative power is important because wars should be shorter when there is an imbalance of power between the two sides. Large totals of military personnel and population should lead to longer wars, while advantages in troop quality and surprise should lead to shorter wars. In addition to international factors, Bennett and Stam account for domestic factors such as issue salience, regime type, and repression. They conduct a quantitative analysis of wars from 1816 to 1985 and find strong support for their theoretical expectations, with military strategy and terrain being the most important predictors of war duration.

Bennett and Stam applied their general model to predict the length of the Iraq War, which began on 19 March 2003. (See Box 11.1 for more details on the Iraq War.) They publicly released a research report on 1 April 2003 predicting the length of the interstate war (see also Bennett and Stam

2006). This provides an ideal opportunity to test their model's ability to make predictions, as the war was still ongoing at the time of their forecast. How long did it last? There are several possible dates for the war's end. We might consider the end to be 9 April, when Baghdad fell; the duration to this point is 22 days, or 0.73 months. This date is coded as the end of the war by the Uppsala Conflict Data Program. A better date to consider is 14 April, when Tikrit was taken and a Pentagon spokesman announced that major combat operations were over. The duration to this point was 27 days, or 0.9 months. The duration through 1 May, when President George W. Bush made a speech aboard the US aircraft carrier *Abraham Lincoln,* was 43 days, or 1.43 months. The COW war data code the end of the war as 2 May 2003. This date has no particular significance in the course of events, however, so the origins of this coding are perplexing.

Bennett and Stam forecast four different scenarios for the Iraq War. Scenario 1 involved the United States employing a maneuver strategy and Iraq using an attrition strategy, with the fighting in open terrain. Given these factors, the predicted duration for the war was 0.88 months. This predicts that the war would end on 14 April. Given that the United States was able to avoid prolonged city fighting, this was the scenario that actually occurred in the war.

Box 11.1

Case in Point: Iraq War

The Iraq War provided a unique opportunity to test political scientists' ability to make real-time predictions about the duration of war. After a coalition of countries led by the United States liberated Kuwait from Iraqi occupation in 1991, disagreements between the United States and Iraq occurred frequently. As part of the cease-fire agreement, UN Security Council Resolution 687 mandated that Iraqi chemical, biological, nuclear, and long-range missile programs be halted and all such weapons destroyed under United Nations control. UN weapons inspectors inside Iraq were able to verify the destruction of a large amount of WMD material, but substantial issues remained unresolved when the inspectors left Iraq in 1998. In addition, no-fly zones were established by the United States, Britain, and France in northern and southern Iraq in an effort to protect the Kurds in the north and the Shiite Muslims in the south from further persecution by the Iraqi government.

Iraq repeatedly interfered with the weapons inspectors and violated the no-fly zones. In retaliation for Iraqi strikes against the Kurds, the United States launched Operation Desert Strike, a one-day air campaign against Iraqi targets in September 1996. In October 1998, the US Congress passed a resolution, supported by President Bill Clinton, calling for regime change in Iraq. In retaliation for Iraq's repeated violations of UN Security Council resolutions, the United States and United Kingdom launched Operation Desert Fox, a four-day air campaign in December 1998.

Continued tensions between the United States and Iraq remained, and American public opinion sharply turned following the attacks by al-Qaida on the World Trade

Center and Pentagon on 11 September 2001. Although Iraq was not connected to the attacks, the George W. Bush administration sought to enforce the 1998 policy of regime change. This led to the controversial invasion of Iraq by the United States, United Kingdom, and Australia on 19 March 2003, after which the Iraq War proceeded quickly. Baghdad fell on 9 April and with the fall of Tikrit, Saddam Hussein's birthplace, on 14 April, major combat operations were over. On 1 May, President Bush gave a "mission accomplished" speech aboard the USS *Abraham Lincoln,* an aircraft carrier returning from the Persian Gulf. As in Afghanistan, however, the United States, the new Iraqi government, and allies faced a strong insurgency campaign in the years following the war. The United States withdrew its remaining forces in December 2011. Although the counterinsurgency campaign following the interstate war was extended, the interstate war itself was quite brief.

Scenario 2 involved a US maneuver strategy, Iraq attrition strategy, but with mixed terrain. At the time of their initial writing in 2003, Bennett and Stam considered this to be the most likely scenario, and it would have arisen if the invading forces had gotten bogged down in a lot of urban, house-to-house fighting, but not enough to lead to an abandonment of the maneuver strategy. In this case, the model predicts a war duration of 2.5 months. So simply getting bogged down in mixed terrain rather than open terrain more than doubles the expected duration of a war.

In scenario 3, the United States employs an attrition strategy, Iraq also uses attrition, and the fighting is in open terrain. In this case, the duration is 10.63 months. The impact of switching from a maneuver strategy to an attrition strategy is enormous, resulting in an expected duration more than twelve times as long. The United States has a long-standing doctrine focused on employing maneuver strategies, and keeping the fighting in open terrain would provide no incentive for the United States to deviate. Thus, this is the most unrealistic scenario of the four, but it is useful for comparison purposes.

Finally, scenario 4 involves the US attrition, Iraq attrition, and mixed terrain. Here, the predicted duration is 12.27 months. This is a much more realistic scenario than scenario 3 because getting bogged down in large-scale urban combat could possibly have pressured American military leaders to switch to an attrition strategy. The change in terrain type extends the war for less than two months more than the previous scenario. Thus, changing strategy has a much larger impact on the duration of the war than changing terrain.

Comparing the predictions with the observed duration, we see that Bennett and Stam's model performed well. Scenario 2, which they initially felt was most likely, overpredicts the duration by about a month. Scenario 1, however, was the actual scenario that played out in the war, and it predicts the war to end on 14 April, which I argue is the best date to consider as the end of the war. But even if we use one of the other possible end dates, Bennett and Stam's prediction was remarkably accurate.

In 2006, Bennett and Stam published the text of their original paper (from 2003) along with additional predictions including, in particular, an additional prediction for war in Iraq. They predicted the length of the second phase of the Iraq War, the insurgency and guerrilla warfare that lasted from 2003 to 2010. For their prediction, they considered the United States to be using an attrition strategy against the insurgents, who were using a punishment strategy, while fighting in mixed terrain. Their prediction resulted in an expected duration of 83 months, which is 6 years and 11 months (Bennett and Stam 2006). Given a start date in May 2003, this suggests that the war would extend until April 2010.

As with their prediction for the interstate war, we can assess how accurate their prediction for the length of the counterinsurgency was. Multinational Force-Iraq ended its patrolling duties and handed responsibility for security in Iraq over to the Iraqi government on 30 June 2009, and on 1 January 2010 the coalition was officially renamed the United States Force-Iraq. The combat mission of US forces in Iraq officially ended on 31 August 2010, and Operation Iraqi Freedom was succeeded by Operation New Dawn. This is probably the best date to use as the end of the American counterinsurgency in Iraq, although final withdrawal of all remaining American forces was not completed until December 2011. While not accurate to a specific month, this is still a good prediction.

Highly related to war duration is another important summary characteristic, war costs. Wars are costly endeavors in terms of lives lost, property destroyed, and money spent. War costs rise as war duration increases. For example, an average of twenty-four thousand soldiers die for each month a war continues (Bennett and Stam 1996). Unfortunately, there has been much less research focused on the costs of war. We will return to examine costs in our discussion of the consequences of war in the next chapter.

War Outcomes

We now turn our attention to explaining **war outcomes**. This is, of course, highly related to war duration because the outcome of a war is not realized until the war ends. But while in the previous section our focus was on explaining how long wars last, we now turn our attention to explaining what factors lead to the difference between victory and defeat in war.

Most scholars have focused on power to explain war outcomes. This represented the conventional wisdom and is implied by the relational definition of power that we discussed in chapter 5 as well as Stoll's (1995) framework presented in Figure 11.1. This traditional, power-based view is illustrated in Figure 11.2. The horizontal axis is State A's share of power within the dyad, while the vertical axis is the probability of victory. The solid line moving from the lower left to the upper right of the figure shows that A's likelihood of winning increases directly as A's relative power increases. The dashed line shows B's probability of victory. It moves directly opposite of the solid line, going from the upper left to the lower right of the figure. Thus, as A's relative power increases, the probability that B wins decreases because B's relative power goes down. The two lines intersect at 0.5, because when each state has exactly half of the power between them, they each have a 50 percent chance of victory.

Stam published *Win, Lose, or Draw,* which revolutionized the study of war outcomes in a variety of ways. In particular, Stam has changed the conventional wisdom expressed in Figure 11.2 that power determines war outcomes. Rather, he argues that we need to account for a variety of domestic and international factors in order to fully explain war outcomes. This is an important departure from realism, which argues that domestic factors do not matter. The importance of domestic factors, however, is highlighted by a number of different examples, such as the Vietnam War, where declining support on the American home front played an important role in the course of the war. Stam argues that we can integrate realist and domestic approaches to the study of war outcomes by focusing on rational choice.

Additionally, while people usually think about war outcomes in terms of either winning or losing, that view is incomplete. Rather than only two,

Figure 11.2 **The Traditional View of the Impact of Power on War Outcomes**

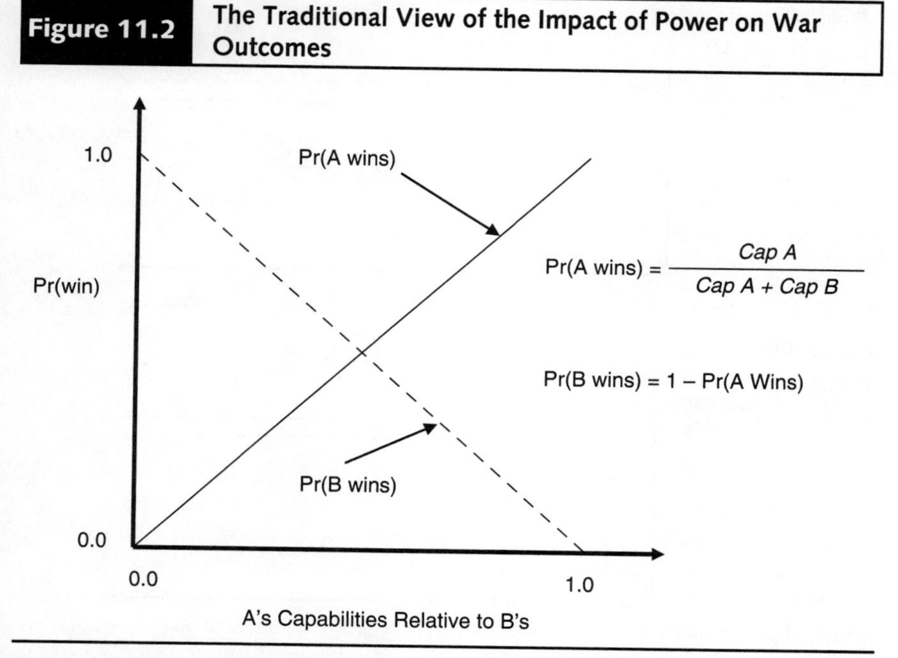

Source: Adapted from Allan C. Stam III, *Win, Lose, or Draw: Domestic Politics and the Crucible of War* (Ann Arbor: University of Michigan Press, 1996), 21.

Stam argues that there are three possible outcomes of a war: win, lose, or draw. For example, who won the Korean War? The United States entered the war seeking to preserve South Korean independence, which was achieved. On the other hand, China entered the war on the opposite side in order to preserve North Korean independence, and their goal was also achieved. At the same time, neither side succeeded in their aims to unify Korea under one ruler (North or South). To wit, neither side won: The Korean War is a classic example of a war that ended in a draw.

The existence of draws highlights an important limitation of the traditional explanation of war outcomes highlighted in Figure 11.2. A central assumption of the traditional view is that the probability that A wins plus the probability that B wins equals one. This allows no possibility for draws. But since we can observe that draws do occur, the probability of a draw in at least some wars is greater than zero. Therefore, the probability that someone wins is over estimated.

Finally, Stam argues that war outcomes are not just driven by power. Instead of simply focusing on power, Stam argues that to explain outcomes one must recognize that war is a form of coercion. This view of **war as coercion** is illustrated in Figure 11.3. The figure depicts the cost-benefit calculations inherent

Figure 11.3 War as Coercion

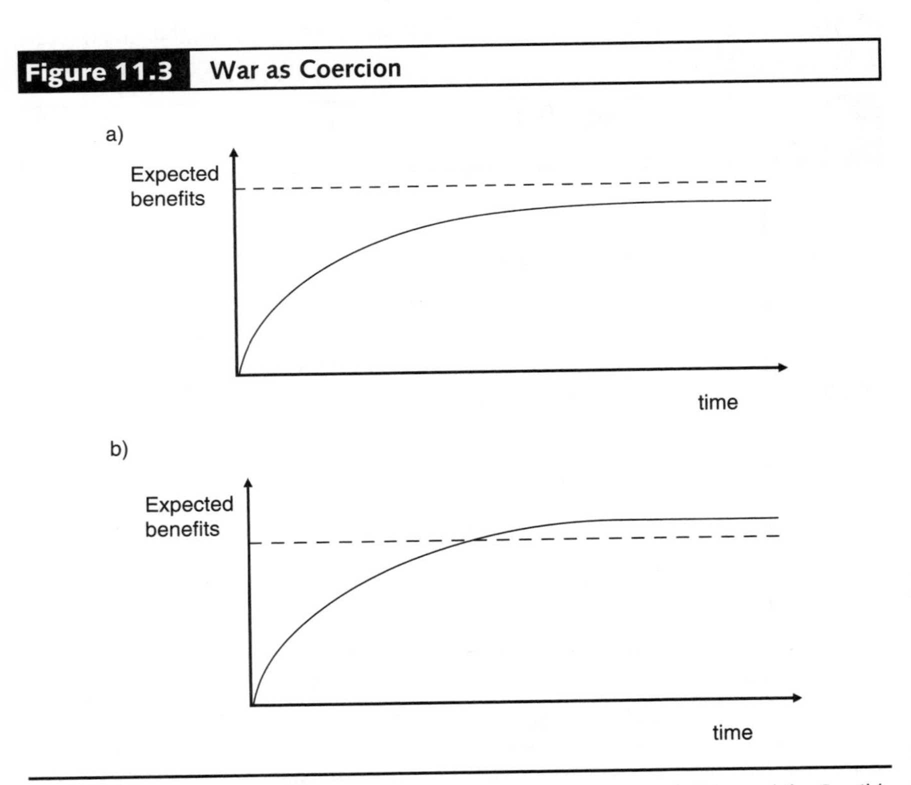

Source: Adapted from Allan C. Stam III, *Win, Lose, or Draw: Domestic Politics and the Crucible of War* (Ann Arbor: University of Michigan Press, 1996), 31–32.

in A's decision about whether to quit the war or continue fighting. The dashed horizontal line in the upper part of each panel is the level of benefits that A expects to receive from the war. These benefits may be in terms of territorial gains, a change in the opponent's regime, policy concessions by the opponent, and so on. The solid curve shows the costs that A has endured as the war continues in length.

In panel a at the top of Figure 11.3, state A's costs always stay below the expected benefits. Thus, A continues to expect a profit from the war, and thus is willing to continue fighting. In contrast, in panel b at the bottom of the figure, State A's costs eventually rise above the expected benefits. Thus, A no longer expects to profit from the war. At this point, A will seek to exit the war; State B has successfully coerced State A.

There are two sides in a war, so the view of war as coercion illustrated in Figure 11.3 is inherently limited. Although the previous discussion focused on B's attempt to coerce A, each side is trying to coerce the other. A more nuanced view recognizes **war as mutual coercion.** A is trying to coerce B at the same time that B is trying to coerce A. This view is illustrated in Figure 11.4. In this figure, A's cost-benefit balance is shown on the vertical axis while B's cost-benefit balance is shown on the horizontal axis.

Figure 11.4 War as Mutual Coercion

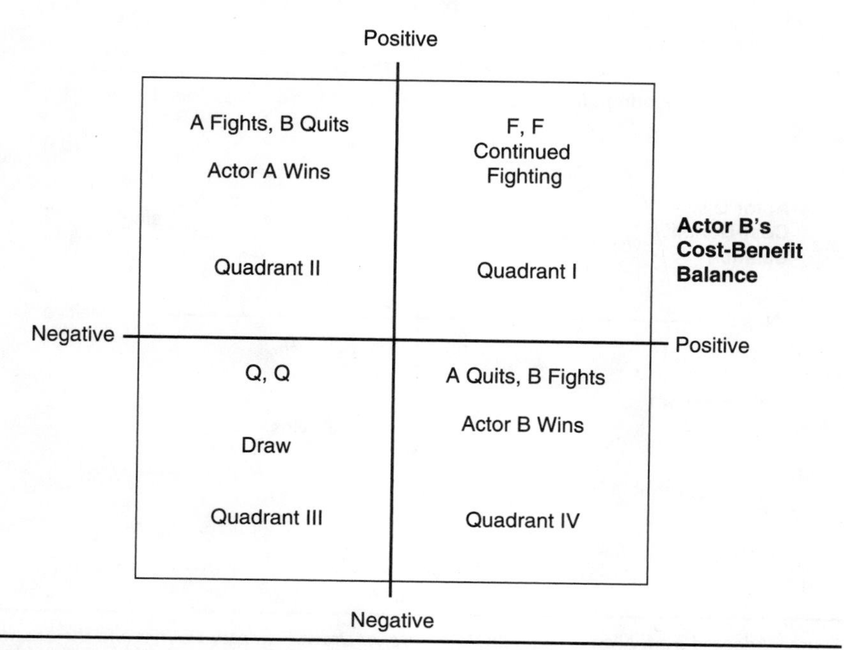

Source: Adapted from Allan C. Stam III, *Win, Lose, or Draw: Domestic Politics and the Crucible of War* (Ann Arbor: University of Michigan Press, 1996), 35.

In quadrant I, both sides have a positive cost-benefit balance and so each will continue fighting the war. In quadrant II, A has a positive cost-benefit balance while B's balance is negative. Thus, B will seek to quit the war and the outcome will be a victory for A. In quadrant III, both sides have a negative cost-benefit balance; both will seek to quit the war so the outcome will be a draw. Finally, in quadrant IV B's balance is positive while A's is negative. A will seek to quit and the war will end with a victory for B.

Figure 11.5 again illustrates the view that war is mutual coercion but uses a polar coordinate system. While in the Cartesian system (like Figure 11.4) the position of any point is indicated by x and y, in the polar coordinate system the position of a point is indicated by the radius r and the angle θ (theta). As θ increases, we move around through quadrant I, II, III, and IV in turn. Thus, we can express the movement between quadrants purely as a function of a single variable.

The reason to use the polar coordinate system is that it simplifies the presentation of probabilities. Figure 11.6 shows the probabilities of the different outcomes of war as we move from one quadrant to the other. The horizontal axis shows θ, the angle from Figure 11.5, in radians. Radians is a

Figure 11.5 A Second Look at War as Mutual Coercion

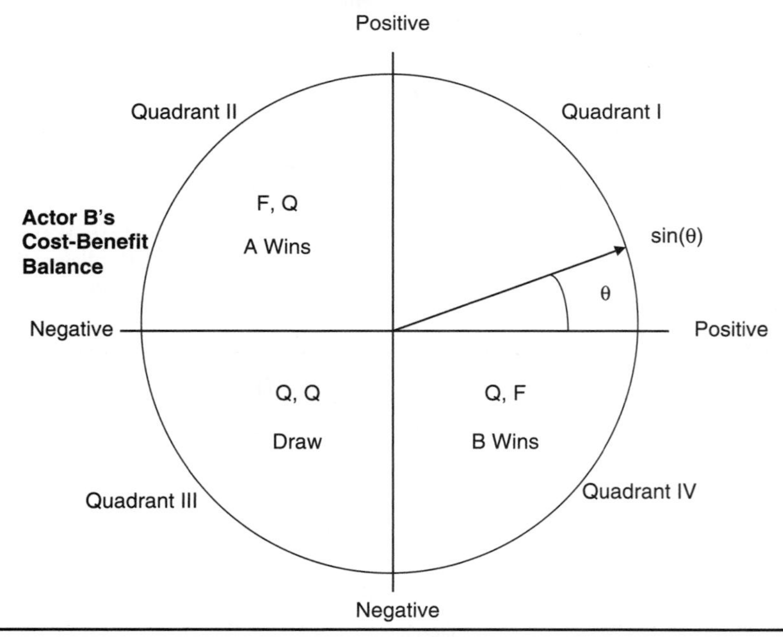

Source: Adapted from Allan C. Stam III, *Win, Lose, or Draw: Domestic Politics and the Crucible of War* (Ann Arbor: University of Michigan Press, 1996), 41.

way of measuring an angle in units of π, rather than degrees. To go around an entire circle is 2π radians, or 360°; half-way around is π radians, or 180°. Stam represents the probabilities of a win, lose, or draw as trigonometric functions of θ.

The insight that Stam's theory provides is the key to remember, rather than the mathematical details behind it. Figure 11.6 provides a much different view of the likelihood of war outcomes than the conventional wisdom represented in Figure 11.2. As θ increases from 0 past $\pi/2$ (from quadrant I into quadrant II), the probability that A wins increases and the probability that B wins goes down. As θ increases between $3\pi/4$ past π to $5\pi/4$ (from quadrant II into quadrant III), however, the probability that either A or B wins goes down simultaneously. Similarly, as θ increases between past $5\pi/4$ to $7\pi/4$ (from quadrant III into quadrant IV), the probability that either A or B wins increases simultaneously. According to the conventional wisdom, the probability that A wins and the probability that B wins must go in completely opposite directions, but once we account for the possibility of draws we see that sometimes they move in the same direction. The probability that B wins peaks in quadrant IV, beyond $7\pi/4$.

Choices Shaping Costs and Benefits

Stam (1996) argues that the key to explaining war outcomes, then, is identifying factors that shape the costs and benefits that shape states' decisions in war. To do so, he identifies factors that shape states' ability to inflict costs, as well as states' ability to absorb costs. In addition, Stam

Figure 11.6 **Probabilities of War Outcomes**

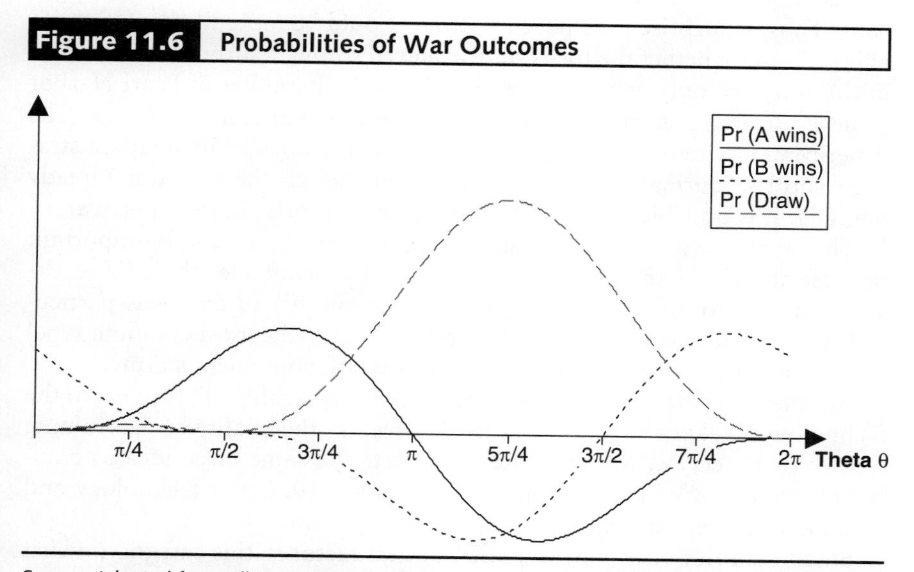

Source: Adapted from Allan C. Stam III, *Win, Lose, or Draw: Domestic Politics and the Crucible of War* (Ann Arbor: University of Michigan Press, 1996), 45.

differentiates these factors depending on whether they are choices that leaders make or constraints that leaders face, as well as distinguishing international, realpolitik factors from domestic ones.

He identifies six different choices that affect the costs and benefits of war participants. The first five of these are realpolitik factors, because they focus on international factors commonly associated with realism. The first realpolitik factor that he identifies is strategy and doctrine. We examined military strategy in chapter 10. Stam analyzed the impact of military strategy by using a typology of maneuver, attrition, and punishment. Coupling this with whether the states are on the strategic offensive or defensive results in eighteen strategy combinations. For example, OMDA (offense maneuver-defense attrition) indicates offense maneuver versus defense attrition; DPOA (defense punishment-offense attrition) indicates defense punishment versus offense attrition. Military strategy has an important effect on war outcomes because it affects the rate and location of costs. Strategy affects outcomes via the ability to raise opponents' costs.

Stam also accounts for contributions by allies. This is important in wars when more than one country fight together because Stam focuses on each country individually. For example, the Crimean War (1853–1856) involved Britain, France, Italy, and Turkey fighting against Russia. When we observe that Turkey fought against Russia and won, it is important to account for the role that its allies played. Understanding ally contribution is important because it disperses costs and it also reduces benefits. For states that fight without any allies, ally contribution is of course zero.

The next factor that Stam identifies is surprise. Surprising the enemy in war is important because it increases the ability to inflict costs on the opponent. Thus, if surprise is achieved, victory should be more likely. This variable indicates whether the focus state achieved strategic surprise at any point in the war, not only at the outset. The Japanese bombing of Pearl Harbor achieved strategic surprise at the outset of World War II in the Pacific. The American amphibious landings at Inchon in September 1950 achieved strategic surprise during the Korean War, even though the war was already ongoing. It is possible for both sides to achieve surprise in the same war.

The next factor Stam examines is mobilization, which is important because the larger the forces that the state has available, the greater the state's ability to inflict costs. Having a large number of troops is particularly important when states use attrition strategies (the most common type) because attrition involves trying to wear down the opponent's army.

Another important realpolitik factor is military quality. This refers to the technology level of the equipment used as well as the quality of the training that soldiers receive. Military quality reflects the same basic idea as combat effectiveness, which we examined in chapter 10. Better technology and training increases rate that costs accrue to opponent.

Stam identifies only one domestic political factor in this category, which is political repression. By political repression, we mean whether political opposition to the government is allowed within a state or whether it is silenced through censorship, imprisonment, or other means. Repression can have an important effect on war outcomes because it allows leaders to hide costs in an attempt to maintain legitimacy with the domestic population.

Resource Constraints Affecting Costs and Benefits

Ten resource constraints that increase or limit the costs of war are identified by Stam (1996). The first six of these are realpolitik factors. The first realpolitik resource constraint that he identifies is the balance of capabilities between the two sides. This is the sole factor that political scientists have traditionally focused on to explain war outcomes, as we examined earlier in the chapter. Relative power is important because it provides states with a greater capacity to absorb industrial losses.

Stam also accounts for relative military capabilities, which is simply calculated by using the military components (personnel and expenditures) of the composite indicator of national capabilities (CINC) score. An advantage in relative military capabilities should make a state more likely to win because it provides the country with a greater capacity to absorb military costs.

A final key constraint factor related to power is population. A larger population provides states with a greater capacity to absorb the costs of manpower losses. One example of the importance of population is found with Russia. For a long time, Russia has been a very large country with a massive population, which has helped it to overcome enormous losses and still emerge victorious against formidable enemies, such as in the French invasion of Russia in 1812 and the German invasion in 1941.

Distance is also an important constraint that states face. As we discussed in chapter 4, it is difficult to project power over long distances. This has a significant impact on war outcomes because as distance increases, it becomes harder to inflict costs on the opponent. Distance is less of an obstacle for stronger states, however, because they have more resources available to provide logistical support to their forces.

Terrain also poses a significant geographic obstacle. Dense terrains such as jungles, mountains, and urban areas are more difficult for armies to operate in, particularly with armored vehicles. On the other hand, open terrain such as open countryside and deserts makes it much easier for armies to maneuver and bring firepower to bear. Increasing the roughness of the terrain reduces the rate at which costs can be imposed on the opponent.

Stam also accounts for the interaction between strategy and terrain, which is important because different strategies work best on different types of terrain. For example, maneuver strategies are best suited to open terrain types, whereas punishment strategies generally depend on rough terrain in order to be effective. States using a strategy that is appropriate for the terrain in question should be more likely to win.

Four domestic political factors are identified by Stam as constraints that states might face. The first of these is democracy. Democracy is important because it increases legitimacy and willingness to incur costs. In addition, democracies tend to select low cost wars to fight. Reiter and Stam (2002) conduct a series of tests to determine why democracies are more likely to win wars. Others have argued that democracies are more likely to win either because they tend to have strong economies and are thus able to overwhelm opponents through massive production of weapons and equipment or because they tend to ally together and fight. Reiter and Stam, however, find that two different factors are the most important ones that make democracies more likely to win. First, armies of democracies tend to encourage soldiers at all levels of the chain of command to take individual initiative, and second, democracies are better at selecting winnable wars to fight.

Stam (1996) also accounts for the interaction between political repression and democracy. This is important because higher levels of political repression and democracy are both expected to make victory more likely, yet democracies tend to have lower levels of political repression than other states. Accordingly, an interaction term allows us to see whether the effect of repression depends on the democracy level of the country.

Issue salience is an important factor because it determines the height of the benefits line. States fighting over high-salience issues such as threats to the state's independence or territorial integrity are highly motivated to win. In contrast, for states fighting over low-salience issues such as disagreements about the opponents' policies, victory is less essential. Thus, wars involving high-salience issues should be more likely to lead to decisive outcomes, whereas wars involving low-salience issues should be more likely to lead to draws.

The final factor that Stam examines is time, meaning the length of the war. Time is important because costs to participants increase monotonically as a war continues. Cumulative costs always go up; they can never go down. Thus, the longer that a war lasts, the more likely it is that the costs will outweigh the benefits. This is true for both sides, of course. Longer war durations should make draws more likely, all else being equal.

Testing Stam's Explanation of War Outcomes

To empirically test his theory, Stam (1996) utilizes a quantitative analysis of interstate wars from 1816 to 1985. He splits large wars such as World War II into a series of smaller wars (Germany versus Poland, Germany versus France, etc.). Doing so allows him to more accurately observe outcomes between different pairs of states; otherwise, for example, it would appear that Germany invaded Poland in 1939 and lost, which is not historically accurate. Stam also devises measures for each of his variables, mostly based on data collected by the Correlates of War project and supplemented by data he gathered from historical sources.

Stam conducts a series of regression analyses to test his theory of war outcomes. He argues that the best model to use is a multinomial logit because draw is a separate category from win and lose rather than a midpoint between them. The results provide strong support for his theory. The model correctly predicts 84 of 88 cases in the sample, for a 95 percent accuracy rate.

In addition, there is a strong agreement between the predicted probability of outcomes and the observed proportion of those outcomes being realized.

To examine Stam's empirical results for particular variables, we begin by looking at the effects of decision makers' choices on war outcomes. Stam finds that strategy has a very large impact. Fighting with a OPDA strategy combination has a predicted probability of victory of about 95 percent. In contrast, OADM has a probability of victory of less than 10 percent. While other strategy combinations are not so decisive, it is clear that strategy has a very large substantive effect on war outcomes. In addition, maneuver and punishment strategies are found to be more effective than an attrition strategy, regardless of whether the state using them is attacking or defending.

As expected, greater levels of alliance contributions make winning or achieving a draw more likely, while making a state less likely to lose. However, ally quality—which captures the quality of allies' military forces—has little effect. Achieving strategic surprise during the war also has little effect on the outcome. Stam examines the effect of mobilization in two different ways and finds that although the relative number of troops has little effect, there is a strong monotonic effect of the absolute number of troops making wins more likely and losses less likely. Finally, military quality (which reflects the technology and training of a country's military) and repression both increase the likelihood of winning and decrease the likelihood of losing.

Turning to Stam's findings regarding the effects of resources and constraints on war outcomes, we start by looking at the effect of military-industrial capacity. Power is very important, and its effect is similar to the expectations of the conventional wisdom. A state with only 1 percent of the relative power in the dyad, however, has about a 20 percent chance of victory, while a state with 99 percent relative power has about an 80 percent chance of winning. Thus, while the conventional wisdom illustrated in Figure 11.2 correctly approximates the probability of winning when relative power is near 50 percent, it does poorly when there is an imbalance of power.

As population of the state increases, Stam finds that winning becomes less likely and draws become more likely. Farther distances make winning less likely and losing more likely on average. However, although the basic effect is the same for powerful states, it takes a much longer distance before the effect is realized. Terrain has the expected effect, as winning is less likely and achieving a draw is more likely in rougher terrain. Higher levels of democracy make wins and draws more likely and losses less likely. Moving to the next variable, we see that high stakes issues tend to have decisive outcomes of either a win or loss, while the three outcomes are nearly equally likely when low stakes issues are being fought over (with a draw the most likely). Finally, time has an important effect because as war duration increases, draws become more and more likely.

We can compare the importance of the different variables for explaining war outcomes in several different ways. We begin by examining the average historical effects for the key variables on the likelihood of winning versus losing, as shown in Figure 11.7. Since the conventional wisdom focuses on power, the effect of each factor is compared relative to the effect of military-industrial capabilities. The most important single factor is an offense maneuver-defense attrition strategy combination, which has an effect four times as strong on average than power. Strategy has a strong effect overall, with four of the six strongest variables being strategy combinations and one of the others being the interaction between strategy and terrain. Alliance contributions are also very important, having an effect more than three times as strong as power.

The other factors in Stam's model all have a weaker average historical effect on the difference between winning and losing than does power. Several strategy combinations, ally quality, repression, issue area, and democracy, all have observable yet relatively weak effects. The effects of distance, population, and mobilization, however, have almost no average historical effect on the likelihood of victory.

Figure 11.8 shows the average historical effects for the different variables on the likelihood of achieving a draw. Once again, strategy combinations have the strongest impacts, with the effect of defense punishment-offense attrition almost two and a half times as strong as military-industrial capabilities.

Figure 11.7 **Average Historical Effects of Various Factors on Winning Versus Losing**

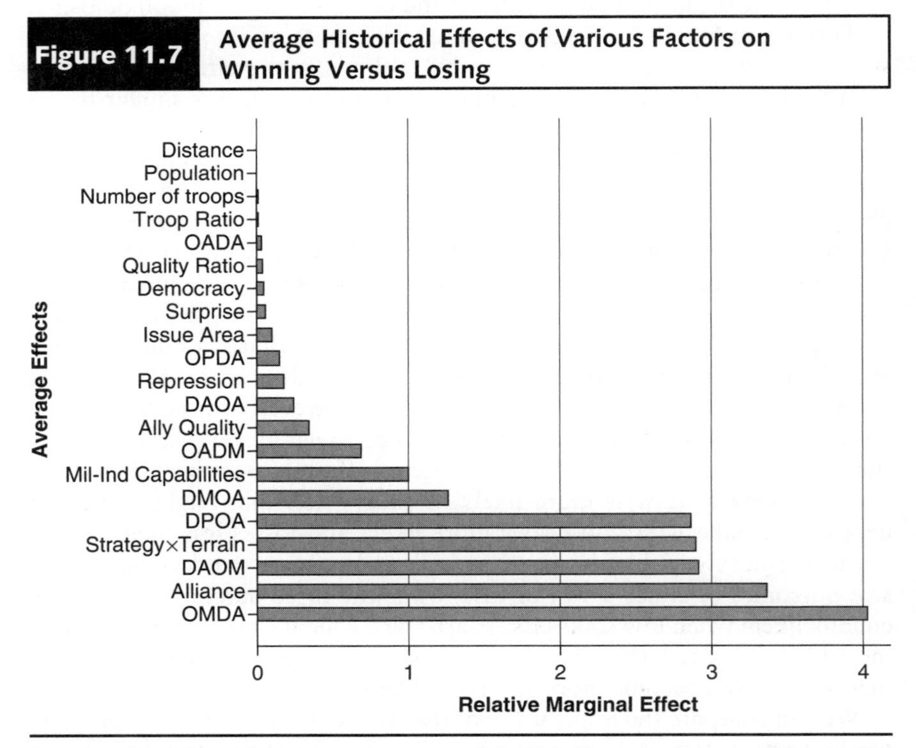

Source: Adapted from Allan C. Stam III, *Win, Lose, or Draw: Domestic Politics and the Crucible of War* (Ann Arbor: University of Michigan Press, 1996), 135.

Figure 11.8 **Average Historical Effects of Various Factors on the Probability of a Draw**

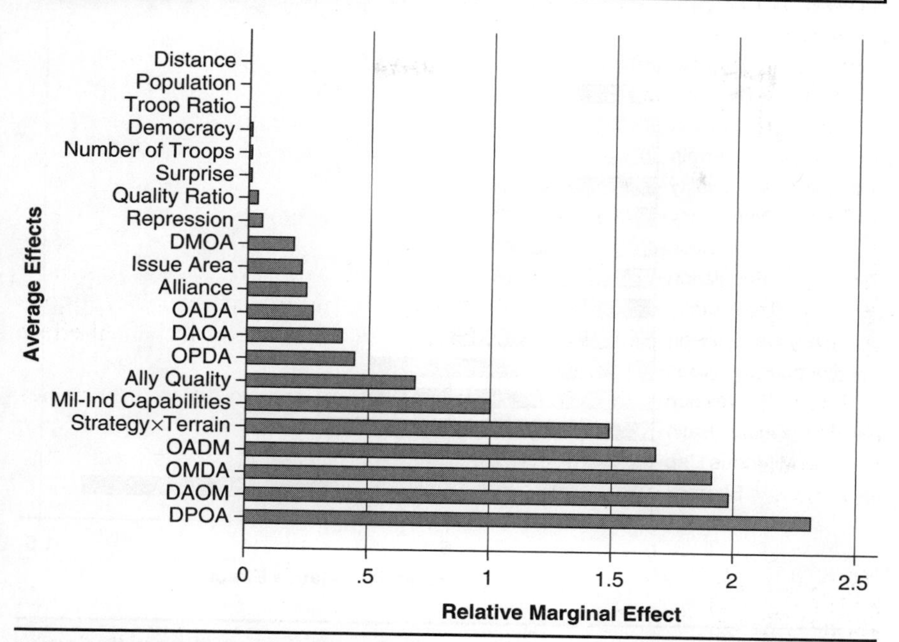

Source: Adapted from Allan C. Stam III, *Win, Lose, or Draw: Domestic Politics and the Crucible of War* (Ann Arbor: University of Michigan Press, 1996), 136.

The interaction between strategy and terrain again has a strong effect, and the effects of the remaining factors are similar in strength as before.

One limitation of focusing on average historical effects is that many variables can vary widely from one war to another but do not tend to vary much on average. For example, distance can vary widely as some states that have fought wars are thousands of miles apart. Most wars, however, occur between states that are either neighbors or very close together, as we saw in chapter 4. Similarly, while some states have extraordinarily large populations, most do not.

To address this limitation, we can look at the potential influence that the different variables can have on war outcomes. Figure 11.9 shows the potential influence of the different variables on the chances of winning. Once again, military strategy is the most important factor, with a potential effect nearly one and a half times as strong as power. The potential effects of quality ratio, repression, number of troops, and distance are all much larger than their historical averages, while surprise has the smallest potential effect.

We examine the potential influence of the variables on the likelihood of observing a draw outcome in Figure 11.10. For the first time, the effect of strategy is not the most important, although it is still more than twice

Figure 11.9 Potential Effects of Various Factors on Winning Versus Losing

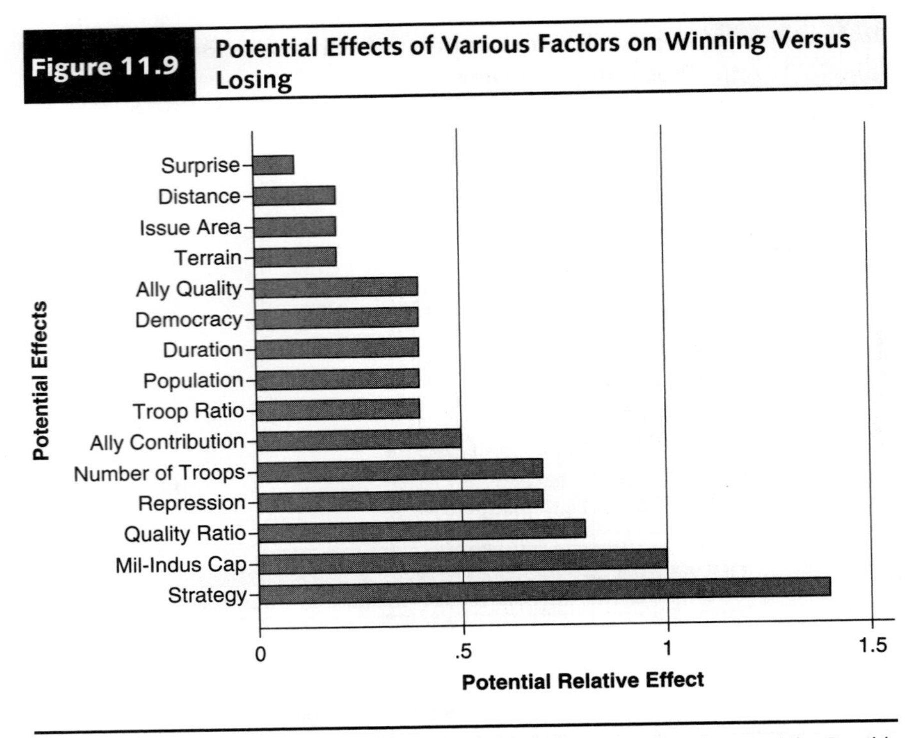

Source: Adapted from Allan C. Stam III, *Win, Lose, or Draw: Domestic Politics and the Crucible of War* (Ann Arbor: University of Michigan Press, 1996), 197.

as strong as the effect of power. Far and away the two variables with the strongest potential effects on the likelihood of observing a draw are population and duration, each of which is nearly five times as important as power. Military-industrial capabilities actually have the second-weakest potential effect of all the variables, with only distance mattering less.

In all, Stam's empirical results show that military strategy is the single most important factor for explaining war outcomes, as his theory suggests. In addition, a number of other factors, both domestic and international, matter for determining the difference between winning, drawing, and losing. This strongly contradicts the traditional power-based view.

Stam's model can be used for forecasting outcomes to wars that arise. These forecasts can shed some light on states' decisions to fight some wars but avoid fighting others. He applies this to two situations that the United States faced in the 1990s, where military intervention was debated. The first situation arose after Iraq invaded Kuwait in August 1990. Ultimately, of course, the United States chose to intervene in what became known as the Gulf War. For the United States facing Iraq in 1991, Stam's model forecasts that the probability of a win was 98 percent, the probability of a loss was 1 percent, and the probability of a draw was 1 percent.

Figure 11.10 **Potential Effects of Various Factors on the Probability of a Draw**

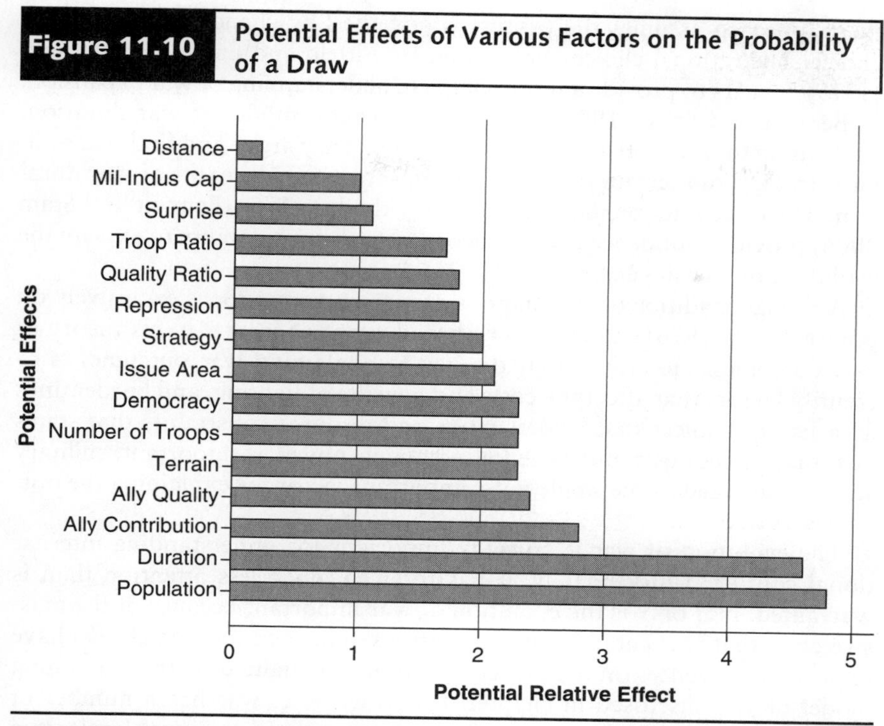

Source: Adapted from Allan C. Stam III, *Win, Lose, or Draw: Domestic Politics and the Crucible of War* (Ann Arbor: University of Michigan Press, 1996), 198.

The second situation that Stam forecasted was broad involvement by the United States using ground forces in the Balkans in the 1990s. The breakup of Yugoslavia led to several wars, and the United States had concerns about Serbian atrocities and sought ways to restore peace to the region. The United States brokered peace at the Dayton Peace Accords in 1995 and fought the Kosovo War against Serbia in 1999. However, the Kosovo War was entirely an aerial campaign. Stam forecasts that the US probability of victory was 0 percent, with the probability of a loss at 24 percent, and a 76 percent chance of a draw. Given these expected outcomes, US reluctance to get involved in a war involving ground forces seems well justified.

Conclusion

The evolution of war is an important subject that is often overlooked in studies of international conflict, and the framework developed by Stoll (1995) provides a useful road map for categorizing our thinking about it. Explaining war expansion is important for understanding why some wars grow to become very large while others remain quite small. Siverson and Starr (1991) explain war expansion by focusing on opportunity and willingness, although their approach assumes the process is governed by chance rather than by choice. The rational choice explanation by Altfeld and Bueno de Mesquita (1979) appears to provide a more complete understanding of war expansion.

Bennett and Stam (1996) developed a general model of war duration, accounting for a variety of different variables. They argue and find that military strategy and terrain are the most important determinants of war duration. Their accurate predictions regarding the Iraq War (Bennett and Stam 2006) provide confidence in their theory, as well as our ability to explain the evolution of war in general.

Although traditional explanations of war outcomes focus exclusively on power, Stam (1996) accounts for a broad range of factors in his theory of war outcomes. He argues that the key to explaining war outcomes is to identify factors that affect the costs and benefits of fighting, and he identifies a variety of choices that leaders make and resource constraints that states face that affect costs and benefits. Although power is important, military strategy emerged as the single most important factor for explaining the outcomes of war.

The evolution of war is crucially important for understanding international conflict. Unfortunately, it has received much less attention than is warranted. Not only is the evolution of war important because of the massive costs that war entails, but expectations about war outcomes also have important influences on the causes of war, as highlighted by the bargaining model of war discussed in chapter 3. Furthermore, war has a number of important consequences at both the domestic and international levels. It is to these consequences, as well as a further consideration of war termination, that we turn our attention in the next chapter.

Key Concepts

War as coercion

War as mutual coercion

War duration

War expansion

War outcome

Warring alliance partners

Warring border states

12 War Termination and Consequences

Diplomats and officers of Germany, their Central Powers allies, and Russia sign the Treaty of Brest-Litovsk, ratifying the exit of Russia from World War I. The origins of this treaty provide important lessons regarding war termination.

Source: Mondadori via Getty Images

The process of war identified by Bremer (1995) and discussed in chapter 1 provides the basic framework for our analysis of international conflict in this book. After examining various causes of conflict and the escalation of disputes to war, we focused on the duration and outcomes of war in the previous chapter. A basic assumption in both topics is that each war will end at some point. But what makes a war end? Why do some states, such as Germany in World War II, continue fighting until its capital is captured and much of the country is destroyed, while other countries, such as Yugoslavia in the Kosovo War, sue for peace before a single enemy soldier has set foot on its soil? Addressing questions such as these is the subject of war termination, the first topic that we examine in this chapter.

Although wars (and disputes) end at some point, the process of war does not stop once a militarized dispute or war ends. Rather, it is cyclical, feeding back into the underlying context, which itself is changed from what it was before the war. The extent and type of these changes are the consequences of war, which is our second main topic in this chapter. For example, World War I lasted from 1914 to 1918, embroiling all of the major powers and a variety of minor powers in combat in Europe, the Middle East, Africa, and the Pacific. The war had far-reaching consequences that shaped the twentieth century and continue to be felt today. The League of Nations, the first international organization of its kind, was formed in the war's aftermath. The Great Depression stemmed in part from the economic devastation brought on by the war, and the rise of Adolf Hitler and the origins of World War II can also be traced to World War I. Finally, military technologies such as the tank and aircraft, as well as doctrinal innovations such as modern-system force employment, were developed during the war.

This chapter begins with an exploration of how war termination can be explained, focusing on bargaining. We look at how war can resolve bargaining problems through the convergence of information and how regime type and commitment problems can affect bargaining and war termination. We then turn to consider the consequences of war, including both national-level and international-level consequences. We take closer looks at the impact of war on regime change within countries and the economic consequences of major wars.

Bargaining and War Termination

Studies of **war termination** seek to answer the basic question, why do wars end? This is closely related to the subjects of war duration and war outcome that we examined in the previous chapter. Yet they are conceptually distinct because, while war duration deals with how long a war lasts (regardless of who wins or how it ends) and war outcome deals with who wins (regardless of how long it lasts or how it ends), war termination focuses on understanding the process through which the war reaches an end (regardless of who wins or how long it lasts).

Explanations of war termination can be grouped into three basic types (Goemans 2000). The first type is focused on one-sided termination. The basic idea is that a state will continue fighting until it no longer sees a benefit from doing so, at which point it will give up. Thus, the war will continue until at least one side is ready to stop fighting; the side that gives up loses, and the side prepared to continue fighting is the victor. Under this explanation, war termination is caused by a state being ready to quit the war: The vanquished makes peace. This is the basic idea of war termination that Stam (1996) assumes in his study of war outcomes that we examined in chapter 11.

There is a key problem, however, with this explanation of war termination. It assumes that the issues that the states are fighting over in a war are fixed. This is illustrated in Figure 11.3 on page 266 in the previous chapter. The benefits line, which represents the amount of benefit that a state expects to gain in the war if its demands are met, is flat. But what if the demands made by one or both sides change during the war? In particular, why does the winner not raise his demands once he realizes his advantage?

This scenario occurred after the Bolshevik revolution in October 1917 during World War I, when the new communist leaders of Russia sought to exit the war immediately, wanting no part of what new Russian leader Vladimir Lenin felt was a capitalist war. Although the Bolsheviks believed that they could end their participation in the war by simply refusing to fight, Germany launched a renewed offensive on 16 February 1918. After the Russians announced that they would agree to German terms on 21 February, Germany raised its demands on 23 February. The Bolsheviks agreed to the new terms on 24 February, and the agreement was formalized in the Treaty of Brest-Litovsk on 3 March 1918, which led to large German territorial gains in the east (Goemans 2000; Wheeler-Bennett 1956).

A more realistic type of explanation focuses on two-sided termination. In this view, both sides must prefer peace over war in order for a war to end. The primary advocate of this view is Wittman (1979, 744), who argues that "an agreement (either explicit or implicit) to end a war cannot be reached unless the agreement makes both sides better off; for each country, the expected utility of continuing the war must be less than the expected utility of the settlement."

Unfortunately, this argument is incomplete. What makes states fighting a war prefer peace over war? And if they prefer peace, then why did they start fighting a war? As Goemans (2000, 9) states, this argument "provides no mechanism that brings such an agreement in reach for both sides and fails to explain why a bargaining space opens up." Additionally, this explanation also ignores the question of whether the war termination agreement can be enforced following the war.

Bargaining and Information Convergence

The third type of explanation of war termination focuses on bargaining. The bargaining model of war has been a prevalent part of the international conflict literature, particularly since Fearon (1995) used a bargaining model to examine rationalist explanations of the causes of war. Bargaining models assume that war results from a breakdown of bargaining. As we examined in chapter 3, Fearon (1995) argued that bargaining failures can arise from three sources: incomplete information with incentives to misrepresent, commitment problems, and indivisible issues.

Commitment problems arise when states cannot credibly promise to cooperate with agreements they reach. At least one side has an incentive to renege on its promise once the other side fulfills its part of the deal. Indivisible issues exist when circumstance dictates that the winner gets everything and the loser gets nothing. Fearon (1995) argues that few, if any, issues are truly indivisible. Even Kashmir, at the heart of Indo-Pakistani conflict, and Jerusalem, central in conflict between Israel and Palestinians, can be and have been divided. Thus, indivisible issues are not an important cause of war, nor are they a key factor in war termination.

Fearon (1995) argues that the most important source of bargaining failure is incomplete information with incentives to misrepresent. Incomplete information can come from several sources. States can be uncertain about each other's willingness to fight because they disagree about the probability of victory or because they disagree about the costs of fighting.

In addition to helping explain the outbreak of conflict, bargaining models are useful explaining war outcomes, duration, and termination (Filson and Werner 2002). The bargaining explanation of war termination is that wars can end only when the minimum terms of settlement of both sides become compatible (Goemans 2000). Thus, when both sides ask for no more than the other side is willing to give up, war termination is possible. Figure 12.1 illustrates war termination from a bargaining perspective. Panel a, at the top of the figure, shows the basic situation where neither side has made any changes to their demands from the start of the war. State A demands a settlement between d_A and 1, while State B only finds settlements between 0 and d_B acceptable. Since $d_A > d_B$, there is no bargaining range in which the states are able to agree on ending the war. In panel b, the situation has changed because State A has lowered its minimal demands. A's ideal point is still the same and it will still only accept settlements between d_A and 1. However, d_A is now lower than d_B, creating a bargaining range in which both sides find the settlement acceptable.

Finally, panel c at the bottom of Figure 12.1 shows the situation in which State B has lowered its minimal demands while State A's has remained the same. However, d_B is now higher than d_A, creating a bargaining range in which both sides find the settlement acceptable. Of course, these are not the only three possible scenarios. Both sides may lower their war aims (by the same amount or different amounts), one side may increase its demands while the other decreases its demands, or both sides may increase their demands.

While there are multiple possibilities, the only way to open up the bargaining space required for war termination is for at least one side to change its war aims. Therefore, the fundamental cause of war termination is a change in the minimal demands of the combatants (Goemans 2000;

Figure 12.1 Changing Demands and War Termination

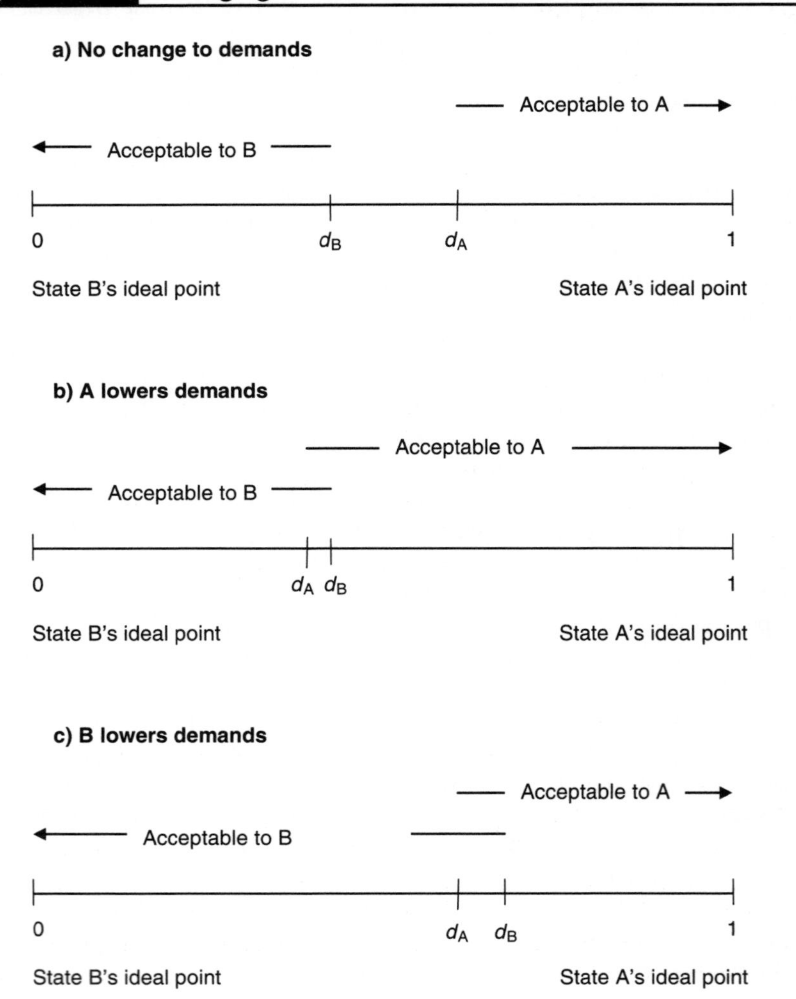

Slantchev 2003; Werner 1998). But what causes these changes in the terms of settlement?

The most basic mechanism is that states raise and lower their demands as a function of new information about the outcome and the costs of war (Slantchev 2003; Werner 1998). The basic logic is that war terminates when the information gap between the two sides closes enough to create a bargaining range (Filson and Werner 2002, 2004; Powell 2004; Slantchev 2003; Smith 1998; Smith and Stam 2004; Wagner 2000). Slantchev (2003) labels this the **principle of convergence.**

Most models assume that the information leading to convergence is revealed through outcomes of battles during the course of the war (e.g., Smith and Stam 2004). Others (e.g., Slantchev 2003) also focus on information revealed by competing offers during intrawar negotiations. In any case, Slantchev (2003, 629) argues that "war results in a relatively quick disclosure of information."

The basic logic of how states adjust their war aims in response to information from the battlefield is provided by the Korean War, which began on 25 June 1950 when North Korea invaded South Korea. The United States intervened on 30 June, followed by other countries fighting under the authority of the United Nations. The initial US demand was a restoration of the prewar border along the 38th parallel. Even this appeared to be an ambitious goal, as North Korean troops continued their advance deep into South Korea, occupying nearly the entire peninsula. By 3 August, the Pusan perimeter was established, halting the North Korean advance. On 15 September, the United States took the offensive with an amphibious landing at Inchon, a port city on the western coast of Korea near Seoul (see Map 12.1).

This took the North Koreans by surprise and completely altered the situation. In particular, things were now looking much better for the United States, and its war aims expanded to include the unification of Korea under South Korean rule (Reiter 2009). US forces began moving north across the 38th parallel on 30 September and captured Pyongyang, the North Korean capital, on 19 October. This move north led China to intervene in the war, and first contact between US and Chinese forces occurred a few days later, on 25 October. The US and South Korean forces were pushed back south again, until the Chinese advance was finally stopped in January 1951. The front line stabilized near the original border by June 1951, and the United States once again lowered its demands to South Korean independence. The front line remained largely stationary until an armistice was finally reached in July 1953, ending the war. As this overview makes clear, the United States adjusted its demands based on changing events over the course of the war.

Ramsay (2008) conducts a quantitative analysis of 379 battles in the twentieth century to test the effect of battlefield events on war termination. He finds only limited support for the idea that information convergence leads to war termination. In particular, "while incomplete information may explain early experiences in war, long conflicts have different explanations"

Map 12.1 Korean War, 1950–1953

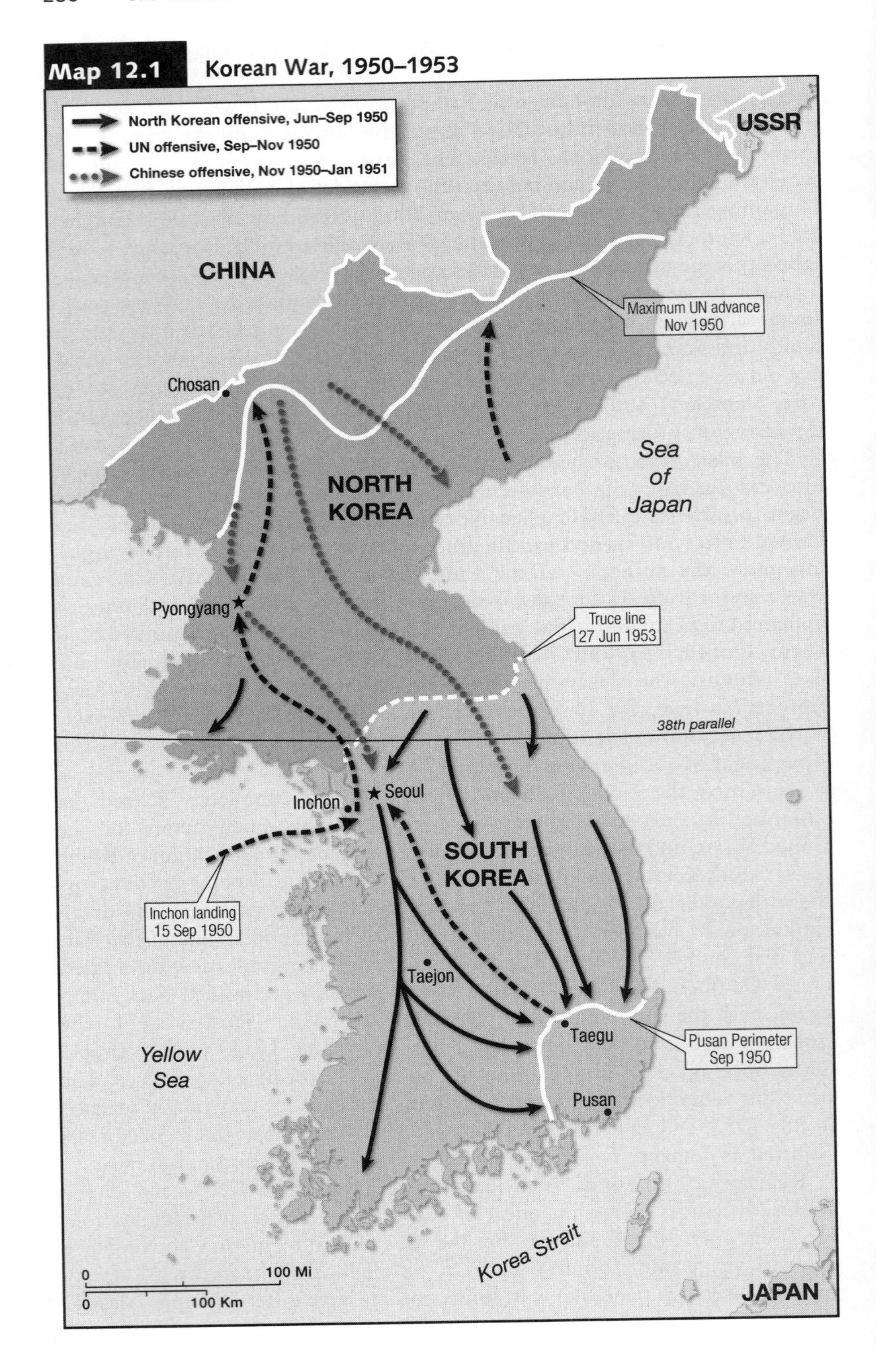

(Ramsay 2008, 872). Fortunately, scholars have established two primary alternative explanations, to which we now turn.

Regime Type and War Termination

One reason for Ramsay's (2008) findings is that some states refuse to lower their war aims—they even sometimes increase their war aims—in response to new information during the war, even when they learn that they will probably lose. Goemans (2000) argues that this is driven in large part by regime type, which plays an important role in determining how states end wars. In particular, he argues that different types of regimes evaluate the expected utility of potential settlements to war differently.

Goemans (2000) classifies regimes according to two criteria, the level of repression and the level of exclusion. Repression refers to the use of measures such as restrictions on the freedoms of speech and assembly, police brutality, and extrajudicial punishment of dissenters by the government in order to stay in power. By the level of exclusion, he means "the proportion of the (productive) population excluded from access to the policy-making process" (Goemans 2000, 37–38). These two variables are closely related, but conceptually distinct.

The first category of regime that Goemans (2000) examines is repressive and exclusionary regimes. These are dictatorships or totalitarian regimes, such as Nazi Germany or the Soviet Union. Because of the high levels of repression and exclusion, the costs for opposition groups to overthrow the regime are very high. For example, Kershaw (2011) shows that, even in the final days of World War II, most Germans were unwilling and unable to oppose Hitler's Nazi regime because of the intense level of repression. Therefore, as long as the war ends in a moderate loss or better, overthrow of the government is unlikely. Only when the war ends in a disastrous loss will overthrow be likely. If the leader is successfully removed, however, the leader is very likely to suffer additional severe punishment such as exile, imprisonment, or death.

The second category that Goemans (2000) uses is semirepressive and moderately exclusionary regimes. These are mixed regimes, which have some characteristics of democracy but also some characteristics of dictatorship. For example, Germany prior to and during World War I had both the Reichstag, an elected parliament with multiple parties and limited power, and also the kaiser, the unelected chief executive who determined foreign policy and was not answerable to the Reichstag. Despite universal male suffrage, the lower classes were effectively excluded from the policy-making process through a variety of provisions (Goemans 2000, 76–81). Because of the lower levels of repression, it is easier for opposition to coordinate attempts to overthrow the regime. Therefore, small losses in war are sufficient for the leaders' loss of power. In addition, leaders are likely to suffer additional severe punishment upon the loss of power.

Goemans's (2000) final category of regime type is nonrepressive and nonexclusionary regimes. These regimes are democracies, such as the United States, Britain, and many other states in the modern world. Because of the lack of political repression in democracies, opposition to the government is cheap. Leaders are expected to lose power following any loss in war. If they do lose power, however, they are unlikely to be punished unless the war results in a catastrophic defeat for the country.

Because of these differences, Goemans (2000) argues that the mixed regimes behave differently in war than other regime types. Leaders of mixed regimes face the worst of both worlds: They are likely to lose power following any war loss (like leaders of democracies) and they are likely to face additional punishment once they lose power (like dictators). Thus, they need to make sure to avoid any loss, whether large or small, because winning is the only thing that will keep them in office. Given this, Goemans (2000) argues that, rather than lowering their demands in response to negative information, leaders of mixed regimes are likely to increase their war aims when things start to go badly in order to show a profit from the war. This can also include switching to a more risky strategy that brings a higher probability of victory that also increases the probability of a disastrous loss, a phenomenon known as **gambling for resurrection** (Downs and Rocke 1994). If the leader recognizes that the war is not going well, then he or she "risks little in the way of further punishment by gambling and may gain the right to stay in office if the gamble is successful" (Downs and Rocke 1994, 375).

To test these expectations, Goemans (2000) conducts a detailed case study of war termination in World War I, examining how Germany, Russia, France, and Great Britain changed their war aims in response to new information. Britain and France (democracies) and Russia (a repressive and exclusionary regime) all lowered their war aims in response to unfavorable new information regarding the likely outcome and costs of the war. For example, with the new German unrestricted submarine warfare that brought heavy losses to British shipping, the continued costly stalemate and unsuccessful offensives on the western front, and the Russian army's collapse, 1917 was a year of mostly negative new information for Great Britain. In response, the British lowered their war aims, in particular by removing the demands for regime change in Germany and the German evacuation of invaded Russian territory (Goemans 2000, chap. 7).

In contrast, Germany—a mixed regime—kept raising its demands in response to negative information about the course of the war. For example, 1916 was a year full of unfavorable happenings for Germany: Its main effort of the year—the offensive at Verdun—failed, and along with the British offensive at the Battle of the Somme, the Germans suffered massive casualties. In addition, Britain implemented conscription for the first time, greatly expanding the size of its army. Despite all of these negative signs for the likelihood of victory and the costs of the war, Germany raised its demands, including the independence of Poland from Russian rule (but under German domination) and a greatly increased monthly tribute from Belgium (Goemans 2000, chap. 4).

Commitment Problems and War Termination

The process of information convergence discussed previously provides the foundation of bargaining explanations of war termination, but it is not able to completely explain war termination. Sometimes states push on for a decisive victory rather than agreeing to a negotiated settlement. As we just examined, Goemans (2000) argues that regime type provides an important explanation of this phenomenon. Others argue that this failure of states to adjust their war aims in response to updated information revealed in the war stems from commitment problems. Reiter (2009) shows that concerns about commitment can lead states to pursue absolute victory. If a belligerent doubts its opponent's ability to credibly commit to a negotiated settlement, a state will push on to pursue decisive victory. This can prove to be effective, as imposed settlements (Quackenbush and Venteicher 2008; Senese and Quackenbush 2003) and foreign imposed regime changes (Lo, Hashimoto, and Reiter 2008) significantly increase durations of peace following conflict.

Walter (1997, 2002) shows that commitment problems play a critical role in the termination of civil wars. Governments and rebel groups in civil wars typically fear that the other will violate an agreement to end the war. For example, if the agreement calls for the rebel group to lay down its arms in exchange for concessions by the government, what is to stop the government from reneging on its concessions once the rebel group is disarmed? These severe commitment problems in civil war explain why civil wars typically last so long and why third-party intervention is important for reaching a peace agreement.

To test these expectations about the impact of commitment problems and information on war termination, Reiter (2009) uses case studies to examine twenty-two war termination decisions across six wars. He finds that when states have commitment credibility concerns and hope for ultimate victory, they make higher war termination demands even in the face of discouraging information. For example, Britain was willing to continue fighting in 1940 despite the fall of France and numerous other setbacks since the start of World War II because the British leadership did not trust Hitler to honor any negotiated settlement and they maintained hope that the United States would enter the war, providing a path to eventual victory (Reiter 2009).

If states face discouraging information from the battlefield coupled with either almost no hope for ultimate victory or prohibitively high costs of continued warfare, however, then they lower their war aims even if they have commitment concerns. The first situation characterizes Finnish decision making in the Winter War (1939–1940). Although initially successful in repelling Soviet attacks, Finland recognized that there was little hope for eventual victory and therefore made concessions to end the war once the Soviets began to gain the upper hand. American war termination decisions in the Korean War following Chinese entry into the war were driven by concerns about the war's high cost, even though the United States was still confident that it could win the war if it had been willing to pay the costs (Reiter 2009).

A few cases fit the basic information argument very well, where states responded to encouraging information from the battlefield by increasing their war termination demands and to discouraging information by lowering their demands. Soviet decision making in the Winter War follows this pattern, as its war aims lowered and raised in response to the flow of information not only from battles, but also the likelihood of intervention by Britain and France (Reiter 2009).

Finally, Reiter (2009) finds that domestic politics has only a very limited role in war termination decision making. Rather, he argues that commitment problems better explain states' refusal to lower demands in response to negative information during war than regime type, calling into question Goemans's (2000) argument. He examines German war termination behavior in World War I and concludes that "credible commitment concerns played an important role in Germany's decision to fight on, certainly more than domestic politics" (Reiter 2009, 184). In addition, he argues that Japan was also a moderately repressive, semiexclusionary regime during World War II. Thus, Goemans's (2000) theory would lead us to expect Japan to respond to battlefield setbacks—which were steady from the Battle of Midway onward—by increasing its war aims or otherwise gambling for resurrection. Instead, "Japan did not increase and if anything decreased its de facto war aims as its military fortunes declined" (Reiter 2009, 199).

In conclusion, bargaining models of war provide strong explanations of war termination. Although information convergence provides an important part of the bargaining dynamic (Slantchev 2003), it is not able to fully explain war termination. Rather, domestic politics (Goemans 2000) and especially commitment problems (Reiter 2009) provide a much more complete explanation, showing why states do not always adjust their war demands in response to information from the battlefield.

Consequences of War

How does war impact the future of international relations? Answering this and related questions requires understanding the consequences of war. By **war consequences**, we mean the ways that a war changes the underlying context in the future relations between states.

There are a number of difficulties in assessing the impact of wars and disputes. As Thompson (1995, 161) points out, there are many reasons for these difficulties: "One is that any given war will have several kinds of effects. The consequences of war may be permanent or temporary. They may be manifested in the short or long term. They may be direct or indirect. They can also be positive or negative."

Additionally, the consequences of a war could be different for different groups, such as participants versus nonparticipants or winners versus losers. These multiple types of effects of war create serious obstacles to study of the consequences of war. How long after a war should one look to find effects? What groups should one focus on? The difficulties in answering these and other questions form part of the reason why the consequences of war have been understudied.

Certainly, some wars have larger consequences than others. World War II, discussed in Box 12.1, had probably the greatest and farthest reaching consequences of any war in history. But it was also the largest war in history. While larger wars tend to have greater consequences, sometimes small wars can have large consequences as well. For example, Bueno de Mesquita (1990b, 28) shows that the Seven Weeks War in 1866 had tremendous consequences, particularly because it "fundamentally changed the international order by providing the foundation for German hegemony on the European continent." This happened despite the small size and short length of the war. He concludes that "great consequences may result from small causes" (Bueno de Mesquita 1990b, 50).

Box 12.1

Case in Point: World War II

World War II, the largest war in history, began on 1 September 1939 when Germany invaded Poland. Britain and France had had enough with Hitler's aggressiveness and declared war two days later. After quickly defeating Poland, Germany turned its attention west. On 9 April 1940, it invaded Denmark and Norway, and one month later, on 10 May 1940, Germany attacked the Netherlands, Belgium, Luxembourg, and France. Italy joined the war as Germany's ally in June 1940. Germany attempted to subdue the United Kingdom with an air campaign in preparation for a possible invasion, but in the ensuing Battle of Britain in the fall of 1940, Britain held on as the only state opposing Germany.

With its western front mostly secure, Germany invaded the Soviet Union on 22 June 1941. The geographic scale of the war expanded greatly when Japan (Germany's ally) attacked the US naval base at Pearl Harbor on 7 December 1941. Japan swept through the south Pacific, taking the Philippines, Malaya, Singapore, the Dutch East Indies, Thailand, Burma, and other locations in a massively successful series of campaigns over the next several months. The United States handed Japan its first defeat at the Battle of Midway in June 1942, sinking four Japanese aircraft carriers. The United States began its campaign to push back Japan by invading the island of Guadalcanal in August 1942.

The Soviets defeated German forces at the Battle of Stalingrad, which lasted from September 1942 to January 1943. Stalingrad is generally considered to be the turning point of World War II, as the Allies gradually pushed German forces back on all sides for the remainder of the war. The western Allies invaded Sicily in July 1943 and then mainland Italy in September 1943. Italy surrendered, but the Allied advance in Italy was slow and arduous. After difficult battles at Anzio and Monte Cassino, the Americans finally captured Rome on 4 June 1944.

Attention had already shifted to France, however, as the Allies invaded Normandy on 6 June 1944, forever known as D-Day. At the same time, the Soviets continued pushing in from the east. Germany launched one final offensive in December 1944 before finally surrendering on 8 May 1945.

In the Pacific, the United States conducted an island hopping campaign, attacking only crucial locations. Battles in the Philippines, Iwo Jima, and Okinawa showed that defeating Japan, while perhaps inevitable, would be incredibly costly. In an

(Continued)

(Continued)

attempt to avoid having to invade Japan itself, the United States dropped atomic bombs on Hiroshima and Nagasaki in August 1945, which led directly to the Japanese surrender.

World War II was the largest war in history, involving over 100 million people serving in military forces from more than thirty countries. Battles and other military operations took place on every continent (other than Antarctica) and ocean in the world. The war resulted in the deaths of over 60 million people (around 24 million military and around 49 million civilians). Given its massive scale, geographic scope, and importance, World War II had probably the greatest and farthest reaching consequences of any war in history.

Thompson (1995) develops a framework to make sense of the consequences of war. In particular, he groups studies of war consequences into two types: those focusing on national-level consequences, and those focusing on international-level consequences. The first category focuses on the various effects of war within countries, while the second category deals with the effects of war on relations between countries.

National-Level Consequences

We begin our more detailed assessment of the consequence of war by looking at national-level consequences. These consequences can be grouped into three primary areas: economic conditions, demographic factors, and sociopolitical conditions (Thompson 1995). War can have a variety of effects on economic conditions within a country. Most obviously, the destruction of land (environmental damage to farm and other land), labor (soldiers and civilians killed and disabled), and capital (infrastructure such as buildings, bridges, and machinery) that happens during war can lead to widespread declines in economic productivity either permanently or temporarily. Other economic consequences for countries in war are less direct. For example, serious mobilization in a small economy such as Israel can be very damaging to the economy if the mobilization lasts for long, since a large percentage of the civilian workforce is mobilized to fight in the war.

One positive effect that war can have on economic conditions within a country is through the acceleration of technological innovation. Given their vast scale and long length, both World War I (Hartcup 1988) and World War II (Hartcup 2000) led to a wide variety of technological innovations. Although preliminary research had already begun before the war, World War II led to a great acceleration of the development of nuclear energy (Rhodes 1986). In addition, electronic computers were invented during World War II, being used most notably for code breaking and development of the atomic bomb. The war also led to great advances in medical technology, particularly through the widespread use of penicillin. The German V-2 rocket developed during the war is not only an early version of modern ballistic missiles, it also paved the way to missions to the moon and elsewhere in space. While war is certainly not necessary for technological innovation, it can certainly provide impetus for innovation to occur more rapidly.

Several scholars have conducted broad, quantitative studies of the impact of war on economic growth within countries. Wheeler (1975, 1980) examined the consequences of war for industrial growth in sixty cases from 1816 to 1965 and found that the effect of war was significant nearly 75 percent of the time. The direction of change, however, varies greatly. In some cases, war has a positive effect on growth, others have a negative effect, and some have only a temporary impact of war on industrial growth. Rasler and Thompson (1985, 1989) focused on the economic growth of major powers from 1700 to 1980. They also find that less than three-quarters of wars exert significant effects on economic growth, with those split nearly evenly between positive and negative impacts. More recently, Koubi (2005) examined the consequences of both interstate and intrastate wars in a broad sample of countries from 1960 to 1989. She found that war has a consistently positive impact on postwar economic performance, which grows even stronger following wars of greater duration and severity.

The second category of domestic consequences of war relates to demographic factors. Most obviously, wars can lead to an increase in mortality rates, particularly of young adult males. The increase in mortality rates is not necessarily confined to military personnel but can also come from civilian casualties as the result of bombings and battles and from increases in premature death due to deteriorating wartime environments. Wars, particularly if they are lengthy, also tend to lead to declines in marriage rates and childbirth, although there can also be a baby boom after the war as families are reunited. All of these factors, driven by the number of people mobilized for the war (and thus not at home in their normal peacetime settings), can cause great strains on family relationships. Between the increased mortality rates and decreased birth rates, war can lead to major shifts in the age and sex structure of populations following a war. These war-induced demographic changes are stronger the larger the war (particularly in terms of number of military personnel mobilized and number of casualties) and the extent to which the home-front quality of life deteriorates, and are smaller the greater the distance between the country and the combat zone (Thompson 1995).

War can also have a large impact on social change within societies (Marwick 1974). One undoubtedly positive consequence of war has related to women's rights. Since war fighting has been a historically male activity, war has led to a variety of changes in gender roles and expectations in society. When large numbers of men went off to fight during the world wars, many women started working outside the home—in factories and other positions previously held by men—for the first time. This impact of women on the war effort during World War I contributed to women's right to vote finally being granted in the United States in 1920. World War II saw even more women entering the work force in the United States and elsewhere. This gave them greater power and voice in society and contributed to a greater level of gender equality.

The third broad category of national-level consequences is changes in the sociopolitical conditions of a country. In particular, war plays a crucial role in state formation and expansion (Rasler and Thompson 1989; Tilly 1975). Indeed, it was crucial for the development of the modern nation-state system as we know it. "War is an important factor in the origin of states and their subsequent expansions in territorial and functional terms" (Thompson 1995, 168). War can lead to permanent raises in state expenditure levels (Peacock and Wiseman 1961), although permanent increases are much more likely if the war is global. Rasler and Thompson (1989) conduct a broad empirical analysis of the impact of war on state expenditures and find that its effect is mostly temporary, although the effect of global wars is abrupt and permanent.

Interstate war can also impact the likelihood of intrastate conflict within a country. States usually attempt to extract more resources (such as conscription, taxes, food, and labor) from the population during war in order to pay for its high costs. Tilly (1978) argues that this can lead to internal conflict if segments of the population resist this extraction of resources and the government reacts with force. Although the details of their arguments vary, similar ideas abound in the literature (Gurr 1988; Stein 1980). Rasler (1986), however, shows that regimes that accommodate demands of various domestic groups during warfare are able to avoid internal conflict.

Even without leading to a civil war, war has important consequences for the survival of political leaders in office. In particular, war defeats increase the probability of regime changes (Bueno de Mesquita, Siverson, and Woller 1992; Chiozza and Goemans 2011). We examine the literature focusing on the relationship between war and regime change in the next section.

Finally, war can have important impacts on societal learning patterns. War participation is costly in terms of lives lost and wealth expended. Additionally, one would expect that greater war costs would create greater reluctance on the part of leaders and societies to repeat the experience. This leads to the idea of war weariness, where countries are expected to be less likely to fight again soon after a war (Farrar 1977; Levy and Morgan 1986). Because societal memories tend to fade with time, war weariness is expected to only have a temporary effect. However, there is little to no empirical support for the war weariness hypothesis (Levy and Morgan 1986; Stoll 1984).

War and Regime Change

A number of studies have examined the impact of war on the likelihood of regime change within a country. The existence of such a connection between war and regime change provided the foundation for Goemans's (2000) argument about war termination that we examined previously. One of the first quantitative studies of the effect of war on regime change was by Bueno de Mesquita, Siverson, and Woller (1992). They looked at all interstate wars from 1816 to 1975 to determine what factors affect the likelihood of violent regime change, such as through revolutions and coups d'état. Their main empirical results are shown in Table 12.1. Winning a war—for either the initiator or the target—greatly decreases the likelihood of violent regime change compared to the initiator losing. A loss by the target state also decreases the likelihood of violent change compared to the initiator losing, but not as much as a victory. Finally, as the number of casualties increase (as measured by the logarithm of battle deaths divided by population), the likelihood of violent regime change increases. We can look at predicted probabilities to assess the strength of this relationship: If the initiator wins, the probability of violent regime change is 0.008, while if the target wins, the probability is 0.11. The probabilities for losses are much higher, 0.22 if the target loses and 0.44 if the initiator loses. Thus, losing a war greatly increases the likelihood of violent regime change, particularly for initiators.

In a follow on study, Bueno de Mesquita and Siverson (1995) seek to further clarify the effect of war on regime change. Their main empirical findings, which come from a survival analysis of the hazard that political leaders face of being removed from office, are shown in Table 12.2. The first variable is the (log of) tenure of the leader in office before the war, which is negative and significant. The second term is an interaction between tenure and democracy, which is positive and significant. Thus, each increase in the prewar tenure for a nondemocratic leader increases their expected duration in office after the war, while for democratic leaders the effect is much smaller. In fact, the effect of prewar tenure in democracies is indistinguishable from zero.

Table 12.1 The Effects of War on Violent Changes of Regime

VARIABLE	COEFFICIENT	STANDARD ERROR
Initiator wins	-1.16**	0.364
Target wins	-1.013**	0.349
Target loses	-0.57*	0.320
Log(battle deaths/population)	0.14*	0.072
Constant	-0.51*	0.306
Chi-square	21.26**	
N	177	

Notes: $*p < 0.05$, $**p < 0.01$. The reference category is initiator loses.

Source: Adapted from Bruce Bueno de Mesquita, Randolph M. Siverson, and Gary Woller, "War and the Fate of Regimes: A Comparative Analysis." *American Political Science Review* 86 (1992), 643.

Table 12.2 Effect of War on Political Survival Time of Leaders

VARIABLE	COEFFICIENT	STANDARD ERROR	HAZARD RATE
TenureL	-0.47**	0.08	0.62
TenureL * Democracy	0.36*	0.16	1.44
(Battle deaths/10K) L	0.07*	0.04	1.07
Win	-0.26*	0.15	0.77
Nonconstitutional overthrow	0.51*	0.19	1.67
Constant	-0.62**	0.20	-
Sigma	1.43	0.08	-
Chi-square (df)	39.8**		
N	191		

Notes: $*p < 0.05$, $**p < 0.01$.

Source: Adapted from Bruce Bueno de Mesquita and Randolph M. Siverson, "War and the Survival of Political Leaders: A Comparative Study of Regime Types and Political Accountability." *American Political Science Review* 89 (1995), 851.

More costly wars (in terms of battle deaths) increase the risk of removal from office following war. On the other hand, winning the war increases the chances of survival in office, exactly as we should expect. They find, however, that this effect is smaller than the effect of prewar tenure for autocratic leaders. Thus, an autocratic leader who has a long tenure in office has a higher probability of maintaining power following a war loss than a democratic leader that wins a war.

Chiozza and Goemans (2011) conduct a broader analysis of the relationship between leaders and international conflict, looking not only at the impact of war on regime change, but also at the role that leaders play in the outbreak of international crises and wars. They find that removal from office, whether forcible or through regular procedures, is always more likely following losses than wins. Democratic leaders, however, are continually at a greater risk for removal from office than nondemocratic leaders. In addition, they find that leaders' decisions about whether or not to initiate conflict are influenced by their expectations about the likelihood of removal from office. If leaders anticipate regular removal from office (such as through elections), then they have little to gain and much to lose from international conflict. In contrast, leaders who anticipate forcible removal from office (such as by coup or revolution) have little to lose and much to gain from international conflict.

International-Level Consequences

While the previous two sections examined the consequences that war can have within countries, war can also have important consequences for relationships between countries. The international-level consequences of a war can be immense and can lead to changes within the dyad and the international system. Thompson (1995) groups these international consequences into five broad categories: geopolitical situations and orientations, relative capabilities, alignment patterns, economic conditions, and rivalries.

War can lead to important changes in countries' geopolitical situations and orientations. Some countries have developed a national aversion to international politics as a result of war. For example, Switzerland and Sweden have been traditionally neutral countries in Europe, avoiding involvement in the two world wars and remaining neutral in the Cold War. However, they were not always neutral. Sweden used to be a regular participant in European wars, including the Thirty Years War (1618–1648), the Great Northern War (1700–1712), and the Napoleonic Wars. But Sweden and Switzerland have both successfully remained neutral throughout wars since 1815.

Wars also often lead to territorial changes. If you compare the map of Europe in 1914 to the map today, the differences are dramatic. Some of the changes came peacefully, such as the breakup of Czechoslovakia into the Czech Republic and Slovakia in 1993. Most territorial changes, however, came as the result of war. Goertz and Diehl (1992) find that territorial changes are likely to lead to a militarized dispute within five years, particularly if the exchange is perceived as illegitimate. If, however, the territorial changes lead to the resolution of territorial issues between the states, then future conflict is much less likely (Gibler 2012).

Others have focused on the consequence of war for the relative capabilities, or power, between states. The outcome of war may reduce or augment a state's relative capabilities. The United States emerged from World War II as the most powerful country in the world by far. Siverson (1980) argues that the extent of a war's consequences is directly related to the changes produced in power relationships between states; the more power is changed by a war, the greater its consequences will be.

War has played a major role in colonization and decolonization in history (Maoz 1990). A number of extrastate wars occurred as European states fought against indigenous peoples to conquer territories, particularly in Africa and Asia, which were then held as European possessions for a number of years. Decolonization occurred in two major waves. Following World War I, Germany lost all of its colonies, Austria-Hungary was broken up into several new states, and Russia lost parts of its empire (particularly Poland and the Baltic States). In the decades following World War II, all of the colonial powers—particularly Britain, France, Portugal, and Belgium—lost nearly all of their remaining colonies. This wave of decolonization was a direct consequence of World War II, although it often took a number of years and a variety of intermediate events before a former colony was finally granted independence.

Wars can also lead to a number of changes in alignment patterns within the international system, including shifts in formal alliances. For example, in World War II the United States, Britain, France, and the Soviet Union were allies fighting against Germany, Italy, and Japan. After the war, Germany, Italy, and Japan all became firm allies of the United States (along with Britain and France), while the Soviet Union became the primary enemy. Unfortunately, there has been little systematic research on these effects.

In addition to the domestic economic impacts that we examined earlier, war can also have a large impact on international economic conditions. In particular, there is strong evidence that war creates inflationary surges in the international economy (Hamilton 1977; Thompson and Zuk 1982). However, there have been very mixed results regarding the consequences of war in other economic areas such as prices and production (Thompson 1995). We examine the economic consequences of war more closely in the next section.

Finally, war can lead to further conflict between the same states (Senese and Quackenbush 2003; Werner 1999). A related consequence is that war can lead to the formation of an international rivalry, which is when states have an ongoing adversarial relationship marked by periodic militarized disputes and wars (Diehl and Goertz 2000; Thompson and Dreyer 2012). We examine recurrent conflict and international rivalry more closely in the next chapter.

The Phoenix Factor

Central to the economic consequences of war are the costs related to it. Organski and Kugler (1977, 1980) examine the costs of major wars. Although there are a variety of different ideas regarding the economic effects of major wars, Organski and Kugler focus on testing the competing expectations of three specific perspectives: the scissors, permanent loss, and the phoenix factor. We consider each in turn.

The influential economist Keynes (1920) argued that the economic consequences of major wars would function like the opening and closing of scissors. The idea here is that the gap in economic fortunes between the winners and losers would increase in the near future following the war; this is the opening of the scissors. The scissors would then close, however, because the economic devastation of the losers would then bring chaos to the entire international economic system. This would cause the winners of the war to fall down to the losers, making the economic gap between the winners and losers disappear.

The second primary perspective is that there would be permanent economic loss to all states following a major war. This idea is most closely associated with Norman Angell (1911; see also Nef 1950), whose ideas we examined previously in chapter 7. The basic idea here is that all nations lose power following a war, but the winners do not lose as much as the losers.

Additionally, unlike Keynes's scissors idea, the gap between winners and losers is expected to continue for a lengthy period.

The third main perspective that Organski and Kugler (1977, 1980) test is their own, the phoenix factor. The basic idea of the **phoenix factor** is that economic consequences of war are temporary. The losers of a major war are expected to lose more economically than the winners. They do not, however, remain down permanently. Rather, like the phoenix of Greek mythology, the losers are expected to rise from the ashes of their defeat and recover relatively quickly. Thus, like the scissors the economic consequences are temporary, but unlike Keynes's expectations, they expect the losers to recover back to the level of the winners rather than the winners falling to the level of the losers.

To test these competing expectations regarding the economic consequences of major wars, Organski and Kugler (1977, 1980) analyze thirty-one cases covering eighteen different countries in World War I and World War II. Each case focuses on the performance of a country's economy in the twenty years following the war. In order to properly test the three perspectives, Organski and Kugler distinguish between different types of actors. First of all, they distinguish between belligerents and nonbelligerents. They also distinguish between winners and losers, as well as between countries that remain active at the end of the war versus those that are occupied. For example, the United States and United Kingdom were both active belligerent winners in both World War I and World War II, while Germany was an active belligerent loser in both wars. France and Netherlands are both classified as occupied belligerent winners, while Czechoslovakia is classified as an occupied belligerent loser in World War II. Finally, Sweden was a nonbelligerent in both wars.

As with power transition theory more generally, Organski and Kugler use GNP to measure power, looking at the economic performance of each country in the two decades before the war and then forecasting the expected baseline economic growth for the country if the war had not occurred. This is illustrated in Figure 12.2, where the dots show the annual observations of GNP for the country before, during, and after the war, the solid curve shows the economic growth trend estimated for the country during the base period, and the dashed curve shows the projected economic performance for the country if there had not been a war. The difference between the normal projected performance and the country's real performance following the war is the war cost.

Organski and Kugler's findings are based on the economic performance of countries following World War I and World War II. Figure 12.3 shows the economic consequences for all countries in their sample. The observations for both wars are combined together in the figure. The solid horizontal line at zero shows the normal growth projection that stems from their forecasting model discussed above. The economic performance of nonbelligerents, shown by the dotted line, is right around the projection.

Figure 12.2 The Costs of Major Wars

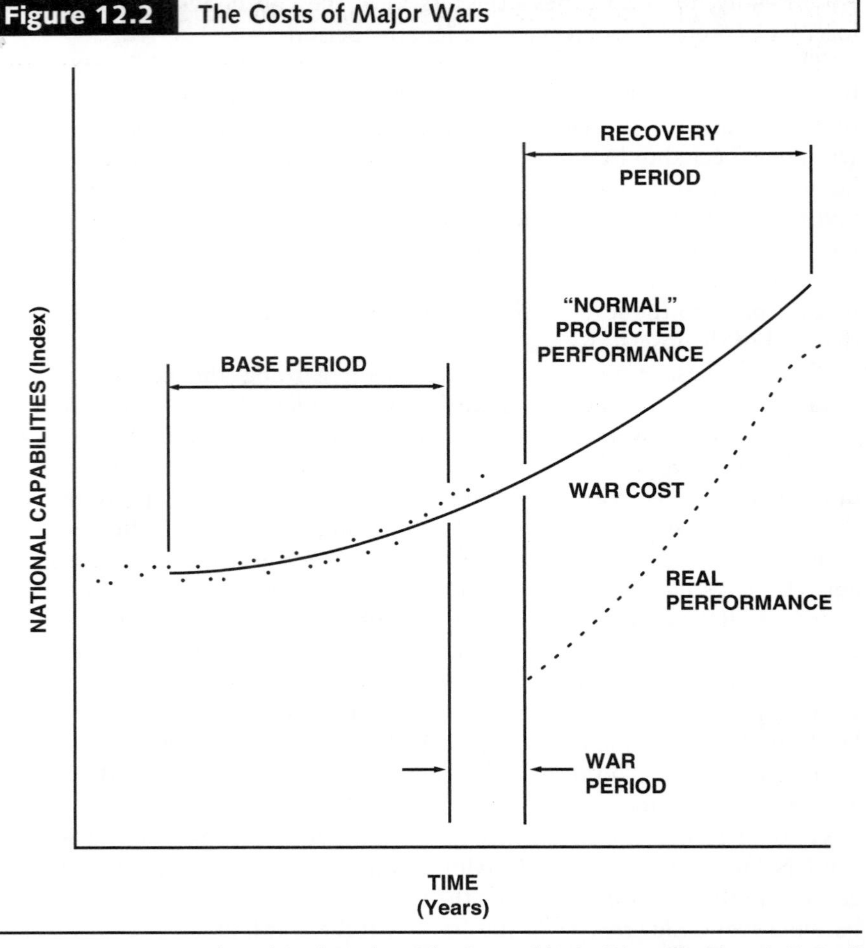

Source: A. F. K. Organski and Jacek Kugler, "The Costs of Major Wars: The Phoenix Factor." *American Political Science Review* 71 (1977), 1350.

Looking at the first two postwar years, we can see that active belligerent winners and nonbelligerents lose about two years of growth, whereas active belligerent losers lose about twenty-one years. This is a sizeable difference of nineteen years. That difference is not sustained, however, as the losers recover rapidly in the subsequent years, catching up to the winners by the eighteenth year following the war. This strongly supports their expectations regarding the phoenix factor.

Organski and Kugler also conduct analyses of World War I and World War II separately. Their findings for the individual wars are complicated by the Great Depression, which affects recovery following World War I and growth projections for World War II. They find that losers lost about four

Figure 12.3 Economic Consequences of War for All Countries in Both World Wars

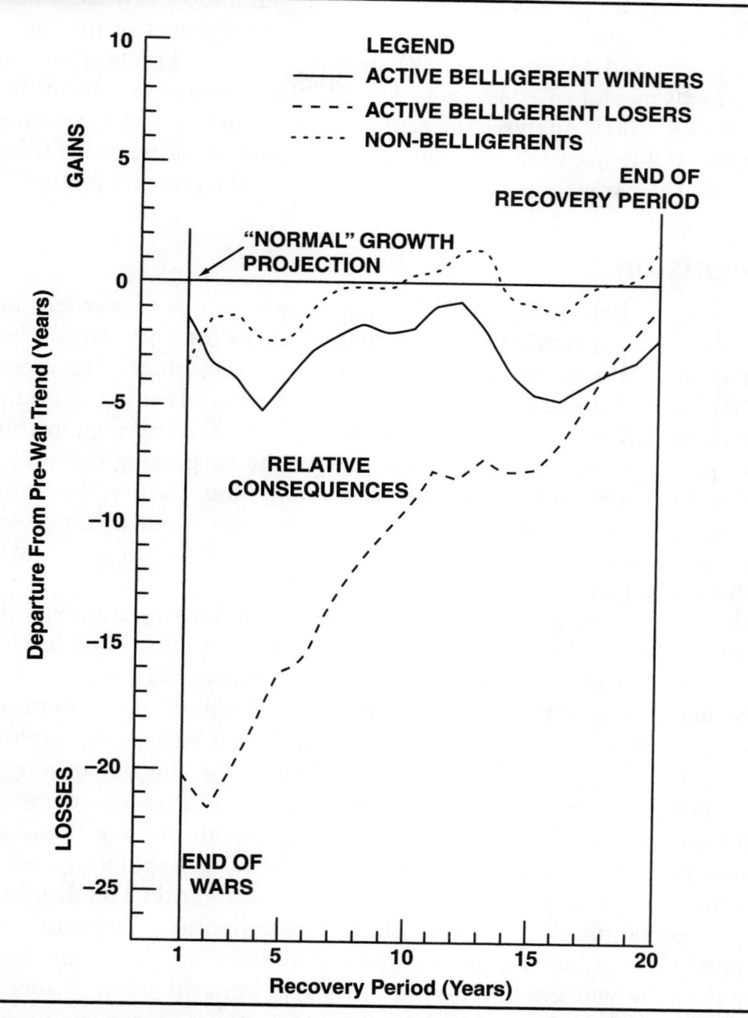

Source: A. F. K. Organski and Jacek Kugler, "The Costs of Major Wars: The Phoenix Factor." *American Political Science Review* 71 (1977), 1360.

times as much as winners following the first war but were able to catch up in only twelve years. Following World War II, the sample of all countries provides some evidence that war losers lose permanently in relation to the winners. The subsample of major powers shows the same phoenix factor trend, however, with the losers catching up to the winners in about sixteen years.

In summary, Organski and Kugler (1977, 1980) find systemic patterns in the economic consequences of major wars. The economies of winners and neutrals are affected only marginally by conflict. In contrast, nations defeated in war suffer intense short-term losses, but in the long run, the effects of war are dissipated. An important limitation of their findings is that Organski and Kugler only look at the consequences of the two world wars. Kugler and Arbetman (1989), however, add the Franco-Prussian War to the analysis and rule out an additional alternative explanation. In addition, Koubi (2005) analyzes the economic performance of a large sample of countries following both interstate and intrastate wars from 1960 to 1989 and finds further empirical evidence supporting the phoenix factor.

Conclusion

Bargaining models of war provide strong explanations of war termination, using the basic logic that war terminates when the information gap between the two sides closes enough to create a bargaining range. The most basic mechanism is that states raise and lower their demands as a function of new information about the outcome and the costs of war. Although information convergence provides an important part of the bargaining dynamic, however, it is not able to fully explain war termination. Rather, domestic politics and especially commitment problems provide a much more complete explanation, showing why states do not always adjust their war demands in response to information from the battlefield.

The consequences of war deal with how the underlying context following a war is changed from what it was before the war. National-level consequences of war can be grouped into three primary areas: economic conditions, demographic factors, and sociopolitical conditions. Examining the impact of war on regime change shows that there is strong evidence that losing a war greatly increases the likelihood of regime change, although there are important differences between democracies and nondemocracies. The international consequences of war can be grouped into five broad categories: geopolitical situations and orientations, relative capabilities, alignment patterns, economic conditions, and rivalries. Organski and Kugler argue that the economic consequences of war are temporary. Empirical evidence supports their argument, as the losers of a major war lose more economically than the winners initially, but they rise from the ashes of their defeat and recover relatively quickly.

Key Concepts

Gambling for resurrection

Phoenix factor

Principle of convergence

War consequences

War termination

13 Recurrent Conflict and Rivalry

A Taiwanese lieutenant colonel shows a satellite photograph of a military base in China's eastern province of Jiangxi. Both sides in the China-Taiwan dyad, which has been plagued by recurrent conflict and rivalry, have built up their arms in recent years.

Source: © Richard Chung/Reuters/Corbis

In the previous chapter, we examined the termination and consequences of war. War has a variety of political, economic, and social consequences at both the domestic and international levels. One consequence that a conflict can have is that it leads to further conflict. Thus, from the new postconflict context, new conflicts of interest may arise or previous ones can remain unresolved, and from these, militarized conflict may result again. For example, World War I ended in the Versailles Treaty, which resulted in a great deal of unhappiness in Germany and elsewhere. Merely two decades later, World War II began, fought between many of the same states—Britain, France, Russia, and the United States versus Germany.

Studies of **recurrent conflict** seek to understand the factors that lead states that have fought in the past to fight again. Usually, this is done by examining the duration of peace following a conflict. Most studies of international conflict assume that different conflicts are independent of one another. For example, Bremer (1992) focuses on contiguity, alliances, and other factors that make dyads more or less dangerous, but there is no consideration of whether the states have fought in the past or how long ago. An alternative way of looking at the history of conflict within a dyad is **rivalry analysis**. The idea of international rivalry is that some pairs of states view each other as enemies, and these rivalry dyads account for a disproportionate share of international conflict.

What factors explain the duration of peace following a conflict? How can we identify international rivalries, and what leads them to form? Answering these and related questions is the purpose of this chapter. We begin by examining explanations of recurrent conflict, focusing in particular on the impact of the type of settlement used to end the previous conflict. We then turn our attention to international rivalry, including approaches to identifying rivalries, theories of rivalry formation, and the relationship between rivalry and international conflict.

Recurrent Conflict

In a classic book, Ikle (1971) wrote that every war must end and that insight could be extended to other forms of interstate conflict as well. Despite the hope that the end of a conflict will bring peace to a dyad, however, these conflicts often recur. This recurrence of conflict is particularly important because subsequent disputes within a dyad tend to be more severe and costly than previous ones. For example, Leng (1983) finds that states tend to use more coercive bargaining strategies in subsequent disputes with the same adversary, and by the third such crisis, war results.

Many historical examples of recurrent conflicts between states are available. One such example is France and Germany/Prussia, who had a long history of conflict following the Napoleonic Wars. The two states fought three wars: the Franco-Prussian War (1870–1871), World War I (1914–1918), and World War II (1939–1945). Moreover, beyond those three wars, France and Germany engaged each other in a total of twenty-two militarized conflicts between 1830 and 1945; only two of the twelve decades in that time frame saw no conflict between the two (the 1890s and the 1900s), whereas France and Germany fought each other at least three times in five different decades (the 1830s, 1870s, 1880s, 1920s, and 1930s). Despite this long history of conflict, however, France and Germany have not fought since World War II; instead, they have become firm allies and members of the European Union.

Another pair of states plagued by frequent conflict is Israel and Syria. From Israel's independence in 1948 (Syria had gained independence in 1946), Syria and Israel immediately began fighting. Most severely, the two sides squared off in the 1948 War of Independence, the Six-Day War in 1967, the Yom Kippur War in 1973, and the War over Lebanon in 1982. Beyond these four wars, Israel and Syria have engaged in a total of fifty-two militarized interstate disputes. There were sixteen disputes between the two sides in the 1950s, and another sixteen in the 1960s; since then, the number has decreased in each decade, to a low of four in the 1990s. While the frequency of conflict between Israel and Syria has decreased, the threat of recurrent conflict between the two sides remains.

Of course not every pair of states is plagued by long histories of conflict like France-Germany and Israel-Syria. For example, the Netherlands and Belgium fought twice early in Belgium's existence as an independent state. Following Belgian independence in 1830, the two engaged in a militarized dispute from August 1831 to February 1832, and another from October 1832 to May 1833. Despite this ominous beginning to their dyadic relationship, they have not opposed one another in a militarized conflict since then—over 180 years of peace.

In addition to the conflict histories of these three dyads, similar dynamics are evident in many others. Clearly, the problem of recurrent conflict is an important one in the study of international conflict. The factors leading to the recurrence of conflict, however, have been subject to a great deal of debate. That is, why do some pairs of states, like France and Germany or Israel and Syria, fight repeatedly, while other pairs, like Belgium and the Netherlands, do not? Alternatively, what factors make peace between states following a conflict more stable, and therefore longer lasting?

Approaches to Explaining Recurrent Conflict

There have been four primary theoretical approaches used for explaining recurrent conflict, focused on bargaining, enduring rivalries, conflict management, and deterrence. We consider each in turn. The first major theoretical perspective on recurrent conflict focuses on bargaining. The bargaining model of war nicely consolidates much of the process of war into a unified theoretical perspective. While some bargaining models focus solely on war outbreak (discussed in chapter 3) or termination (discussed in chapter 12), Filson and Werner (2002, 2004) show that the onset, duration, and outcome of war are all interrelated. In addition, bargaining models also shed light on recurrent conflict (Walter 2002; Werner 1999). Bargaining explanations of recurrent conflict have made predictions about the effect of outcomes, settlements, changes in relative power, and third-party involvement on durations of peace following conflict.

The importance of outcomes for explaining recurrent conflict stems from the bargaining model's assumption that power is central to explaining international conflict. For example, Blainey (1988, 293) argues that "[w]ars usually begin when two nations disagree on their relative strength, and wars usually cease when the fighting nations agree on their relative strength." He then explains stability following decisive wars in terms of power:

> Any factor which increases the likelihood that nations will agree on their relative power is a potential cause of peace. One powerful cause of peace is a decisive war, for war provides the most widely-accepted measure of power. [However,] even a decisive war cannot have permanent influence, for victory is invariably a wasting asset. (294)

Bargaining theorists argue that the mode of settlement used to end a dispute is not particularly important in determining the duration of peace following a conflict (Werner 1999; Werner and Yuen 2005). Rather, they argue that incentives to renegotiate the settlement (however arrived at) drive recurrent conflict. This argument stems from the bargaining model's implication that the terms of settlement agreed to reflect the belligerents' mutual expectations about the consequences of continued fighting (Werner 1998). Thus, as long as those expectations remain the same, peace should last.

However, if those expectations change, that is, if either belligerent anticipates that it would fare better in a new conflict than the last, then they have an incentive to demand renegotiation. This may lead to a new conflict. Werner (1999) argues that the primary indicator of these incentives is shifts in power: As the dyadic balance of power shifts in one state's favor, that state has an increased incentive to challenge the settlement and a recurrence of conflict becomes more likely.

The final key prediction of the bargaining perspective focuses on third-party involvement. The bargaining model identifies commitment problems as a primary reason that bargaining breaks down, and thus as an important cause of war (Fearon 1995; Powell 2002). Walter (1997, 2002) argues that third-party involvement—either in the form of security guarantees or intervention with troops on the ground to enforce peace agreements—provides a way through which commitment problems can be overcome. Thus, third-party involvement is expected to be an important way to increase the stability of peace following a conflict. Walter (1997, 2002) examines the effect of third-party involvement in the settlement of civil wars, although the bargaining perspective expects them to be pacifying following interstate conflicts as well (Fortna 2004c).

The second major theoretical perspective focuses on enduring rivalries. The enduring rivalries literature (e.g., Colaresi, Rasler, and Thompson 2007; Diehl 1998; Diehl and Goertz 2000; Goertz and Diehl 1995; Hensel 1996, 1999; Maoz and Mor 2002; Vasquez 1996) focuses on pairs of states that view each other as enemies and fight with some regularity. Much of the rivalry literature has been devoted to identifying rivalries and examining them as a whole, rather than the stability of peace following individual conflicts, and we examine this in detail later in the chapter.

Rivalry theorists agree with bargaining theorists that outcomes play a central role in explaining recurrent conflict. For example, Hensel (1994) looks at patterns of recurrent interstate disputes by examining the effects that outcomes have on future conflict. His results, looking solely at contiguous Latin American dyads, are mixed. Hensel finds that 93 percent of disputes characterized by a decisive outcome (i.e., a clear victory or yield by one side) are followed by a later dispute, compared to only 85 percent of stalemate and 73 percent of compromise outcomes. He also finds that the postdispute stability of decisive outcomes, however, is (on average) over two years longer than compromise outcomes (12.24 years to 10.20), and over six years longer than stalemate outcomes (12.24 years to 5.94 years).

The third primary theoretical perspective has focused on conflict management. This literature (e.g., Dixon and Senese 2002; Frazier and Dixon 2006; Gartner and Bercovitch 2006; Regan and Stam 2000) has examined the effects of factors such as third-party intervention, UN peacekeeping missions, and regime type on the likelihood of reaching negotiated resolutions to conflict and stable peace following international conflicts. The primary forms of conflict management that this literature focuses on are third-party involvement and peacekeeping. Third-party involvement is expected to lead to more stable peace following conflicts. Similarly, peacekeeping is also expected to be more stable. Gartner and Bercovitch (2006) find two contrasting influences at work in the relationship between mediation and postdispute peace. First, they contend that the effects of mediators are positively related with stability; clearly, this is the purpose of mediators' involvement in the first place. The second important influence at work in this relationship, though, is a selection effect that is negatively related to stability. Conflicts that attract the involvement of mediators tend to be conflicts that are inherently difficult to settle peacefully. If the sides in a dispute could peacefully resolve a dispute by themselves, there would be no need for a mediator. Beardsley et al. (2006) advance a similar argument, but also consider whether the principle actors engage in another crisis within the five-year period following a dispute.

Frazier and Dixon (2006) explore the utility of varying conflict management techniques and the contexts in which they are most likely to succeed. Their study focuses on the efficacy of different mediators—intergovernmental organizations (IGOs), states, and coalitions—and the conflict management techniques available to them, including verbal offers, mediation attempts, adjudication, and military intervention. Overall, they find that IGOs are the most effective managers while military interventions are the most effective conflict management type. Frazier and Dixon are interested in which combination of mediator and technique is most likely to bring about a negotiated settlement.

The conflict management literature assumes that negotiated settlements are inherently desirable compared to other settlement types, and thus that negotiated settlements lead to greater durations of peace after a dispute (e.g., Bercovitch, Anagnoson, and Wille 1991; Butterworth 1978; Dixon and Senese 2002; Frazier and Dixon 2006). Accordingly, imposed settlements are expected to produce less stable relations following conflict. For the same reasons, victor-imposed regime changes are expected to be less stable.

The final theoretical perspective focuses on deterrence to explain recurrent conflict (e.g., Peterson and Quackenbush 2011; Quackenbush 2010b; Quackenbush 2014; Quackenbush and Venteicher 2008; Senese and Quackenbush 2003). This perspective has identified the mode of settlement of a conflict as a primary determinant of the duration of peace following. Maoz (1984) provides a preliminary treatment of the impact of conflict (dispute and war) settlements on the outbreak of future engagements. Utilizing logistic regression and simple measures of central tendency to examine 163 dispute settlements between 1816 and 1976, his findings suggest that imposed settlements are more stable than negotiated ones.

Relatedly, Fortna (2003, 2004a) develops a theory of agreements, arguing that cease-fire agreements can help maintain peace by altering the incentives for war and peace, reducing uncertainty, and helping to prevent or manage accidents that could lead to war. Additionally, the stronger these agreements are, the more durable the peace following them is. Like the mode of settlement, her analysis highlights agreement strength as another important political factor characterizing the termination of the previous conflict.

A final argument explained by the deterrence approach to recurrent conflict is that conflicts terminated with a victor-imposed regime change are followed by longer durations of peace than other conflicts (Lo, Hashimoto, and Reiter 2008; Quackenbush 2014; Werner 1999). By establishing a new regime in the defeated state, the victor is able to create more friendly relations between the states, thereby making enforcement of the settlement easier (Thomson, Meyer, and Briggs 1945). This argument also augments the primary argument of the deterrence perspective that imposed settlements are more pacifying than other types. We now turn our attention to a fuller exploration of the effect of settlement on recurrent conflict.

Settlements and Recurrent Conflict

On 1 September 1939, Germany invaded Poland. With the British and French declarations of war on Germany two days later, World War II was under way. Less than twenty-one years earlier, World War I had ended.

Adolf Hitler, the leader of Germany, had precipitated a series of crises that culminated in the second war by remilitarizing the Rhineland in 1936, annexing Austria in the Anschluss of March 1938, and then demanding the Sudetenland from Czechoslovakia (Weinberg 1994, 1998; Wright 2007). At Munich in September 1938, Britain and France agreed to give the Sudetenland to Germany, after which British Prime Minister Neville Chamberlain announced, "I believe it is peace for our time." Nonetheless, they were at war less than a year later.

Given this history of conflict, Allied leaders in World War II wanted to ensure that the peace following the war would last. At a press conference in Casablanca, where he was meeting with Prime Minister Winston Churchill and members of the Combined Chiefs of Staff, US President Franklin D. Roosevelt announced on 24 January 1943 that the Allies would demand nothing less than the unconditional surrender of Germany, Italy, and Japan (O'Connor 1971). The unconditional surrender policy served several purposes, such as reassuring the Soviet Union that the United States and Britain were committed to opening a second front against Germany, preventing bickering within the Allies that could potentially split the alliance, and unifying American public opinion (Chase 1955; O'Connor 1971). It is also clear, however, that Roosevelt believed that unconditional surrender was the surest path to postwar peace. As he went on to state at the Casablanca press conference, unconditional surrender "means a reasonable assurance of world peace" (Department of State 1968, 727).[1]

The unconditional surrender policy was widely criticized as one of the "great mistakes" of World War II (Baldwin 1950). Many critics believed that the policy galvanized resistance in the Axis powers and thereby prolonged the war (e.g., Baldwin 1950, 1954; Fuller 1949; Wilmot 1952; Woodward 1970). Yet although the quest for total victory (which these critics take for granted) certainly lengthened the war, there is no evidence that the unconditional surrender policy lengthened it further (Kecskemeti 1958). Other critics suggested that the United States and its allies should have ended the war in a negotiated peace instead (e.g., Grenfell 1953). Indeed, one critic alleged that the unconditional surrender policy provided the roots of World War III (Miksche 1952).

Did the unconditional surrender policy create the foundation of a stable peace after the war, as Roosevelt claimed, or did it foster instability, as critics claimed (e.g., Grenfell 1953; Kalijarvi 1948; Miksche 1952)? Or more generally, do imposed settlements or negotiated settlements lead to the most stable peace following a conflict? Before exploring the answer to this question, it is important to establish what we mean by the different methods of settlement. The militarized interstate dispute (MID) data have a variable identifying settlement type, which includes four categories of settlements: negotiated, imposed, no settlement, and unclear. A **negotiated settlement** is identified as

[1]Roosevelt's notes for the press conference state that "unconditional surrender by them [Germany, Italy, and Japan] means a reasonable assurance of world peace, for generations" (Department of State 1968, 837), which indicates that the ensuing peace was expected to be long lasting.

"some type of agreement (formal or informal), the lack of any unconditional surrender or giving up on concessions, and the absence of any attempt of external imposition of a settlement." An **imposed settlement** is defined as "an agreement that has been forced upon another state by means of overwhelming authority and without invitation." The no settlement category denotes "the lack of any formal or informal effort which successfully resolves or terminates the dispute." And finally, the unclear grouping includes cases where "the historical sources present either a conflicting or opaque interpretation of dispute termination" (Jones, Bremer, and Singer 1996, 181).

The ending of a conflict is also characterized by the outcome. The outcome of a dispute is the military result, whereas the settlement of a dispute is political in nature. There are three types of conflict outcomes—decisive, compromise, and stalemate. Decisive outcomes are characterized by a victory—"the favorable outcome achieved by one state through the use of militarized action that imposes military defeat upon the opponent"—or a yield, in which "one state capitulates by offering concessions that appease the demands of another state before . . . any substantial tactical gains on the battlefield." Compromise outcomes occur when "each side in the dispute agrees to give up some demands or make concessions to the other." Finally, disputes where no "decisive changes in the predispute status quo" occur are stalemates (Ghosn, Palmer, and Bremer 2004, 137).

Senese and Quackenbush (2003) and Quackenbush (2014) argue that in the context of relations following conflicts, states are satisfied with the status quo in predictable ways, depending on the method of settlement associated with the preceding dispute. For instance, winners of disputes that impose settlements are likely to be quite satisfied with the postsettlement status quo, while losers of those disputes (states that had settlements imposed on them) are likely to be dissatisfied with the new status quo. Further, in the case of negotiated settlements, each state is likely to be somewhat satisfied with the new status quo (they did, after all, agree to it), but they are also likely to be somewhat dissatisfied with it as well (they did not get everything that they wanted). And finally, following disputes that end without a settlement, both states are likely to be dissatisfied with the status quo.

States' dissatisfaction with the postsettlement state of affairs provides an incentive to challenge the status quo. To prevent these challenges, states use deterrence in an attempt to keep dissatisfied former opponents from violating the terms of the settlement. To explore the dynamics of these deterrence relationships following dispute settlements, Senese and Quackenbush (2003) apply perfect deterrence theory, which we examined in chapter 8. To demonstrate how this application is made, we now turn to a discussion of deterrent relationships following various modes of settlement.

Settlement Type, Deterrence, and Recurrent Conflict

The winner of a dispute who unilaterally imposes a settlement is establishing a new status quo (Pillar 1983). Being dissatisfied with this state of affairs, the recipient of an imposed settlement has an incentive to challenge this new status quo following the settlement. The winning state, seeking to defend the status quo that it established through the imposed settlement, will attempt to deter its former opponent from challenging the terms of the settlement. Whereas the recipient must be deterred from challenging the settlement, there is no corresponding need to deter the settlement imposer. Dyadic relations following imposed settlements are therefore situations of unilateral deterrence.

An example will help clarify this relationship between imposed settlements and unilateral deterrence. On 2 April 1982, Argentina invaded the British-controlled Falkland Islands (which Argentina claims as its own, calling the islands the Malvinas). Great Britain responded to Argentina's seizure of the Falklands by retaking the islands, and Argentina surrendered on 14 June 1982. A settlement that essentially reinstated the status quo (British control of the Falklands) was imposed, and Britain, as the victor, was satisfied, but Argentina was not. Following the war, Britain adjusted its defense policy, including greater emphasis on power projection capability (Gibran 1998), in recognition of the need to deter further Argentinean attacks. However, Argentina had, and continues to have, no need to deter British aggression; with possession of the Falklands in hand, Great Britain lacks an incentive to challenge the status quo.

To examine these types of relations following imposed settlements, Senese and Quackenbush (2003) apply the unilateral deterrence game of perfect deterrence theory (Zagare and Kilgour 2000), which we discussed in detail in chapter 8. For this application, the recipient of the imposed settlement—given its dissatisfaction with the new status quo—is Challenger, and the state which imposed the settlement—being satisfied—is Defender.

Unlike imposed settlements, negotiated settlements require the two sides to agree on the postdispute status quo. Having played a role in its determination, each state is somewhat satisfied with the new status quo. Because of the compromises inherent in a negotiated resolution, however, neither state received all that it wanted. Therefore, each state also has an incentive to challenge the terms of the settlement. Further, since each side is better off with the status quo than with the other side's unilateral alteration of the status quo, it is in each state's interest to attempt to deter the other from initiating a challenge. Hence, the two sides must resort to mutual deterrence in an attempt to maintain the settlement.

One example of the relationship between negotiated settlements and mutual deterrence is the Korean War. North Korea attacked South Korea on 24 June 1950. After spanning the length of the peninsula (first moving down to the Pusan perimeter and then moving up almost to the Yalu River), the front lines stabilized around the 38th parallel for the next two years (Toland 1991). On 27 July 1953, the armistice agreement—a negotiated settlement—was signed. Each side was somewhat satisfied; both North Korea and South Korea had retained their independence. Each was still dissatisfied, however, since neither had accomplished reunification. Indeed, the settlement has been called "a substitute for victory" (Foot 1990), forcing the two sides into a mutual deterrence relationship in an attempt to maintain the settlement.

Senese and Quackenbush (2003) apply the mutual deterrence game of perfect deterrence theory (Zagare and Kilgour 2000) to examine such conditions following negotiated settlements. We examined unilateral deterrence and mutual deterrence in chapter 8. In contrast to unilateral deterrence, in mutual deterrence situations following negotiated settlements, even if both states have perfectly credible threats, stability (i.e., the survival of the status quo) cannot be ensured, whereas in unilateral deterrence situations following imposed settlements, the status quo is the *only* rational outcome if Defender's threat is sufficiently credible.

Thus, the fundamental difference between relations following imposed and negotiated settlements is that unilateral deterrence is required to maintain an imposed settlement while mutual deterrence is required to maintain a negotiated settlement. This is significant because perfect deterrence theory demonstrates that unilateral deterrence is more stable than mutual deterrence and that the enhanced stability of unilateral deterrence is even more pronounced at high levels of credibility, as we saw in chapter 8. This is true because in unilateral deterrence, the defender can ensure the stability of the status quo by having a highly credible threat, whereas in mutual deterrence, conflict always remains possible even if both states have perfectly credible threats. Therefore, Senese and Quackenbush (2003) expect that imposed dispute settlements are more stable than negotiated dispute settlements.

An important similarity between relations following negotiated settlements and those following no settlement is that both represent situations of mutual deterrence. In the latter case, not having come to any resolution of the previous dispute, both sides are likely to be dissatisfied with significant aspects of the new status quo. They each have incentives to challenge those aspects of the status quo that they least prefer while defending those aspects that they most prefer. As instances of mutual deterrence, these relations following no settlement—like those following negotiated settlements—are less stable than relations subsequent to imposed settlements. But although situations of negotiated settlements and no settlement are both less stable than imposed resolutions, they are not equally so. Relations following disputes that end with no settlement are even less stable than those following negotiated settlements, for even though both are situations of mutual deterrence, states are likely to be more satisfied with the status quo following a negotiated settlement than they are when no settlement occurred. And the higher each state's evaluation of the status quo, the more likely deterrence is to be successful (Zagare and Kilgour 2000, 116–7). Hence, the status quo is more likely to survive following a negotiated settlement than following no settlement. Accordingly, Senese and Quackenbush (2003) also expect that dyadic relations following disputes with a settlement (imposed or negotiated) are more stable than relations following disputes with no settlement.

Empirical Findings

Given the argument laid out above, Senese and Quackenbush (2003) expect the longest durations of peace to be associated with imposed dispute settlements and the shortest to follow disputes absent a traceable settlement, with periods of peace subsequent to negotiated resolutions falling somewhere in between. Senese and Quackenbush (2003) initially tested their expectations using data between 1816 and 1992. Here, we focus on an update of their empirical tests, using peace periods following 2,974 dyadic militarized interstate disputes between 1816 and 2001 (Quackenbush 2014).

We start by looking solely at the impact of settlement type on peace duration, as shown in the log-rank tests of Table 13.1. Although 353.08 terminations were statistically expected during peace periods following imposed settlements, only 237 are observed, a significant dissimilarity. For peace periods subsequent to militarized disputes characterized by no settlement, only 1,341.87 failures are expected, while fully 1,466 are observed, again a significant difference. Finally, the disparity between the observed (427) and expected (435.05) terminations for negotiated peace durations is very small and statistically insignificant. Taken together, these findings suggest that peace periods following imposed resolutions are the least, and those following no settlement the most, likely to be followed by another dispute (through 2001), with negotiated peace durations positioned somewhere in between. This ordering of peace termination propensities across the three settlement categories is supportive of their expectations.

The analyses reported in Table 13.2 examine the effects of the covariates on the duration of peace following interstate disputes. In model 1, the reported parameter estimates for the negotiated and no settlement dummy variables are both positive and highly significant, indicating an important reduction in the survival time across peace periods following

Table 13.1 Log-Rank Tests for Equality of Survivor Functions

SETTLEMENT TYPE	EVENTS OBSERVED	EVENTS EXPECTED	χ^2	P-VALUE
Imposed	237	353.08	46.84	<0.001
No settlement	1,466	1,341.87	31.82	<0.001
Negotiated	427	435.05	0.19	0.663

Notes: N = 2130, *df* = 1.

The events expected for each settlement type are obtained by assuming the proportion of events (failures) is the same for each group (type of settlement), and multiplying that proportion by the number at risk in each group.

Source: Compiled by author.

these types of conflict settlements when compared with peace periods following imposed settlements. These findings are in line with the hypotheses deduced from the application of perfect deterrence theory. Further, the greater magnitude of the no settlement coefficient is also in line with expectations, as it suggests the presence of shorter peace periods following these types of settlements compared to those following negotiated resolutions.

Model 2 in Table 13.2 displays the results from a Cox estimation that adds consideration of regime type, change in relative power, geographic proximity, decisive outcomes, and war to that of the core settlement variables. Based on these findings for the model with controls, negotiated

Table 13.2 Cox Model Results for the Effect of Settlement Type on Peace Duration

VARIABLE		MODEL 1	MODEL 2	MODEL 3
Negotiated settlement	β	0.383***	0.361***	0.352***
	Se_β	0.076	0.087	0.086
No settlement		0.493***	0.513***	0.507***
		0.073	0.087	0.087
Minimum democracy			-0.020***	-0.020***
			0.006	0.006
Change in relative power			1.782**	-1.774
			0.603	2.630
Change in relative power * *ln*(Time)				1.046
				0.840
Geographic contiguity			0.775***	0.776***
			0.079	0.080
Decisive outcome			0.144*	0.284
			0.061	0.229
Decisive outcome * *ln*(Time)				-0.039
				0.056
War			-0.105	0.933*
			0.099	0.435
War * *ln*(Time)				-0.281*
				0.113
Log-likelihood		-15,389.6	-15,221.0	-15,210.3
N		36,570	36,570	36,570

Notes: $*p < 0.05$, $**p < 0.01$, $***p < 0.001$

Imposed settlement is the reference category for both models. Standard errors are adjusted for clustering on the dyad.

Source: Compiled by author.

settlements again produce significantly shorter periods of postdispute peace than do imposed settlements. In fact, the risk of peace termination is 1.4 times higher in the presence of a negotiated settlement compared to an imposed resolution. Similarly, disputes characterized by no settlement are followed by significantly less lengthy durations of stability than are imposed settlements, with the risk of termination fully 1.7 times higher subsequent to no settlement. These findings are in accord with those of the base model and again support the hypothesis that negotiated settlements produce longer periods of peace than do no settlements.

One key characteristic of all proportional hazards models is the assumption that the effects of the covariates are constant over time. In these specifications, this would require the effects of settlement type to be the same in the first year following a settlement as they are in the tenth year, twentieth year, and so on. It is, however, quite plausible to expect that the effects of specific dispute settlements might wane over time. That is to say, as the memory of the previous conflict recedes, the precise mode by which past grievances were resolved could well become less important to present and future relations. If this were the case, then, the relationships between the settlement indicators and peace period termination would indeed be nonproportional. Estimation of proportional hazards models in the presence of such nonproportionality could produce biased results (Box-Steffensmeier and Zorn 2001).

Quackenbush (2014) tests for nonproportional hazards using the appropriate test. Model 3 controls for these nonproportional hazards by reanalyzing Model 2 and incorporating interactions with the natural log of time for each of the problematic variables. Most importantly, the coefficients for the negotiated and no settlement variables remain positive and highly significant, indicating that imposed settlements remain the most stable type of settlement even after controlling for nonproportional hazards. Furthermore, analyses run with no settlement as the base category confirm that the difference between negotiated settlements and cases with no settlement is statistically significant. Thus, each of the theoretical expectations is strongly supported.

Several of the parameter estimates for the effects of the control variables are also statistically significant. As the democracy level of the least democratic side of a dyad increases, the duration of postdispute peace also increases, a finding that fits nicely with those chronicling the pacifying nature of democratic norms and institutions on the foreign policy decisions of leaders.

While Werner (1999) argues that changes in relative power are key indicators of incentives to renegotiate a settlement, and previous studies (Werner 1999; Werner and Yuen 2005) have found them to significantly increase the hazard of conflict, the finding here shows them to have no effect. The negative coefficient on change in relative power indicates that initially, larger changes in power are pacifying. The positive coefficient on the interaction between these changes and time indicates that as the duration of peace increases, this negative effect moves closer to zero and then eventually becomes positive. Neither coefficient, however, can be statistically distinguished from zero effect. Accordingly, it appears that incentives to renegotiate do not drive recurrent militarized disputes, or at least that changes in relative power do not identify such incentives.

The results clearly reveal the penchant for geographically nearby former foes to renew their antagonisms. Peace periods following conflicts between contiguous disputants are significantly shorter than those succeeding disputes between noncontiguous states, with a risk of peace termination 2.2 times greater. The findings, however, also indicate that the decisiveness of the previous dispute's outcome has no effect on how long it takes for a pair to renew hostilities. This differs from Hensel's (1994, 1996, 1999) work showing disparate effects across outcome types for pairs of states. The absence of a settlement control in his model may help to explain these differing results for the outcome variable.

Finally, from the results for the war indicator, it appears that peace periods following wars are initially less stable than those following short-of-war disputes. The negative coefficient on the interaction of war with time, however, indicates that as the duration of peace increases, this difference between wars and other disputes becomes increasingly smaller. Importantly, the settlement variables remain substantively unchanged when other important variables are controlled for.

While the results shown in Table 13.2 clue one in to the direction and significance of the covariate effects, examining survival curves is intuitively pleasing and useful for assessing the substantive significance of settlement type. Figure 13.1 shows the estimated survival curves for settlement types based on the results of model 3. The curve for imposed settlement is consistently the highest curve, indicating that they have the highest probably of survival (i.e., uninterrupted peace) at any interval of time following a dispute. The least likely cases to survive are those characterized by no settlement, while negotiated settlements are in between. From this figure, one can also estimate the duration of peace following a dispute to be the time interval where the probability of survival is 50 percent. Thus, the predicted duration of peace following imposed settlements is about 260 months (over 21 years). This is more than twice as long as negotiated settlements, with a predicted duration of about 120 months (10 years), and over three times as long as the no settlement category, with a predicted duration of about 80 months (less than 7 years). Overall, the theoretical expectations of Senese and Quackenbush (2003) and Quackenbush (2014) are strongly supported.

As we discussed previously, bargaining and rivalry explanations argue that outcomes, such as victory or stalemate, are the key to explaining recurrent conflict, not dispute settlements. Nonetheless, Quackenbush and Venteicher (2008) find that settlements are far more important for explaining recurrent conflict than outcomes are, as expected by Senese and Quackenbush (2003). Quackenbush (2010b) examines the interaction

Figure 13.1 Survival Curves by Settlement Type

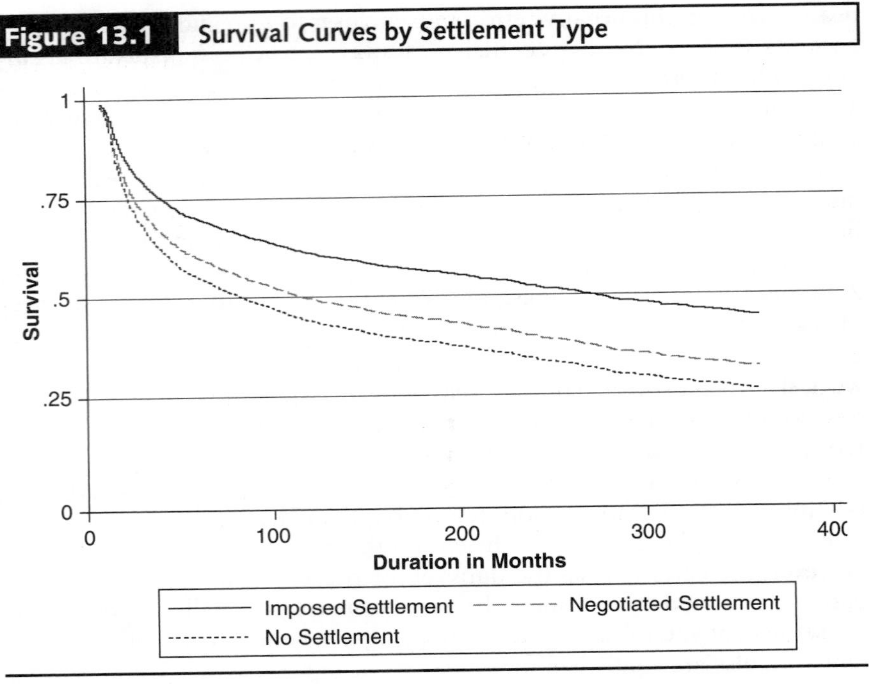

Note: These survival curves are based on model 3.

Source: Compiled by author.

between territorial issues and dispute settlements. Once again, he finds that imposed settlements are more stable than other types of settlement. In addition, territorial issues not only increase the probability of recurrent conflict, but this effect does not wane over time.

Given the consistent evidence that imposed settlements are the most pacifying, it is important to explore factors leading to the varying stability of imposed settlements themselves. An initial step in this direction is taken by Peterson and Quackenbush (2011), who find that following imposed settlements, mutual reliance on trade for income leads to longer durations of dyadic peace. This suggests that imposers can prevent renewed hostilities by balancing vulnerability and trade gains between themselves and their trade partner. Conversely, trade appears to have no effect on stability following other types of settlement.

International Rivalry

Studies of international rivalry have also focused on understanding this tendency of dyads to engage in repeated conflict, focusing squarely on the observation that "a small group of dyads is responsible for a disproportionately large number of conflicts and wars" (Maoz and Mor 2002, 3).

This challenges the typical assumption that international conflicts are independent across time. Gartzke and Simon (1999) argue that it is possible that pairs of states can fight repeatedly even though the separate conflicts are independent, although Colaresi and Thompson (2002) find strong evidence that previous crises make subsequent crises more likely. An **international rivalry** is seen as a pair of states that are competing with and trying to outdo each other.

The idea of rivalries between states in the international system can be compared with rivalries in sports. In baseball, there is a bitter rivalry between the New York Yankees and the Boston Red Sox. The two teams are in the same division and have a decades-long history of intense competition. Similarly, Real Madrid and Barcelona have an intense rivalry in Spanish soccer, which extends into broader society, not just sport.

A clear example of a rivalry in international politics is the relationship between the United States and the Soviet Union during the Cold War. Despite fears of World War III, the two countries never directly opposed each other in a war. Instead, they engaged in a series of crises, formed opposing alliance blocs, and fought a series of wars (in Korea, Hungary, Vietnam, and Afghanistan) in attempts to keep states in their umbrella of influence. Additionally, the Cold War spanned not only military and strategic issues but filtered into competition over the international economy, technology (as illustrated most famously by the space race), sports competitions such as the Olympics, and virtually all aspects of society.

Identifying International Rivalries

Although some rivalries—such as the United States and Soviet Union—are obvious, identifying rivalries more generally has been subject to a great deal of attention and debate (e.g., Bennett 1996, 1997a, 1998; Diehl 1998; Goertz and Diehl 1995; Huth, Bennett, and Gelpi 1992; Wayman 2000). Most scholars have taken a **dispute density approach** to rivalry, where rivalries are identified by the number of militarized disputes they have been involved in, usually within a specified period of time.

For example, Wayman (2000) focuses on three criteria to identify enduring rivalries: severity, durability, and continuity. For severity, Wayman argues that a dyad must have at least five reciprocated MIDs. Further, each MID must last a minimum of thirty days, and multiyear disputes are counted as one dispute per year. The durability criteria is that there must be at least twenty-five years between the outbreak of the first dispute and the termination of the last one. Finally, for continuity, Wayman argues that the rivalry cannot have a gap of more than ten years between disputes. Dyads that meet each of these three criteria are considered enduring rivalries.

Although severity, durability, and continuity are important concepts underlying any measure of rivalry, other scholars identify them in different ways. Some also classify different categories of rivalry, not just enduring ones. An important step in this direction was taken by Diehl and Goertz (2000), who define three types of rivalry: enduring, proto, and isolated.

Enduring rivalries are those dyads experiencing at least six disputes over a time period lasting at least twenty years, representing the most serious of conflicts. Isolated rivalries are those dyads that experience two or fewer disputes, making their rivalry much briefer and less intense. Finally, proto-rivalries include dyads that have too many disputes to be isolated but do not qualify as enduring either (Diehl and Goertz 2000, 45).

Similarly, Maoz and Mor (2002, 231–2) define two types of rivalry, although their criteria are different. Enduring rivalries are dyads with at least six militarized disputes and no gap of more than thirty years between disputes, while proto-rivalries have either between three and five disputes or more than six disputes, but a gap of more than thirty years between two of them.

In contrast, Klein, Goertz, and Diehl (2006, 337–8) eliminate different categories of rivalry and simply define rivalries as dyads with three or more militarized interstate disputes between them over the period 1816–2001. If there is a period of forty to fifty years between disputes within the dyad, then they are coded as separate rivalries.

Clearly, the key differences between these different measures of rivalry are (1) how many disputes over what span of time are used to define an enduring rivalry, and (2) whether less conflictual dyads are considered rivalries or not. Nonetheless, each of these measures employing the dispute-density approach identify rivalries by examining how many militarized disputes dyads have engaged in.

A second approach to identifying international rivalries is the **diplomatic history approach**, which identifies rivalries by examining diplomatic histories and determining which states view each other as sufficiently threatening competitors to qualify as enemies. This is a more subjective approach than the dispute-density approach, as it requires scholars to make judgments about which states view each other as enemies at different periods of time. This is done by focusing on questions such as "who do armies and navies practice against in their simulated battle training? Which enemies are acknowledged when it comes time to justify military and intelligence budgets? Whether the states actually clash physically is beside the point . . . What matters is how decision makers decide which actor(s) is (are) most threatening at any point in time" (Thompson and Dreyer 2012, 12). Thompson, along with coauthors, uses this approach to define **strategic rivalry** (e.g., Colaresi, Rasler, and Thompson 2007; Thompson 2001), although in subsequent work, these are referred to as international rivalries (Thompson and Dreyer 2012).

There are a lot of differences between these two approaches. Even within the dispute-density approach, there are many differences between different measures of rivalry. Colaresi, Rasler, and Thompson (2007, 56) examine six different measures of rivalry and find that "of some 355 dyads presented as rivalry candidates, there are only 23 cases (6.5 percent) on which all 6 agree in some respect." These cases of agreement are listed in Table 13.3. There is general agreement that each of these pairs of states were in a rivalry, although there is still disagreement over the dates of those rivalries. For example, the six measures each have different dates for the France-Germany rivalry, such as 1816–1955, 1830–1945, and 1866–1955.

The dispute-density and diplomatic history approaches to identifying international rivalry each have different advantages and disadvantages. Since the first approach simply requires counting how many conflicts states fought over given periods of time, it is quick and does not require subjective judgments beyond the basic criteria to apply. In contrast, the second approach requires analysts to make judgment calls when examining the historical record. Dispute-density approaches, however, identify "rivalries," such as Ecuador and the United States, between states that do not seem to have considered each other enemies or otherwise behave like rivals. Another example of an implausible rivalry is Canada and the United States, which is counted as a rivalry by Klein, Goertz, and Diehl (2006) from 1974 to 1997 because of a handful of militarized disputes regarding fishing vessels.

Ultimately, the key advantage of the diplomatic history approach is that rivalries are identified independently of their history of conflict. If one wants to argue that rivalries make conflict between states more likely, then one should not use the observation that states fight a lot in order to identify that they are rivals. Otherwise, the result is circular reasoning where dyads are more likely to fight because they fight a lot. This is similar to our discussion of measures of power in chapter 5; if the outcome of conflicts is used to determine power, as in the relational definition, then power cannot be used to explain or predict the outcome of conflicts.

Table 13.3 Consensus Rivalries Agreed on by Six Different Measures

Afghanistan-Pakistan	Greece-Turkey
Argentina-Chile	India-Pakistan
Britain-Germany	Israel-Jordan
Britain-United States	Israel-Syria
China-India	Italy-Turkey
China-Japan	Japan-Russia
China-Russia	North Korea-South Korea
China-United States	Mexico-United States
Ecuador-Peru	Russia-United States
Egypt-Israel	Spain-United States
Ethiopia-Somalia	Turkey-Russia
France-Germany	

Source: Michael P. Colaresi, Karen Rasler, and William R. Thompson. *Strategic Rivalries in World Politics* (Cambridge: Cambridge University Press, 2007), 57.

Explaining Rivalry Formation

Much recent research has focused on analyzing the dynamics of rivalry. Several scholars have addressed the origin of rivalries (e.g., Maoz and Mor 2002; Stinnett and Diehl 2001; Valeriano 2013), although there are competing views to answer the question of how rivalries form. Two prominent theoretical models of rivalry formation are the punctuated equilibrium model and the evolutionary model.

Diehl and Goertz (2000) introduce the punctuated equilibrium model to explain rivalry formation. Figure 13.2 illustrates the punctuated equilibrium model. "The punctuated equilibrium model predicts rapid change at the beginning and end of enduring rivalries" (Diehl and Goertz 2000, 139) because of political shocks. Diehl and Goertz focus on system level political shocks, involving the two world wars, periods of frequent territorial change, and periods of intense changes in the power distribution in the system. They also argue that state level political shocks—national independence, civil war, regime change, and democratization—can have important effects.

Diehl and Goertz (2000) argue that once a rivalry is formed because of a political shock, the severity of the rivalry locks in to a new equilibrium at a certain level. The level of severity that characterizes the underlying relationship in a rivalry is the basic rivalry level and is illustrated by the height of the line in Figure 13.2. Different rivalries will lock in to different levels of severity, and so the basic rivalry level will be higher for some rivalries and lower for others. Diehl and Goertz (2000), however, argue that there is great stability in conflict behavior within a rivalry; different periods of peace and conflict are random

Figure 13.2 **Punctuated Equilibrium Model of Rivalry**

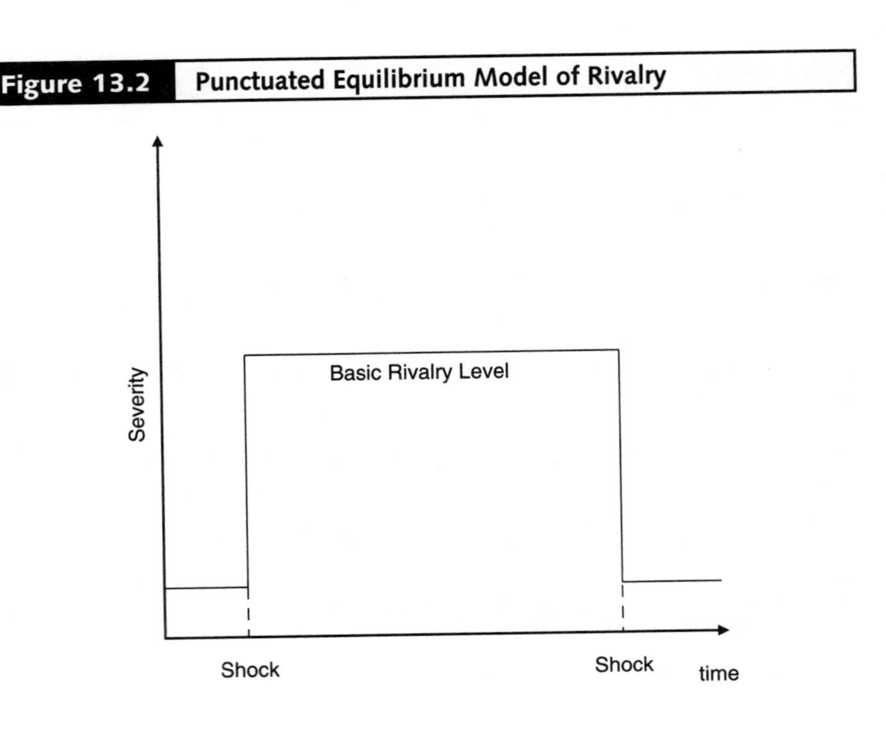

variations around this level, with no "trend toward more conflictual or more peaceful relations" (165). Wayman (2000) also agrees that overall, there is not a lot of fluctuation in conflict during enduring rivalries.

An alternative explanation of rivalry formation and behavior is the evolutionary model introduced by Hensel (1996, 1999). The evolutionary approach to rivalry "is based on the premise that rivalry—rather than being inevitable or predetermined by structural conditions—is a dyadic phenomenon and changes over time" (Hensel 1999, 183). Thus, rivalries change over time in response to interactions between the rival states. In particular, disputes involving territorial issues or ending in stalemate outcomes tend to lead to further conflict and entrenchment of rivalry, while disputes involving nonterritorial issues or ending in decisive outcomes make further conflict less likely, moving away from enduring rivalry.

In addition, Hensel (1996, 1999) argues that conflict patterns vary systematically over the rivalry process. He identifies three phases of rivalry (early, intermediate, and advanced), and finds strong empirical evidence that further disputes become more and more likely as a rivalry advances from one phase to the next. This of course reveals a much different pattern of rivalry behavior than the punctuated equilibrium model's focus on great stability in conflict behavior around the basic rivalry level.

Wayman (2000) examines the origins of rivalry in a more inductive way by trying to uncover patterns that are common in many rivalries. First of all, geographic contiguity is crucially important. Contiguity creates persistent opportunity to attack, and is common to most rivalries. Given the strong effects of contiguity and territory that we examined in chapter 4, this is not surprising. There are also a number of rivalries that are born feuding. Box 13.1 discusses the rivalry between India and Pakistan. Other prominent

Box 13.1

Case in Point: India and Pakistan

India and Pakistan provide a clear example of a rivalry that was born feuding. India had been the most prized colony of the British Empire. Upon decolonization in August 1947, India was split into two states: The areas with a Muslim majority became Pakistan, while the rest (with a Hindu majority) became India. The ruler of Kashmir, on the border between India and Pakistan, joined India even though the majority of the population was Muslim. This decision was hotly contested. The First Kashmir War between India and Pakistan lasted from October 1947 to January 1949 and ended in a stalemate, with Pakistan in control of about one third of Kashmir.

A line of control was established between the two sectors, although India and Pakistan continued to claim all of Kashmir as their own. This is shown on Map 13.1. Nonetheless, the area remained relatively peaceful for nearly two decades. In August 1965, however, Pakistan invaded Indian-controlled Kashmir, launching the Second Kashmir War. Fighting continued through September, with Pakistan gaining a small victory against India.

(Continued)

(Continued)

Map 13.1 India and Pakistan Rivalry Over Kashmir

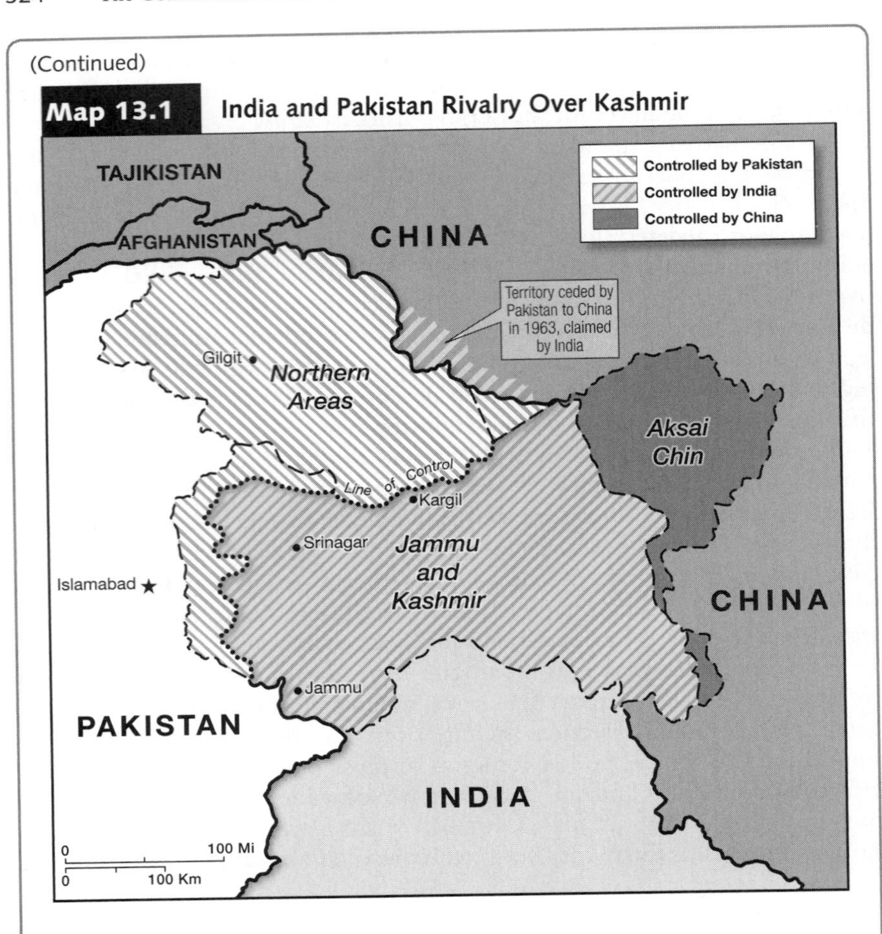

At the time of decolonization, Pakistan was divided into two parts, East and West Pakistan, on either side of India. Although they were both Muslim, significant cultural and language differences—in addition to geography—divided East and West Pakistan. East Pakistan began rebelling against the government of West Pakistan in the late 1960s, and in December 1971, India went to war to support East Pakistan's independence. In the resulting Bangladesh War, India defeated Pakistan in only two weeks, and East Pakistan gained independence as Bangladesh.

In the ensuing years, India and Pakistan both developed and deployed nuclear weapons. The two sides battled again over Kashmir in 1999. In the Kargil War, Pakistan infiltrated troops past the line of control. In the two-month war from May through July 1999, India defeated Pakistan and reestablished its positions along the line of control. The Kargil War was the first war between nuclear powers, and although they have not had another war since, tensions remain, as illustrated most prominently by the attacks in Mumbai, India, by Islamic terrorists from Pakistan in November 2008.

rivalries born feuding include Syria-Israel and North Korea-South Korea. Finally, Wayman finds that ethnic issues are important in the formation of rivalries. These can happen when the dominant ethnic group in one state is a minority—particularly an oppressed minority—in the other.

Thompson and Dreyer's (2012) data identify four types of issues in rivalry: spatial, positional, ideological, and interventionary. In spatial rivalries, the states contest the control of territory. Positional rivalries center on competition over relative shares of influence and prestige within a region or the global system. In ideological rivalries, the states contest the relative virtues of different political, economic, social, or religious belief systems. Finally, in interventionary rivalries states intervene in the internal affairs of other states in order to reduce threat or acquire leverage in the other state's decision making. These categories are not mutually exclusive as a rivalry might involve multiple issue types, not just one.

Other research has focused on the termination of enduring rivalries. Dispute-density measures of rivalry use some arbitrary amount of time after the last militarized dispute within a rivalry to identify the end date of a rivalry. Goertz and Diehl (2000) use a period of ten years, which is fairly typical. Bennett (1996, 1997a) argues that using a fixed number of years to code the end of rivalry is problematic, and instead focuses on the settlement of issues at stake within a rivalry to identify rivalry termination. Bennett (1998) goes on to test three models of rivalry duration, focused on security, domestic politics, and political shocks.

Bennett (1998) finds that domestic political factors and issue salience are the most likely causes of rivalry termination. His findings on political shocks and security concerns, however, are sensitive to the way that they are measured. Colaresi (2001) and Cornwell and Colaresi (2002) analyze rivalry termination using strategic rivalries and find similar results. Furthermore, Bennett (1998) finds that rivalries are more likely to end the longer that they last, which contradicts theories of rivalry.

Rivalries and International Conflict

Many studies have used rivalry as a mechanism for case selection. For example, Diehl (1985) examines arms races within enduring rivalries and Huth and Russett (1993) use rivalries to identify cases of general deterrence in their study. A number of other studies have examined the dynamics of rivalry evolution (e.g., Colaresi, Rasler, and Thompson 2007; Gibler 1997; Goertz, Jones, and Diehl 2005; Maoz and Mor 2002; Tir and Diehl 2002) to see how conflict changes over the course of a rivalry.

Our concern here is on the more direct connection between rivalries and international conflict. Table 13.4 compares conflict within rivalry. Diehl and Goertz (2012) identify 881 isolated conflicts, which is nearly 73 percent of all dyadic conflicts. These cases account for less than 32 percent of militarized disputes and have a probability of war of 0.18. Nearly 15 percent of conflicts are proto-rivalries, and they account for over 17 percent of disputes. Enduring rivalries are even less common (less than 13 percent), but they account for more than half of all disputes.

Diehl and Goertz (2012) separate enduring rivalries into two subcategories. There are 105 rivalries that have between 6 and 13 disputes, which is 8.7 percent of cases. These rivalries account for about 23 percent of all disputes and have a probability of war of 0.23. The final category is enduring rivalries with more than 13 disputes, which make up less than 4 percent of cases (47 total). This small number of dyads accounts for about 29 percent of all militarized disputes. Furthermore, the probability of war at some point within one of these most conflictual rivalries is 0.64, which is more than three times the probability of war overall (0.18).

It is important to understand what these empirical results regarding the relationship between rivalry and international conflict tell us. This is overwhelming evidence that a small number of dyads account for a disproportionate share of militarized disputes and wars. However, we cannot conclude that being in a rivalry drove these dyads to fight so much due to the fact that they are identified as rivals because they had engaged in more militarized interstate disputes than other pairs of states did.

Conclusion

The process of war does not end once a dispute or war ends. Rather, it is cyclical, as a conflict can have a variety of consequences that shape the underlying context within a dyad. One of those consequences is that new conflicts and rivalry may arise. There have been different approaches to explaining recurrent conflict between states. Examining the impact of settlement type on recurrent conflict reveals that very strong logic and evidence

Table 13.4 Rivalry Context and the Frequency of Interstate Conflict, 1816–2001

RIVALRY TYPE	NUMBER OF RIVALRIES	DISPUTE FREQUENCY	PROBABILITY OF WAR (N WITH WAR)
Isolated Conflict	881 (72.7%)	1,217 (31.8%)	0.13 (118)
Proto-	178 (14.7%)	654 (17.1%)	0.26 (46)
Enduring	152 (12.6%)	1,960 (51.2%)	0.36 (54)
6–13 Disputes	105 (8.7%)	862 (22.5%)	0.23 (24)
>13 Disputes	47 (3.9%)	1,098 (28.7%)	0.64 (30)
All	1,211 (100%)	3,831 (100%)	0.18 (218)

Source: Paul F. Diehl and Gary Goertz, "The Rivalry Process: How Rivalries are Sustained and Terminated." In *What Do We Know About War?*, 2nd ed., ed. John A. Vasquez (Lanham, MD: Rowman and Littlefield, 2012), 87.

shows that imposed settlements increase the stability of relations between states following conflict.

A highly related subject deals with international rivalry. There are a variety of measures of rivalry, falling into either the dispute-density or diplomatic history approach. The punctuated equilibrium and evolutionary models are two of the primary theories for explaining the dynamics of rivalry formation and behavior. The relationship between rivalry and international conflict is strong, as a small number of dyads account for a large proportion of all militarized interstate disputes and wars.

As with the previous topics we have examined in this book, such as territory, alliances, deterrence, and war outcomes, we have only been able to examine the tip of the iceberg with regard to recurrent conflict and international rivalry in this chapter. Nonetheless, the tendency for states to fight repeatedly is an important subject to examine in order to gain a basic understanding of international conflict.

Key Concepts

Diplomatic history approach

Dispute density approach

Imposed settlement

International rivalry

Negotiated settlement

Recurrent conflict

Rivalry analysis

Strategic rivalry

Part IV
Conclusions

14 What Have We Learned About War?

Italian reinforcements march on their way to the front in June 1918. The massive costs of World War I a century ago led to a great increase in research seeking to understand international conflict. What have we learned about war since then?

Source: De Agostini/Getty Images

In this book, we have examined the entire process of war, including a variety of factors explaining the causes of conflict, the escalation of disputes to war, and war's conduct and aftermath. Contiguous states are more likely to fight while jointly democratic dyads are less likely to. Disputes over territorial issues are particularly likely to escalate to war. Military strategy plays an important role in shaping war outcomes. These are just a few of the many trends we have examined regarding international conflict. But in the end, what have we learned about war? What does it all mean? Can all of these different theories and trends be tied together into one overall explanation of international conflict or are we forever limited to separate understandings of separate factors?

In this chapter, we seek to answer these questions. We begin by discussing the cumulation of knowledge, focusing on the distinction between additive and integrative cumulation. This highlights the importance of theory, which we cover in the final section.

The Cumulation of Knowledge

We have examined a wide variety of topics in this book in order to make our introduction to international conflict as comprehensive as possible. At the same time, we have been able to only skim the surface of research on international conflict. Numerous books and journal articles are published each year on all aspects of international conflict. Although we have reviewed many of these studies in this book, there are complete topics—such as the relationship between public opinion and conflict, the origins and effects of peacekeeping operations, and the connection between conflict and respect for human rights—that have received little if any attention here because it is simply not possible to cover everything.

The amount of research on international conflict has grown tremendously over time. Cusack (1995) examines the number of quantitative studies of war that were published between the mid-1960s and early 1990s. He does this by focusing on the average number of articles published per year in seven top journals (*American Political Science Review, American Journal of Political Science, International Interactions, International Studies*

Quarterly, Journal of Conflict Resolution, Journal of Peace Research, and *World Politics*). He finds that the amount of quantitative research increased dramatically during that period, from two to three articles per year between 1965 and 1974, to about eight per year in the late 1970s, around sixteen to eighteen articles per year in the 1980s, and twenty-four per year in the 1990s. How has this growth in research improved our understanding of international conflict?

Answering this question requires us to determine how we can assess the progress toward better and more complete understanding of international conflict made by the field. Scholars disagree about how best to measure progress in the field (Bennett and Stam 2004; Cusack 1995; Most and Starr 1989). Some see the addition of new findings over time as the key toward progress. In this view, the growth of scientific research on international conflict in the past fifty years has undoubtedly resulted in tremendous progress. For example, nearly all of the empirical findings we reviewed in this book have been made during that time, most of them within the past two decades.

Others, however, argue that more is required than simply adding more studies to those that came before. Zinnes (1976) characterizes these approaches to progress by distinguishing between integrative and additive cumulation. **Additive cumulation** occurs when a study adds information to an existing body of knowledge. For example, when studies add new findings about the effect of democracy on international conflict, this is additive cumulation. **Integrative cumulation** is marked by the tying together and explanation of research findings. For example, when Bueno de Mesquita et al. (2003) developed selectorate theory, which not only provided an explanation for the democratic peace but also a wide variety of other effects of democracy, this is integrative cumulation. Furthermore, Zinnes argues that additive cumulation is a precondition for integrative, but is not a substitute for it.

There has been a great deal of additive cumulation of knowledge on international conflict over the years. Yet there has not been as much integrative cumulation (Cusack 1995; Most and Starr 1989; Zinnes 1976). In various reviews of the field over the years (e.g., Deutsch and Senghaas 1973; Simowitz and Price 1986; Sullivan and Siverson 1981), international relations has repeatedly been criticized for lack of integrative cumulation. How is integrative cumulation to be achieved?

One important approach to advancing the field is to develop a synthesis of research findings to establish which empirical patterns hold over a broad range of research and which ones do not. Geller and Singer (1998) take this approach in their book *Nations at War: A Scientific Study of International Conflict*. They review over five hundred empirical studies on international conflict. They group research by the level of analysis, focusing on war-prone states, war-prone dyads, war-prone regions, and war-prone systems. Table 14.1 shows the empirical patterns that they find consistent support for in the literature.

Table 14.1 Consistent Empirical Patterns in International Conflict

ONSET (OCCURRENCE/INITIATION) OF WAR
Level of analysis: State
Power status (major power)
Power cycle (critical point if major power)
Alliance (alliance member)
Borders (number of borders)
Level of analysis: Dyad
Contiguity/proximity (common border/distance)
Political systems (absence of joint democracies)
Economic development (absence of joint advanced economies)
Static capability balance (parity)
Dynamic capability balance (unstable: shift/transition)
Alliance (unbalanced external alliance-tie)
Enduring rivalry
Level of analysis: Region
Contagion/diffusion (presence of ongoing regional war)
Level of analysis: System
Polarity (weak unipolarity/declining leader)
Unstable hierarchy
Number of borders
Frequency of civil/revolutionary wars
SERIOUSNESS (MAGNITUDE/DURATION/SEVERITY) OF WAR
Level of analysis: State
Power status (major power)
Level of analysis: System
Alliance (high polarization)

Source: Adapted from Daniel S. Geller and J. David Singer, *Nations at War: A Scientific Study of International Conflict* (Cambridge: Cambridge University Press, 1998), 27–28.

Most of the empirical patterns relate to the onset of war and are worded in terms of what characteristic increases the probability of war onset, occurrence, or initiation. Thus, for political systems, the absence of joint democracies increases the likelihood of war, as we saw in chapter 7. Only two of the empirical patterns that they found relate to the seriousness of war (in terms of magnitude, duration, or severity). They conclude with case studies of the outbreak of two wars, the Iran-Iraq War in 1980 and World War I in 1914, showing how different empirical patterns can be applied to the particular case.

A second approach is taken by Bennett and Stam (2004) in their book *The Behavioral Origins of War*. This approach is **comparative hypothesis testing**, which is based on the idea that "many theories or hypotheses may simultaneously be correct" (Bennett and Stam 2004, ix), but that "some theories have more empirical relevance than others" (x). Accordingly, they evaluate the relative strength of sixteen different explanations of international conflict (using between one and six variables for each explanation) looking at all dyad years from 1816 to 1992.

They find that some factors (such as contiguity, a war equilibrium from Bueno de Mesquita and Lalman's International Interaction Game, and dyadic democracy) are consistently important for explaining war. Other factors, such as nuclear weapons, are found to not matter very much. No single factor, however, is found to dominate the others: "If there were one single story to take from this book, it would be that there is no single story of war" (Bennett and Stam 2004, 201).

The works of Geller and Singer (1998) and Bennett and Stam (2004) are undoubtedly useful for establishing which empirical regularities we can have confidence in because (1) they are consistent across a broad range of studies, and (2) they are still present when put against other explanations through comparative hypothesis testing. One limitation of these approaches, however, is that they are both entirely empirical in nature, yet explanation is not spontaneously generated. We cannot just roll around in the data and magically find explanations of international conflict. Rather, explanation derives from the explicit construction of theories. Cusack (1995) argues that the key reason for the lack of integrative cumulation is that there is a theoretical deficit in the study of war.

The Importance of Theory

In our discussion of each topic in this book, we focused a great deal of attention on theories. We have done so for a couple of key reasons. First, theory establishes the logic of relationships between variables, and we have been concerned with evaluating the logic and evidence behind explanations of international conflict. Second, theory is the key step toward integrative cumulation.

In the first chapter, we examined the basics of scientific theories. There are different important parts of a theory, including assumptions and hypotheses. Theories specify relationships between independent variables—factors that we think will provide us with all or part of our explanation—and dependent variables—the phenomena we are trying to explain. Theory can start either inductively or deductively. Since scientific explanation is based on both logic and evidence, both inductive and deductive reasoning are necessary.

We also examined the criteria for a good theory, especially logical consistency, accuracy, and falsifiability. We expect theories to be able to provide prediction, explanation, organizing power, and heuristic power. These latter two are particularly important for integrative cumulation. Organizing power refers to the ability to bring together information, while heuristic power deals with the fruitfulness of a theory, the ability of the theory to raise new questions. Does the theory lead anywhere?

Before further considering the role of theory in providing integrative cumulation, it is important to consider the **ranges of theory**. Similar to the levels of analysis, theories can be targeted at different scales, in terms of how broad or narrow the range of phenomena the theory covers is. Grand theory is overarching, using the interrelationship of several variables to explain a wide range of phenomena. Middle-range theory focuses on explaining a limited range of phenomena. Narrow-range theory deals with the attempt to establish a relationship between two or more variables. This is really just hypothesis testing, not theory, because there is little if any attempt to fully lay out the logic behind the relationship.

The two most prominent attempts at grand theory in the study of international politics are realism and liberalism. For example, in his *Theory of International Politics,* Waltz (1979) seeks to establish how the entire international system works. Similarly, major works of liberalism such as Keohane's (1984) *After Hegemony* seek to establish how the international system works. While realism and liberalism disagree about the key aspects of the international system and how it works (Baldwin 1993; Keohane 1986), they are both grand theories that are trying to explain a very wide range of phenomena. Unfortunately, they both are limited in their ability to explain international politics.

Most modern theoretical work in international conflict focuses on middle-range theory. The result is that there are many islands of theory in the study of international politics. Some islands are separated by what the theorists are trying to explain. Thus, there is the international conflict island, the international political economy island, the international organizations island, and so on. But within each of these broad islands of theory, there are numerous smaller islands, divided in large part by the approach to explanation and the key factor(s) that are expected to provide the explanation. Within international conflict research, there is the realist island, the democratic peace island, the alliances island, the war outcomes island, the recurrent conflict island, and many others.

Bremer's (1995) process of war model that we have used as a framework in this book provides a conceptual tool for bridging the various islands of theory within the study of international conflict. Theories looking at the causes of conflict, the escalation of disputes to war, war outcomes, and the consequences of war are all looking at different stages along the same overall process.

The bargaining model of war provides a promising path toward greater integrative cumulation. Fearon (1995), Powell (1999), and others use bargaining models to explain the onset of war. Goemans (2000), Reiter (2009), and others use it to explain war termination. Filson and Werner (2002) develop an integrated bargaining model to explain the onset, duration, and outcome of war. Thus, bargaining models are already paving a path toward greater integrative cumulation in terms of explaining multiple stages of the process of war within a unified theoretical framework. They have not, however, gone as far in terms of explaining how different factors such as territory, democracy, military strategy, and so forth all fit together to help explain those different stages.

Selectorate theory (Bueno de Mesquita, et al. 2003) has made important strides toward integrative cumulation by showing how regime type affects a wide range of phenomena, not only in international politics, but in domestic politics as well. In addition to explaining the democratic peace (Bueno de Mesquita et al. 1999, 2003, 2004), selectorate theory has also been used to explain topics as widely varied as foreign aid (Bueno de Mesquita and Smith 2009), environmental politics (Bernauer and Koubi 2009), government finance (Bueno de Mesquita and Smith 2010), human rights (Bueno de Mesquita et al. 2005), and the general behavior of politicians in office (Bueno de Mesquita and Smith 2011). While we still do not know how many other factors related to international conflict fit into selectorate theory, it is nonetheless impressive that a theory is able to explain such a wide range of phenomena within a single, logically consistent, overarching theoretical structure.

The ultimate level of integrative cumulation would be the development of a single theoretical structure that could explain all of international politics. Although that is an important goal to strive for, few if any academic disciplines have achieved that level of integration. A more realistic goal in the near term is the continued building of bridges connecting various islands of theory, both across stages of the process of war and incorporating multiple factors that have been found to be important within each stage.

Key Concepts

Additive cumulation

Comparative hypothesis testing

Integrative cumulation

Ranges of theory

Appendix

International Conflict Since 1815

There have been wars throughout history. Chapter 2 provided an overview of the history of international conflict, and we used various wars and crises as examples in other chapters. Modern political science commonly focuses on international relations since the Napoleonic Wars, which ended in 1815. What are the major international conflicts since then? A major resource for all wars is *Resort to War: A Data Guide to Inter-state, Extra-state, Intra-state, and Non-state Wars, 1816–2007,* by Meredith Reid Sarkees and Frank Whelon Wayman (2010), which is the official Correlates of War (COW) data book for the latest version of the war data and has summaries of every war.

Three other important resources for overviews of wars in history are *Warfare and Armed Conflicts: A Statistical Encyclopedia of Casualty and Other Figures, 1494–2007,* by Michael Clodfelter (2008), *The Harper Encyclopedia of Military History: From 3500 BC to the Present,* by R. Ernest Dupuy and Trevor N. Dupuy (1993), and *Wars and Peace Treaties, 1816–1991,* by Erik Goldstein (1992).

Two other works that are useful for examining some lesser-known wars are volume 2 of *Wars of the Americas: A Chronology of Armed Conflict in the Western Hemisphere, 1492 to the Present,* by David F. Marley (2008) and *Wars in the Third World since 1945,* by Guy Arnold (1995). Each of these books covers only a subset of all wars.

We first examine every interstate war (as defined by the Correlates of War data) from 1816–2007, providing the primary opponents, dates, COW war number, summary, and recommended sources for each. These recommended sources are good places to start research about particular conflicts, enabling those interested to go further in their study of the conflict. We then turn our attention to examining some major crises and disputes, providing the same information for each.

Interstate Wars

Franco-Spanish War
Primary opponents: France v. Spain
Dates: 7 April–13 November 1823

COW war number: 1
Summary: After a military revolt in Spain in 1920, France invaded to restore the traditional powers of Spanish King Ferdinand VII.
Recommended sources: Sarkees and Wayman (2010, 78); Goldstein (1992, 3–4).

First Russo-Turkish War
Primary opponents: Russia v. Ottoman Empire (Turkey)
Dates: 26 April 1828–14 September 1829
COW war number: 4
Summary: Russia invaded the Ottoman Empire and successfully annexed territory around the mouth of the Danube River and in the Caucasus.
Recommended sources: Sarkees and Wayman (2010, 78–79); Goldstein (1992, 22–23); Clodfelter (2008, 207–8).

Mexican-American War
Primary opponents: United States v. Mexico
Dates: 25 April 1846–14 September 1847
COW war number: 7
Summary: Mexico was upset that the United States annexed Texas in 1845. In addition, the two sides disagreed about the proper location of the border. The war began in April 1846, and after initial battles along the Rio Grande, the United States invaded Mexico. After the fall of Mexico City in September 1847, Mexico sued for peace. As a result of the war, the United States annexed much of what is now the southwestern United States.
Recommended sources: Clary (2009); Wheelan (2007); Marley (2008, 747–74); Goldstein (1992, 205–7).

Austro-Sardinian War
Primary opponents: Sardinia, Tuscany, and Modena v. Austria
Dates: 24 March 1848–30 March 1849
COW war number: 10
Summary: Sardinia/Piedmont attacked Austria in an attempt to eliminate Austrian rule of Lombardy and Venetia. Tuscany and Modena fought alongside Sardinia, but the war ended in March 1849 with an Austrian victory.
Recommended sources: Sarkees and Wayman (2010, 80–81); Goldstein (1992, 14–16).

First Schleswig-Holstein War
Primary opponents: Prussia v. Denmark
Dates: 10 April 1848–10 July 1849
COW war number: 13
Summary: Prussia fought against Denmark after the German-speaking majorities in Schleswig and Holstein revolted against Danish rule. Prussian troops drove the Danish from the two duchies, but ultimately the issue was unresolved.

Recommended sources: Sarkees and Wayman (2010, 81–82); Goldstein (1992, 6–8).

War of the Roman Republic
Primary opponents: Two Sicilies, France, and Austria v. Papal States
Dates: 30 April–2 July 1849
COW war number: 16
Summary: Italian liberals declared a Roman Republic in February 1849. France sent an army to restore the Pope to power. In the resulting War of the Roman Republic, France was joined by the Kingdom of Two Sicilies and Austria, and defeated the burgeoning republic from April to July 1849.
Recommended sources: Sarkees and Wayman (2010, 82–83); Clodfelter (2008, 190–1).

La Plata War
Primary opponents: Brazil v. Argentina
Dates: 19 July 1851–3 February 1852
COW war number: 19
Summary: Argentina besieged Montevideo, the capital of Uruguay, for nine years. In July 1851, Brazil intervened in order to maintain Uruguay's independence (which later became a state in 1882). Brazil went on to invade Argentina, winning the decisive battle near Buenos Aires.
Recommended sources: Marley (2008, 742–4); Sarkees and Wayman (2010, 83–84).

Crimean War
Primary opponents: France, Ottoman Empire (Turkey), Sardinia/Piedmont (Italy), and United Kingdom v. Russia
Dates: 23 October 1853–1 March 1856
COW war number: 22
Summary: Russia sought to push the Turks out of Europe and expand its access to the Mediterranean and attacked the Ottoman Empire on 23 October 1853. The United Kingdom and France joined the war to defend the Ottoman Empire in 1854, with Sardinia/Piedmont joining the following year. Russia sued for peace on 1 March 1856.
Recommended sources: Figes (2010); Royle (2000); Goldstein (1992, 27–29).

Anglo-Persian War
Primary opponents: United Kingdom v. Persia (Iran)
Dates: 25 October 1856–5 April 1857
COW war number: 25
Summary: Persian forces seized British-controlled Herat in Afghanistan. Britain invaded Persia in a combined naval and army action and Persia sued for peace.
Recommended sources: Sarkees and Wayman (2010, 85–86); Goldstein (1992, 76–77).

War of Italian Unification
Primary opponents: Austria v. Sardinia/Piedmont and France
Dates: 29 April–12 July 1859
COW war number: 28
Summary: France fought alongside Sardinia/Piedmont and they defeated Austria from April to July 1859. Austria ceded Lombardy to Sardinia (through France), and in 1860, Modena, Parma, and Tuscany were incorporated into Sardinia/Italy. While foreign influence had been successfully reduced, independent states in southern Italy stood in the way of unification.
Recommended sources: Goldstein (1992, 17–18); Clodfelter (2008, 195).

First Spanish-Moroccan War
Primary opponents: Spain v. Morocco
Dates: 22 October 1859–25 March 1860
COW war number: 31
Summary: Spain attacked Morocco to expand Spanish influence in the country. Spain prevailed in a matter of months.
Recommended sources: Sarkees and Wayman (2010, 85–86); Clodfelter (2008, 214).

Italian-Roman War
Primary opponents: Sardinia/Piedmont v. Papal States
Dates: 11 September–29 September 1860
COW war number: 34
Summary: Sardinia/Piedmont defeated the Papal States to move a step closer to Italian unification.
Recommended sources: Sarkees and Wayman (2010, 87–88); Clodfelter (2008, 196).

Neapolitan War
Primary opponents: Sardinia/Piedmont v. Two Sicilies
Dates: 15 October 1860–13 February 1861
COW war number: 37
Summary: Sardinia/Piedmont defeated the Kingdom of Two Sicilies. Apart from Venetia, still under Austrian rule, Italy was finally unified from north to south.
Recommended sources: Sarkees and Wayman (2010, 89); Clodfelter (2008, 196).

Franco-Mexican War
Primary opponents: France v. Mexico
Dates: 16 April 1862–5 February 1867
COW war number: 40
Summary: France invaded Mexico to support Mexican conservatives against the liberal government of Juarez. Despite occupying Mexico City for five years, fighting continued. France eventually withdrew under pressure from the United States.
Recommended sources: Marley (2008, 824–44); Goldstein (1992, 207–8).

Ecuadorean-Colombian War

Primary opponents: Colombia v. Ecuador

Dates: 22 November–6 December 1863

COW war number: 43

Summary: Ecuador invaded Colombia but was swiftly and decisively defeated.

Recommended sources: Sarkees and Wayman (2010, 91); Clodfelter (2008, 339).

Second Schleswig-Holstein War

Primary opponents: Prussia and Austria v. Denmark

Dates: 1 February–20 July 1864

COW war number: 46

Summary: Austria and Prussia took the two southernmost duchies of Denmark, with Prussia gaining control of Schleswig and Austria gaining control of Holstein.

Recommended sources: Embree (2006); Goldstein (1992, 8–10).

Lopez War [War of the Triple Alliance]

Primary opponents: Paraguay v. Brazil and Argentina

Dates: 12 November 1864–1 March 1870

COW war number: 49

Summary: Paraguay's leader, Francisco Solano Lopez, wanted to incorporate Uruguay and a portion of Brazil into Paraguay. Paraguay invaded Brazil, and later Argentina. Through six years of fighting, Paraguay lost half of its population before Lopez was finally captured and killed, ending the war. Uruguay fought alongside Brazil and Argentina but was not yet an independent state.

Recommended sources: Leuchars (2002); Marley (2008, 854–65); Goldstein (1992, 194–6).

Naval War

Primary opponents: Spain v. Peru and Chile

Dates: 25 September 1865–9 May 1866

COW war number: 52

Summary: Spain sent a fleet to the Pacific in an attempt to reassert control of its former colony Peru. Chile and Peru eventually defeated the Spanish fleet in a naval campaign.

Recommended sources: Marley (2008, 849–54); Goldstein (1992, 193–4).

Seven Weeks War [Austro-Prussian War]

Primary opponents: Prussia, Italy, and allies v. Austria and allies

Dates: 15 June–26 July 1866

COW war number: 55

Summary: Bismarck sought war with Austria to establish Prussia as the leading German state. Prussia invaded Austria and, taking advantage of superior rifles and tactics, decisively defeated the Austrian forces. Following the culminating Battle of Königgrätz, Austria sued for peace.

Recommended sources: Wawro (1996); Barry (2010); Clodfelter (2008, 197–8).

Franco-Prussian War
Primary opponents: France v. Prussia and allies
Dates: 19 July 1870–26 February 1871
COW war number: 58
Summary: Prussia and minor German states invaded France following the French declaration of war on 15 July 1870. They quickly defeated the French army at the Battle of Sedan on 1 September 1870 and then laid siege to Paris beginning on 19 September. Germany was unified on 18 January 1871 before the war finally ended on 26 February 1871.
Recommended sources: Howard (2001); Wawro (2003); Clodfelter (2008, 199–202).

First Central American War
Primary opponents: Guatemala v. El Salvador
Dates: 27 March–25 April 1876
COW war number: 60
Summary: Guatemala invaded El Salvador following a change of government in El Salvador. Guatemala prevailed quickly.
Recommended sources: Sarkees and Wayman (2010, 98–9); Clodfelter (2008, 329–30).

Second Russo-Turkish War
Primary opponents: Russia v. Ottoman Empire (Turkey)
Dates: 24 April 1877–31 January 1878
COW war number: 61
Summary: Russia invaded the Ottoman Empire to decrease its territorial control of the Balkans. By the end of the war, Russia occupied almost all of the Ottoman Balkans.
Recommended sources: Goldstein (1992, 32–33); Clodfelter (2008, 209–10).

War of the Pacific
Primary opponents: Chile v. Bolivia and Peru
Dates: 14 February 1879–11 December 1883
COW war number: 64
Summary: Chile invaded Bolivia and Peru to seize disputed territory. Early battles were at sea, followed by a series of land battles. Chile won, greatly expanding its territory in the north.
Recommended sources: Marley (2008, 879–89); Goldstein (1992, 196–8).

Conquest of Egypt
Primary opponents: United Kingdom v. Egypt
Dates: 11 July–15 September 1882
COW war number: 65

Summary: Britain invaded Egypt following claims that Europeans—in Egypt because of the Suez Canal—were being massacred. Britain won and took control of Egypt as a colony.

Recommended sources: Sarkees and Wayman (2010, 102–3); Clodfelter (2008, 217).

Sino-French War

Primary opponents: France v. China

Dates: 15 June 1884–19 June 1885

CO*W war number:* 67

Summary: France and China went to war over the growth of France's Indochina colony. France was eventually able to force China to accept the growth of its colony.

Recommended sources: Sarkees and Wayman (2010, 104–5); Goldstein (1992, 103–4).

Second Central American War

Primary opponents: Guatemala v. El Salvador

Dates: 28 March–15 April 1885

CO*W war number:* 70

Summary: Guatemala attacked El Salvador in order to pave the way for a Union of Central America. Although initially successful, the death of Guatemalan President Barrios brought a swift end to the war.

Recommended sources: Sarkees and Wayman (2010, 105–6); Clodfelter (2008, 330).

First Sino-Japanese War

Primary opponents: Japan v. China

Dates: 25 July 1894–30 March 1895

CO*W war number:* 73

Summary: Japan fought China over influence in Korea. Japan won, acquiring Formosa and other concessions.

Recommended sources: Paine (2003); Goldstein (1992, 104–6).

Greco-Turkish War

Primary opponents: Greece v. Ottoman Empire (Turkey)

Dates: 15 February–19 May 1897

CO*W war number:* 76

Summary: Greece intervened in a civil war between Crete and the Ottoman Empire, fighting alongside the rebellion. Ottoman forces then invaded Greece before a cease-fire was signed.

Recommended sources: Sarkees and Wayman (2010, 107–8); Goldstein (1992, 35–36).

Spanish-American War

Primary opponents: United States v. Spain

Dates: 22 April–12 August 1898

CO*W war number:* 79

Summary: Tensions had been rising over Spain's repression in their colony of Cuba and then boiled over after the US battleship *Maine* was sunk in Havana harbor on 15 February 1898. Rallying around the cry of "Remember the *Maine*," the United States went to war on 22 April 1898. Following a series of naval battles, the United States invaded the Philippines, Puerto Rico, and Cuba. An armistice was signed on 12 August, and the United States annexed Puerto Rico and the Philippines and gained a naval base at Guantanamo Bay, Cuba, while Cuba was granted independence.
Recommended sources: O'Toole (1984); Trask (1981); Marley (2008, 907–19).

Boxer Rebellion
Primary opponents: United States, United Kingdom, France, Russia, and Japan v. China
Dates: 17 June–14 August 1900
COW *war number:* 82
Summary: The United States, United Kingdom, France, Russia, and Japan fought to put down a rebellion against their influence in China by the Boxers, a group of Chinese martial artists. China fought alongside the Boxers. The foreign powers emerged victorious.
Recommended sources: Silbey (2013); Preston (1999); Sarkees and Wayman (2010, 110–2); Goldstein (1992, 106–7).

Sino-Russian War
Primary opponents: Russia v. China
Dates: 17 July–10 October 1900
COW *war number:* 83
Summary: In this war related to the Boxer Rebellion, Russia launched a separate invasion of Manchuria. Russia won decisively, gaining control of most of Manchuria.
Recommended sources: Sarkees and Wayman (2010, 112–3); Clodfelter (2008, 382).

Russo-Japanese War
Primary opponents: Japan v. Russia
Dates: 8 February 1904–15 September 1905
COW *war number:* 85
Summary: Following wars with China in the previous decade, Russia and Japan had increased their influence in Manchuria and Korea, respectively. Each sought to expand their influence in the area, and Russia sought to contain Japan's influence. Japan launched a surprise attack on the Russian fleet at Port Arthur, followed up by a siege of Port Arthur, land battles in Korea and Manchuria, and a series of naval engagements. The war resulted in a decisive Japanese victory, after which Korea was occupied by Japan.
Recommended sources: Connaughton (2003); Warner and Warner (2002); Goldstein (1992, 107–9).

Third Central American War

Primary opponents: Guatemala v. El Salvador and Honduras

Dates: 27 May–20 July 1906

COW war number: 88

Summary: Guatemala invaded El Salvador following El Salvador's attempt to overthrow the Guatemalan president. Honduras fought alongside El Salvador. Guatemala won, but the war ended due to pressure from the United States and Mexico.

Recommended sources: Sarkees and Wayman (2010, 115–6); Clodfelter (2008, 400).

Fourth Central American War

Primary opponents: Nicaragua v. El Salvador and Honduras

Dates: 19 February–23 April 1907

COW war number: 91

Summary: Nicaragua attacked Honduras and then El Salvador in order to promote regime change and regional unity. The war ended under pressure from the United States and Mexico.

Recommended sources: Sarkees and Wayman (2010, 116–7); Goldstein (1992, 210–1); Clodfelter (2008, 400).

Second Spanish-Moroccan War

Primary opponents: Spain v. Morocco

Dates: 7 July 1909–23 March 1910

COW war number: 94

Summary: Spain attacked Morocco following the killing of four Spanish rail workers. Although they faced a series of early defeats, Spanish forces were eventually able to prevail, adding territory to Spain's Moroccan possessions.

Recommended sources: Sarkees and Wayman (2010, 117–8); Clodfelter (2008, 379).

Italian-Turkish War

Primary opponents: Italy v. Ottoman Empire (Turkey)

Dates: 29 September 1911–18 October 1912

COW war number: 97

Summary: Italy attacked and seized Libya from the Ottoman Empire in a combined land and naval campaign.

Recommended sources: Sarkees and Wayman (2010, 118–9); Goldstein (1992, 36–38).

First Balkan War

Primary opponents: Serbia, Bulgaria, and Greece v. Ottoman Empire (Turkey)

Dates: 17 October 1912–19 April 1913

COW war number: 100

Summary: Encouraged by Ottoman weaknesses exposed by the Italian-Turkish War, Serbia, Bulgaria, and Greece attacked the Ottoman Empire to

seize additional territory in the Balkans. They won, leaving the Ottoman Empire with only a small portion of European territory near Istanbul.
Recommended sources: Hall (2000); Sarkees and Wayman (2010, 119–20); Goldstein (1992, 38–40).

Second Balkan War
Primary opponents: Bulgaria v. Greece, Romania, Serbia, and Ottoman Empire (Turkey)
Dates: 30 June–30 July 1913
COW *war number:* 103
Summary: Bulgaria attacked its former allies Greece and Serbia because of disagreements over territory they had just won from the Ottoman Empire. Romania and the Ottoman Empire entered the war against Bulgaria. Bulgaria lost, losing some territory to each of the other states.
Recommended sources: Hall (2000); Sarkees and Wayman (2010, 120–1); Goldstein (1992, 40–41).

World War I
Primary opponents: Germany, Austria-Hungary, and allies v. Britain, France, Russia and allies
Dates: 29 July 1914–11 November 1918
COW *war number:* 106
Summary: The war came with the July Crisis of 1914, which was sparked by the assassination of Archduke Franz Ferdinand, heir to the Austrian throne, by a Serbian nationalist. The war lasted from August 1914 until November 1918, and was dominated—particularly on the decisive western front—by trench warfare and stalemate. Attempts to restore mobility to the battlefield in the face of massive firepower, particularly from artillery and machine guns, led to the emergence of modern warfare.
Recommended sources: Keegan (1999); Hart (2013); Hastings (2013); Clodfelter (2008, 415–64); Goldstein (1992, 42–50).

Estonian War of Liberation
Primary opponents: Russia v. Estonia and Finland
Dates: 22 November 1918–3 January 1920
COW *war number:* 107
Summary: Russia tried to reestablish control over Estonia after World War I. Finland supported Estonia, and the Estonians were able to drive Soviet forces out of Estonia.
Recommended sources: Sarkees and Wayman (2010, 124–5).

Latvian War of Liberation
Primary opponents: Russia and Germany v. Latvia, Estonia, and Germany
Dates: 2 December 1918–1 February 1920
COW *war number:* 108

Summary: German forces remained in Latvia following World War I. Russia invaded Latvia in an attempt to regain control of Latvia. After an unsuccessful coup d'état, Germany switched sides to fight against the Latvians. Estonia entered the war alongside Latvia, and they were eventually able to prevail.
Recommended sources: Sarkees and Wayman (2010, 125–6).

Russo-Polish War
Primary opponents: Russia v. Poland
Dates: 14 February 1919–18 October 1920
COW war number: 109
Summary: Poland gained independence from Russia in World War I. After the war, both Polish and Russian troops fought to gain more of each other's territory. Poland was able to prevail, gaining territory.
Recommended sources: Davies (1972); Borzecki (2008); Goldstein (1992, 50–51).

Hungarian Adversaries War
Primary opponents: Romania and Czechoslovakia v. Hungary
Dates: 16 April–14 August 1919
COW war number: 112
Summary: Romania and Czechoslovakia attacked Hungary in order to gain additional territory following World War I and the breakup of Austria-Hungary. They succeeded, and Hungary lost much of its territory.
Recommended sources: Sarkees and Wayman (2010, 128–9); Clodfelter (2008, 370).

Second Greco-Turkish War
Primary opponents: Greece v. Turkey
Dates: 5 May 1919–11 October 1922
COW war number: 115
Summary: Greece occupied areas around Smyrna in Turkey. They began advancing further inland. Although Greece was successful at first, Turkey eventually was able to counterattack and drive Greek forces out of Turkey.
Recommended sources: Sarkees and Wayman (2010, 129–30); Goldstein (1992, 51–52).

Franco-Turkish War
Primary opponents: France v. Turkey
Dates: 1 November 1919–20 October 1921
COW war number: 116
Summary: France occupied Istanbul and invaded southern Turkey to expand its new protectorate of Syria. Fierce Turkish resistance eventually led France to withdraw.
Recommended sources: Sarkees and Wayman (2010, 130–1).

Lithuanian-Polish War
Primary opponents: Poland v. Lithuania
Dates: 15 July–1 December 1920
COW *war number:* 117
Summary: Poland and Lithuania, newly independent following World War I, disputed control of the city of Vilna. Fighting began between the two sides over the city, but Poland was able to prevail and gained control of the city.
Recommended sources: Sarkees and Wayman (2010, 131–2); Clodfelter (2008, 370).

Manchurian War
Primary opponents: Soviet Union v. China
Dates: 17 August–3 December 1929
COW *war number:* 118
Summary: The Soviet Union and China disputed the status of a Soviet railroad crossing Chinese territory in Manchuria. Soviet troops invaded Manchuria, and the two sides came to an agreement concerning cooperation on the railroad.
Recommended sources: Sarkees and Wayman (2010, 132–3); Clodfelter (2008, 390).

Second Sino-Japanese War
Primary opponents: Japan v. China
Dates: 19 December 1931–22 May 1933
COW *war number:* 121
Summary: Japan sought to expand its influence in Manchuria while China was distracted in civil war. Japan defeated China and established a puppet state, called Manchukuo, in Manchuria.
Recommended sources: Sarkees and Wayman (2010, 133–5); Goldstein (1992, 62–65).

Chaco War
Primary opponents: Bolivia v. Paraguay
Dates: 15 June 1932–12 June 1935
COW *war number:* 124
Summary: Bolivia attacked Paraguay to resolve competing territorial claims to the Chaco region. Although initially unprepared for war, Paraguay recovered and won, gaining territory from Bolivia.
Recommended sources: Farcau (1996); Marley (2008, 989–1002); Goldstein (1992, 199–200).

Saudi-Yemeni War
Primary opponents: Saudi Arabia v. Yemen
Dates: 20 March–13 May 1934
COW *war number:* 125

Summary: Yemen and Saudi Arabia disputed the territory of Asir. Saudi Arabia attacked and won decisively, forcing Yemen to renounce most of its claims to Asir.
Recommended sources: Sarkees and Wayman (2010, 136–7); Clodfelter (2008, 376).

Italo-Ethiopian War [Conquest of Ethiopia]
Primary opponents: Italy v. Ethiopia
Dates: 5 October 1935–9 May 1936
COW war number: 127
Summary: Italy invaded Ethiopia from its colony in Italian Somaliland. In a relatively short campaign from October 1935 to May 1936, Italy conquered Ethiopia and made it a colony.
Recommended sources: Goldstein (1992, 164–6); Clodfelter (2008, 381–2).

Third Sino-Japanese War
Primary opponents: Japan v. China
Dates: 7 July 1937–6 December 1941
COW war number: 130
Summary: In July 1937, Japan invaded the rest of China, launching the Third Sino-Japanese War. Japan captured Beijing and much of China's urban areas and coastal provinces. The brutal campaign was highlighted by a variety of abuses of the Chinese population, especially the Rape of Nanking, where over 600,000 civilians were killed in 1937 alone. Nonetheless, China continued fighting, and after Japan's attack on Pearl Harbor in 1941, this became the Chinese front of World War II.
Recommended sources: Sarkees and Wayman (2010, 138–9); Goldstein (1992, 62–65); Clodfelter (2008, 391–5).

Changkufeng War
Primary opponents: Japan v. Soviet Union
Dates: 29 July–11 August 1938
COW war number: 133
Summary: Japan attacked Soviet forces controlling Changkufeng hill on the border between the Soviet Union and Japan's Manchukuo puppet. Faced with a strong Soviet defense, Japanese forces withdrew.
Recommended sources: Coox (1977); Goldstein (1992, 65–66).

Nomonhan War
Primary opponents: Japan v. Soviet Union and Mongolia
Dates: 11 May–16 September 1939
COW war number: 136
Summary: Japan attacked Soviet and Mongolian forces around the Nomonhan Bridge, at the border between Mongolia and Manchukuo.

Despite initial Japanese gains, the tide turned after Soviet reinforcements arrived, and Japan was driven back to the border.
Recommended sources: Goldman (2012); Goldstein (1992, 66–67).

World War II
Primary opponents: Germany, Japan, Italy, and allies v. Britain, France, United States, Soviet Union, and allies
Dates: 1 September 1939–14 August 1945
COW war number: 139
Summary: Germany invaded Poland, and Britain and France declared war two days later. Germany conquered most of Europe in the first two years of war. The war expanded greatly when Japan attacked the United States at Pearl Harbor. Japan swept through the South Pacific and Southeast Asia in a massively successful series of campaigns over the next several months. Following the Battle of Midway in June 1942, the United States began its campaign to push back Japan. The Soviets defeated German forces at the Battle of Stalingrad in January 1943, after which the Allies began to push German forces back on all sides before Germany finally surrendered on 8 May 1945. In the Pacific, the United States conducted an island hopping campaign, attacking only crucial locations. After costly battles in the Philippines, Iwo Jima, and Okinawa, the United States dropped atomic bombs on Hiroshima and Nagasaki in August 1945, which led the Japanese to surrender.
Recommended sources: Weinberg (1994); Murray and Millet (2000); Keegan (1990); Clodfelter (2008, 464–570); Goldstein (1992, 53–71).

Winter War [Russo-Finnish War]
Primary opponents: Soviet Union v. Finland
Dates: 30 November 1939–12 March 1940
COW war number: 142
Summary: The Soviet Union invaded Finland to seize contested territory. Despite being badly outnumbered, Finland fought well, stopping the initial Soviet attacks. Eventually, the Soviets were able to break through Finnish lines, and Finland sued for peace.
Recommended sources: Sander (2013); Trotter (1991); Goldstein (1992, 57–59).

Franco-Thai War
Primary opponents: Thailand v. France
Dates: 1 December 1940–28 January 1941
COW war number: 145
Summary: Thailand attacked French Indochina to take disputed territory, taking advantage of France's defeat to Germany. Thailand won, taking the disputed provinces.
Recommended sources: Sarkees and Wayman (2010, 144–5); Clodfelter (2008, 396).

First Kashmir War [First Indo-Pakistani War]

Primary opponents: India v. Pakistan

Dates: 26 October 1947–1 January 1949

COW war number: 147

Summary: The First Kashmir War between India and Pakistan lasted from October 1947 to January 1949 and ended in a stalemate, with Pakistan in control of about one third of Kashmir.

Recommended sources: Goldstein (1992, 85–86); Arnold (1995, 275–81).

Palestine War [Arab-Israeli War of 1948–1949]

Primary opponents: Jordan, Iraq, Egypt, Lebanon, and Syria v. Israel

Dates: 15 May 1948–7 January 1949

COW war number: 148

Summary: Israel became an independent state on 14 May 1948, and the Arab states attacked the next day. The Palestine War lasted from May 1948 to January 1949, as Israel was able to prevail against a coalition of Egypt, Syria, Jordan, Lebanon, and Iraq, gaining control of almost 80 percent of the Palestine mandate.

Recommended sources: Morris (2008); Herzog and Gazit (2005); Goldstein (1992, 122–5); Arnold (1995, 307–15).

Korean War

Primary opponents: North Korea and China v. South Korea, United States, and allies

Dates: 24 June 1950–27 July 1953

COW war number: 151

Summary: The Korean War began in June 1950 when North Korea invaded South Korea. The United States, along with other countries fighting under the authority of the United Nations, intervened. After US forces began moving north across the 38th parallel in October, China intervened. The front line stabilized near the original border by June 1951, and remained largely stationary until an armistice was finally reached in July 1953, ending the war.

Recommended sources: Halberstam (2007); Hastings (1987); Goldstein (1992, 111–2); Arnold (1995, 151–8).

Off-shore Islands War

Primary opponents: China v. Taiwan

Dates: 3 September 1954–23 April 1955

COW war number: 153

Summary: China attacked Taiwan's hold on islands off the coast of China. Taiwan evacuated all of the islands except for Quemoy and Matsu.

Recommended sources: Sarkees and Wayman (2010, 148–9); Clodfelter (2008, 673).

Sinai War [Suez War]
Primary opponents: Israel, United Kingdom, and France v. Egypt
Dates: 29 October–6 November 1956
COW war number: 155
Summary: Egypt nationalized the Suez Canal, which had been operated by Britain and France, in July 1956. To punish Egypt, Britain and France developed a plan alongside Israel to invade. Israel invaded the Sinai Peninsula, quickly routing the Egyptian troops. The United Kingdom and France began air strikes and landed troops to seize the canal. After the war ended, a United Nations peacekeeping force was deployed to the Sinai.
Recommended sources: Turner (2006); Herzog and Gazit (2005); Goldstein (1992, 125–7); Arnold (1995, 315–20).

Russo-Hungarian War [Soviet Invasion of Hungary]
Primary opponents: Soviet Union v. Hungary
Dates: 4–14 November 1956
COW war number: 156
Summary: Hungarian leader Imre Nagy tried to liberalize Hungary. Soviet troops invaded Hungary, and Hungarian forces were defeated in heavy fighting.
Recommended sources: Gati (2006); Sarkees and Wayman (2010, 150–2).

Ifni War
Primary opponents: Morocco v. Spain and France
Dates: 21 November 1957–10 April 1958
COW war number: 158
Summary: After regaining its independence from France, Morocco sought to regain Ifni and other Spanish possessions. Morocco attacked Ifni, but Spain, joined by France, was able to drive back the Moroccan forces.
Recommended sources: Sarkees and Wayman (2010, 152–3); Clodfelter (2008, 589).

Taiwan Straits War
Primary opponents: China v. Taiwan
Dates: 23 August–23 November 1958
COW war number: 159
Summary: The unresolved status of Quemoy and Matsu led to the Taiwan Straits War in 1958, which ended in stalemate after the US Navy assisted in supplying Taiwan.
Recommended sources: Sarkees and Wayman (2010, 153–4); Clodfelter (2008, 673–4).

Sino-Indian War [War in Assam]
Primary opponents: China v. India

Dates: 20 October–22 November 1962
COW war number: 160
Summary: In an attempt to resolve disagreement about the location of their border, China attacked India. India fought poorly, and China was able to capture much of the disputed territory.
Recommended sources: Goldstein (1992, 86–88); Arnold (1995, 271–2).

Vietnam War
Primary opponents: North Vietnam v. South Vietnam, United States, and allies
Dates: 7 February 1965–30 April 1975
COW war number: 163
Summary: The South Vietnamese government was very unpopular and faced an insurgency supported by North Vietnam. After years of providing aid and military advisors, the United States became directly involved in 1965, beginning the Vietnam War. Fighting ground battles in the south against Viet Cong insurgents and the North Vietnamese Army and an air campaign in the north, the United States struggled to make progress. The war became immensely unpopular on the American home front, and the United States withdrew from the war in January 1973. In January 1975, North Vietnam invaded South Vietnam, and by the end of April had won.
Recommended sources: Prados (2009); Schulzinger (1997); Daddis (2014); Clodfelter (2008, 712–63).

Second Kashmir War [Second Indo-Pakistani War]
Primary opponents: Pakistan v. India
Dates: 5 August–23 September 1965
COW war number: 166
Summary: Pakistan invaded Indian-controlled Kashmir. Fighting continued through September, with Pakistan gaining a small victory against India.
Recommended sources: Goldstein (1992, 88–89); Arnold (1995, 275–81).

Six-Day War
Primary opponents: Israel v. Egypt, Jordan, and Syria
Dates: 5–10 June 1967
COW war number: 169
Summary: Fearing an attack by its Arab neighbors, Israel preemptively attacked Egypt, Jordan, and Syria. In a lightning campaign during 5–10 June 1967, Israel decisively defeated the Arab states. Following their victory, Israel occupied the Gaza Strip, Sinai Peninsula, West Bank, and Golan Heights.
Recommended sources: Oren (2002); Herzog and Gazit (2005); Goldstein (1992, 127–8); Arnold (1995, 320–5).

Second Laotian War

Primary opponents: Vietnam v. Laos, Thailand, and United States

Dates: 13 January 1968–17 April 1973

COW war number: 170

Summary: North Vietnam intervened in the civil war in Laos, fighting against the Laotian government. The United States and Thailand fought alongside the Laotian government. This war was connected to the Vietnam War; following the withdrawal of the United States from the Vietnam War, American forces withdrew from Laos, ending the war.

Recommended sources: Sarkees and Wayman (2010, 158–9); Clodfelter (2008, 660–2).

War of Attrition

Primary opponents: Egypt v. Israel

Dates: 6 March 1969–7 August 1970

COW war number: 172

Summary: Egypt launched the War of Attrition to challenge Israeli control of the Sinai following the Six-Day War. The war was characterized by artillery duels, air strikes, periodic land raids, and general stalemate.

Recommended sources: Bar-Siman-Tov (1980); Korn (1992); Herzog and Gazit (2005).

Football War

Primary opponents: El Salvador v. Honduras

Dates: 14–18 July 1969

COW war number: 175

Summary: Honduran fans traveling home after a World Cup qualifying match in El Salvador were attacked and beaten. In retaliation, Hondurans began attacking and killing Salvadorans living there. El Salvador invaded Honduras and won, gaining a pledge by Honduras to protect Salvadorans living in the country.

Recommended sources: Goldstein (1992, 211–2); Arnold (1995, 299–301).

War of the Communist Coalition

Primary opponents: North Vietnam v. Cambodia, South Vietnam, and United States

Dates: 23 March 1970–2 July 1971

COW war number: 176

Summary: North Vietnam intervened in Cambodia to overthrow the government. The United States and South Vietnam fought alongside the Cambodian government. North Vietnam withdrew its forces, but fighting continued as a civil war.

Recommended sources: Sarkees and Wayman (2010, 161–2).

Bangladesh War [Third Indo-Pakistani War]
Primary opponents: Pakistan v. India
Dates: 3–17 December 1971
COW war number: 178
Summary: East Pakistan began rebelling against the government of West Pakistan in the late 1960s, and in December 1971, India went to war to support East Pakistan's independence. In the resulting Bangladesh War, India defeated Pakistan in only two weeks, and East Pakistan gained independence as Bangladesh.
Recommended sources: Raghavan (2013); Goldstein (1992, 89–91); Arnold (1995, 275–81).

Yom Kippur War [October War]
Primary opponents: Egypt, Syria, Jordan, Iraq, and Saudi Arabia v. Israel
Dates: 6–24 October 1973
COW war number: 181
Summary: The Arab states remained dissatisfied with Israel's control of the territories occupied since the Six-Day War and decided to retake the territory by force. On 6 October 1973, Egypt and Syria launched a coordinated surprise attack on Israel. Initially, the Arab forces succeeded in pushing into the Sinai Peninsula and Golan Heights before Israel was able to gain the upper hand through a series of counterattacks. Israel won another decisive victory, crossing to the west side of the Suez Canal before agreeing to a cease-fire on 24 October, ending the war.
Recommended sources: Rabinovich (2004); Herzog and Gazit (2005); Goldstein (1992, 128–30); Arnold (1995, 325–30).

Turco-Cypriot War
Primary opponents: Turkey v. Cyprus
Dates: 20 July–16 August 1974
COW war number: 184
Summary: Turkey invaded Cyprus after Greece rejected its demand to withdraw Greek forces from the island. Turkey gained control of the northern 40 percent of the island as an enclave for Turkish Cypriots.
Recommended sources: Sarkees and Wayman (2010, 164–5); Clodfelter (2008, 580).

War Over Angola
Primary opponents: South Africa and Democratic Republic of the Congo v. Angola
Cuba
Dates: 23 October 1975–12 February 1976
COW war number: 186
Summary: Angola was involved in a nonstate war between different factions (it became a system member in November 1975). South

Africa invaded in support of the National Union for the Total Independence of Angola (UNITA) and the National Front for the Liberation of Angola (FNLA). Cuban forces opposed them, assisting the People's Movement for the Liberation of Angola (MPLA). By February 1976, Cuban and Angolan forces had driven South African troops from Angola. This was followed by continued civil war until 1991.

Recommended sources: Sarkees and Wayman (2010, 166–7); Goldstein (1992, 177–80).

Second Ogaden War

Primary opponents: Somalia v. Ethiopia and Cuba

Dates: 23 July 1977–9 March 1978

COW war number: 187

Summary: Somalia invaded Ethiopia to support the Western Somali Liberation Front (WSLF), a group fighting for independence of the Ogaden region of Ethiopia. After initial success, further Somali advances were repulsed by Ethiopian and Cuban forces. Somalia withdrew in March 1978, but fighting continued as a civil war.

Recommended sources: Sarkees and Wayman (2010, 170–1); Goldstein (1992, 170–1); Arnold (1995, 208–16).

Vietnamese-Cambodian Border War

Primary opponents: Cambodia v. Vietnam

Dates: 24 September 1977–8 January 1979

COW war number: 189

Summary: Cambodia attacked Vietnam over disputed territory along their border. Vietnam counterattacked and won decisively, installing a new government in Cambodia. An insurgency against the new government continued until 1989.

Recommended sources: Sarkees and Wayman (2010, 168–70); Goldstein (1992, 118–9).

Ugandan-Tanzanian War

Primary opponents: Uganda and Libya v. Tanzania

Dates: 28 October 1978–11 April 1979

COW war number: 190

Summary: Uganda attacked Tanzania in an attempt to capture a large amount of land and gain access to the Indian Ocean. Tanzania pushed Ugandan troops out and moved into Uganda. Libya sent troops to help Uganda, but Tanzania still emerged victorious.

Recommended sources: Goldstein (1992, 182–3); Arnold (1995, 235–9).

Sino-Vietnamese Punitive War

Primary opponents: China v. Vietnam

Dates: 17 February–16 March 1979

COW war number: 193

Summary: China attacked Vietnam in retaliation for the expulsion of ethnic Chinese from Vietnam. After a short campaign, China withdrew its forces, declaring that Vietnam had been sufficiently punished.
Recommended sources: Goldstein (1992, 119–21); Arnold (1995, 272–5).

Iran-Iraq War
Primary opponents: Iraq v. Iran
Dates: 22 September 1980–20 August 1988
COW *war number:* 199
Summary: The Iran-Iraq War was initiated by Iraq in September 1980. Although Iraqi forces were initially successful, Iran counterattacked in January 1981, and by May 1982 had driven Iraqi forces out of Iranian territory. The war bogged down in a long stalemate, finally ending in August 1988.
Recommended sources: Hiro (1991); Goldstein (1992, 133–5).

Falklands War
Primary opponents: Argentina v. United Kingdom
Dates: 25 March–15 June 1982
COW *war number:* 202
Summary: The Falklands War occurred from March through June 1982 when Argentina seized the British-controlled Falkland Islands. The United Kingdom fought back, and after a difficult naval and air campaign, successfully retook the islands.
Recommended sources: Hastings and Jenkins (1984); Middlebrook (1987, 1989); Marley (2008, 1059–77).

Lebanon War [War over Lebanon]
Primary opponents: Israel v. Syria
Dates: 21 April–15 September 1982
COW *war number:* 205
Summary: The Palestine Liberation Organization (PLO) had moved into Lebanon after being expelled from Jordan. Israel invaded in April 1982 to remove the PLO from Lebanon. Syria, who occupied part of Lebanon, fought against Israel. Although Israel dominated a series of aerial battles with Syria, in the end the war was a stalemate.
Recommended sources: Herzog and Gazit (2005); Arnold (1995, 334–40).

War over the Aouzou Strip
Primary opponents: Chad v. Libya
Dates: 15 November 1986–11 September 1987
COW *war number:* 207
Summary: After intervening in a civil war in Chad, Libya occupied a piece of territory called the Aouzou Strip. Chad attacked Libyan forces there and won, seizing a number of Libyan bases.

Recommended sources: Goldstein (1992, 180–2); Arnold (1995, 192–202).

Sino-Vietnamese Border War
Primary opponents: China v. Vietnam
Dates: 5 January–6 February 1987
COW war number: 208
Summary: China attacked Vietnam along their disputed border. Vietnam fought back, and the war ended in a draw.
Recommended sources: Sarkees and Wayman (2010, 175–6); Arnold (1995, 272–5).

Gulf War
Primary opponents: Iraq v. Kuwait, United States, Saudi Arabia, United Kingdom, France, and others
Dates: 2 August 1990–11 April 1991
COW war number: 211
Summary: Iraq invaded Kuwait on 2 August 1990. The United States led a coalition of the United Kingdom, France, Saudi Arabia, and other Arab states to liberate Kuwait. The campaign, called Operation Desert Storm, began on 16 January 1991 with a six-week air campaign preceding a short, decisive ground offensive beginning on 24 February.
Recommended sources: Atkinson (1993); Goldstein (1992, 135–8); Arnold (1995, 144–51).

War of Bosnian Independence
Primary opponents: Yugoslavia v. Bosnia and Croatia
Dates: 7 April–5 June 1992
COW war number: 215
Summary: Bosnia, a former republic of Yugoslavia, became independent. Yugoslavia fought alongside Bosnian Serbs against Bosnian Muslims, Bosnian Croats, and Croatia. Yugoslavia withdrew its forces, marking the end of the interstate war, but fighting continued as a civil war.
Recommended sources: Sarkees and Wayman (2010, 177–8).

Azeri-Armenian War
Primary opponents: Armenia v. Azerbaijan
Dates: 6 February 1993–12 May 1994
COW war number: 216
Summary: Armenia and Azerbaijan, former Soviet republics, disagreed about the Nagorno-Karabakh territory. Armenia attacked Azerbaijan and eventually won.
Recommended sources: Sarkees and Wayman (2010, 178–80); Arnold (1995, 263–6).

Cenepa Valley War
Primary opponents: Ecuador v. Peru
Dates: 9 January–27 February 1995
COW war number: 217

Summary: Ecuador attacked Peru in an effort to resolve long-standing disagreements regarding the location of their border. Six weeks of fighting ended in a stalemate.
Recommended sources: Sarkees and Wayman (2010, 180–1).

Badme Border War
Primary opponents: Eritrea v. Ethiopia
Dates: 6 May 1998–12 December 2000
COW *war number:* 219
Summary: After Eritrea gained its independence from Ethiopia in 1993, the border between them was not resolved. Eritrea attacked contested territory under Ethiopian control. Heavy fighting for more than two years ended in a stalemate.
Recommended sources: Sarkees and Wayman (2010, 181).

Kosovo War
Primary opponents: United States, France, Germany, United Kingdom, and allies v. Yugoslavia
Dates: 24 March–10 June 1999
COW *war number:* 221
Summary: The Kosovo War was an air campaign from March through June 1999 by the United States and its NATO allies against Yugoslavia (essentially just Serbia) to end the Serbian campaign of ethnic cleansing in Kosovo.
Recommended sources: Daalder and O'Hanlon (2001); Lambeth (2001); Sarkees and Wayman (2010, 181–4).

Kargil War
Primary opponents: Pakistan v. India
Dates: 8 May–17 July 1999
COW *war number:* 223
Summary: Pakistan infiltrated troops past the line of control. India defeated Pakistan and reestablished its positions along the line of control.
Recommended sources: Tellis, Fair, and Medby (2001); Lavoy (2009); Sarkees and Wayman (2010, 184).

Afghanistan War [Invasion of Afghanistan]
Primary opponents: United States, United Kingdom, and allies v. Afghanistan
Dates: 7 October–22 December 2001
COW *war number:* 225
Summary: Afghanistan had provided a haven for al-Qaida, and so was an obvious target of retaliation for the 11 September 2001 terrorist attacks on the United States. The war was launched on 7 October 2001 by the United States, United Kingdom, and allies. The interstate portion of the war was a decisive success, as the Taliban government

was removed from power, and the war ended on 22 December 2001. It was followed, however, by an insurgency against the United States and the new Afghan government, which is still ongoing.

Recommended sources: Jones (2009); Sarkees and Wayman (2010, 184–6).

Iraq War [Invasion of Iraq]

Primary opponents: United States, United Kingdom, and allies v. Iraq

Dates: 19 March–2 May 2003

COW *war number:* 227

Summary: The United States, United Kingdom, and Australia invaded Iraq on 19 March 2003. Saddam Hussein's government was removed from power in a quick and decisive campaign. The United States, the new Iraqi government, and allies, however, faced a strong insurgency campaign in the years following the war.

Recommended sources: Gordon and Trainor (2006); Harvey (2012); Sarkees and Wayman (2010, 186–7).

Major Crises and Disputes

Fashoda Crisis (1898)

Primary opponents: Britain v. France

Summary: France sent a small force into southern Sudan, part of the British colony of Egypt. Britain responded with a flotilla of gunboats, meeting the French forces near Fashoda. The two sides reached an agreement over colonial borders without going to war.

Recommended sources: MacMillan (2013).

First Moroccan Crisis (1905)

Primary opponents: France v. Germany

Summary: Kaiser Wilhelm II of Germany visited Morocco and pledged German support for Moroccan independence. This was a threat to French influence in the area. France mobilized, but the crisis was resolved short of war.

Recommended sources: MacMillan (2013).

Bosnian Crisis (1908–1909)

Primary opponents: Austria-Hungary and Germany v. Serbia, Russia, and Turkey

Summary: Austria-Hungary announced the annexation of Bosnia-Herzegovina. Serbia, Russia, and Turkey protested but then backed down in the face of German support of Austria-Hungary.

Recommended sources: MacMillan (2013).

Agadir (Second Moroccan) Crisis (1911)

Primary opponents: France and Britain v. Germany

Summary: France deployed a substantial number of troops in the interior of Morocco, breaking their previous agreement with Germany.

Germany sent a gunboat to the port of Agadir. Britain supported France, and Morocco became a French colony.
Recommended sources: MacMillan (2013).

Rhineland Crisis (1936)
Primary opponents: Germany v. France, United Kingdom, and Belgium
Summary: Germany moved its army west of the Rhine River, territory that had been demilitarized in the Versailles Treaty that ended World War I. France, Britain, and Belgium did no more than complain.
Recommended sources: Wright (2007); Brecher and Wilkenfeld (1997, 242–4).

Munich (Sudetenland) Crisis (1938)
Primary opponents: Germany v. France, United Kingdom, and Czechoslovakia
Summary: Hitler demanded that the Sudetenland, territory of Czechoslovakia mostly comprised of ethnic Germans, be united with Germany. In a meeting at Munich, France and the United Kingdom agreed to cede the Sudetenland to Germany.
Recommended sources: Faber (2008); Wright (2007); Brecher and Wilkenfeld (1997, 228–30).

Berlin Blockade (1948–1949)
Primary opponents: Soviet Union v. United States, United Kingdom, and France
Summary: Berlin, divided into occupation zones among the four powers after World War II, was in the midst of the Soviet occupation zone. The crisis began on 24 June 1948 when the Soviet Union blockaded all Western transportation into and out of Berlin. The Western powers responded by airlifting supplies to the city. The blockade ended on 12 May 1949, and the division of Germany into two states was further solidified.
Recommended sources: Harrington (2012); Parrish (1998); Brecher and Wilkenfeld (1997, 342–3).

Cuban Missile Crisis (1962)
Primary opponents: United States v. Soviet Union and Cuba
Summary: On a routine reconnaissance flight, the United States discovered that the Soviets were installing nuclear missiles in Cuba. The ensuing Cuban Missile Crisis was probably the closest the world has come to nuclear war. War, however, was successfully avoided as the Soviets backed down in the face of an American naval quarantine (a polite term for a blockade).
Recommended sources: Allison and Zelikow (1999); Dobbs (2008); Munton and Welch (2011); Brecher and Wilkenfeld (1997, 352–4).

Sino-Soviet Border Crisis (1969)
Primary opponents: China v. Soviet Union

Summary: Fighting between the two states broke out between March and August 1969 over disputed territory along the Ussuri River. Casualties were below the level required for a war.

Recommended sources: Goldstein (1992, 116–8); Brecher and Wilkenfeld (1997, 554–5); Clodfelter (2008, 675–6).

First Gulf of Syrte Crisis (1981)

Primary opponents: United States v. Libya

Summary: Although it was traditionally viewed as international waters, Libya under Muammar Qaddafi claimed the entire Gulf of Syrte as its territorial waters. Under the Reagan administration, the US Navy challenged these claims by holding naval exercises in the Gulf, leading to a crisis between the United States and Libya in August 1981. Two Libyan aircraft were shot down by American aircraft.

Recommended sources: Arnold (1995, 129–30); Brecher and Wilkenfeld (1997, 463–4).

Osirak Crisis (1981)

Primary opponents: Iraq v. Israel

Summary: Iraq was developing a nuclear reactor at Osirak. On 7 June 1981, Israel destroyed the reactor with an air strike before it became operational.

Recommended sources: Claire (2004); Nakdimon (1988); Brecher and Wilkenfeld (1997, 292–4).

Grenada Invasion (1983)

Primary opponents: United States and Caribbean states v. Grenada

Summary: In October 1983, a military coup overthrew the government of Grenada, an island state in the Caribbean. In response, the United States invaded Grenada and restored the constitutional government to power.

Recommended sources: Adkin (1989); Marley (2008, 1077–80); Arnold (1995, 176–7).

Second Gulf of Syrte Crisis (1986)

Primary opponents: United States v. Libya

Summary: The United States again held naval exercises in the Gulf of Syrte in March–April 1986. The United States responded to Libyan attempts to shoot down American aircraft as well as Libyan support for terrorism, by conducting air raids on Tripoli and Benghazi, the two largest Libyan cities.

Recommended sources: Arnold (1995, 129–30); Brecher and Wilkenfeld (1997, 475–7).

Panama Invasion (1989–90)

Primary opponents: United States v. Panama

Summary: The United States invaded Panama in December 1989 to remove Manuel Noriega from power and bring him to trial on charges of direct involvement in drug trafficking from South America to the United States. In a quick campaign from 20 December 1989 through 3 January 1990, called Operation Just Cause, the United States succeeded in its objectives.

Recommended sources: Marley (2008, 1080–3); Arnold (1995, 178–81).

Taiwan Missile Crisis (1995–1996)

Primary opponents: China v. Taiwan

Summary: China test-launched a series of missiles over Taiwan and conducted military maneuvers off of its coast in an attempt to influence Taiwan's first democratic presidential election.

Recommended sources: ICB data viewer (http://www.cidcm.umd.edu/icb/dataviewer).

Operation Desert Strike (1996)

Primary opponents: United States v. Iraq

Summary: Iraq intervened in the Kurdish civil war, moving forces into the northern zone prohibited by UN Security Council resolutions. The United States launched a brief bombing campaign against Iraqi military positions in September 1996.

Recommended sources: ICB data viewer (http://www.cidcm.umd.edu/icb/dataviewer).

Operation Desert Fox (1998)

Primary opponents: United States and United Kingdom v. Iraq

Summary: After Iraq demanded that UN weapons inspectors be withdrawn from the country, the United States and United Kingdom launched a series of air strikes against Iraq.

Recommended sources: ICB data viewer (http://www.cidcm.umd.edu/icb/dataviewer).

References

Achen, Christopher H. 1986. *Statistical Analysis of Quasi-Experiments.* Berkeley: University of California Press.

Achen, Christopher H., and Duncan Snidal. 1989. "Rational Deterrence Theory and Comparative Case Studies." *World Politics* 41(2): 143–69.

Adkin, Mark. 1989. *Urgent Fury: The Battle for Grenada.* Lexington, MA: Lexington Books.

Alger, John I. 1982. *The Quest for Victory: The History of the Principles of War.* Westport, CT: Greenwood.

Allison, Graham T. 1971. *Essence of Decision: Explaining the Cuban Missile Crisis.* Boston: Little, Brown.

Allison, Graham T., and Philip Zelikow. 1999. *Essence of Decision: Explaining the Cuban Missile Crisis.* 2nd ed. New York: Longman.

Altfeld, Michael, and Bruce Bueno de Mesquita. 1979. "Choosing Sides in War." *International Studies Quarterly* 23(1): 87–112.

Angell, Norman. 1911. *The Great Illusion: A Study of the Relation of Military Power in Nations to their Economic and Social Advantage.* New York: Putnam.

Antal, John F. 1992. "Maneuver versus Attrition: A Historical Perspective." *Military Review* 72 (10): 21–33.

Arnold, Guy. 1995. *Wars in the Third World since 1945.* 2nd ed. London: Cassell.

Asal, Victor, and Kyle Beardsley. 2007. "Proliferation and International Crisis Behavior." *Journal of Peace Research* 44(2): 139–55.

Asher, Jerry, and Eric M. Hammel. 1987. *Duel for the Golan: The 100-hour Battle that Saved Israel.* New York: Morrow.

Atkinson, Rick. 1993. *Crusade: The Untold Story of the Persian Gulf War.* Boston: Houghton Mifflin.

Axelrod, Robert. 1984. *The Evolution of Cooperation.* New York: Basic Books.

Babst, Dean V. 1972. "A Force for Peace." *Industrial Research* 14: 55–58.

Baldwin, David A., ed. 1993. *Neorealism and Neoliberalism: The Contemporary Debate.* New York: Columbia University Press.

Baldwin, Hanson. 1950. *Great Mistakes of the War.* New York: Harper.

———. 1954. "Churchill Was Right." *The Atlantic* 194(1): 23–32.

Barbieri, Katherine. 1996. "Economic Interdependence: A Path to Peace or a Source of Conflict?" *Journal of Peace Research* 33(1): 29–49.

Barry, Quinton. 2010. *Road to Königgrätz: Helmuth von Moltke and the Austro-Prussian War 1866.* Solihull, UK: Helion.

Bar-Siman-Tov, Yaacov. 1980. *The Israeli-Egyptian War of Attrition, 1969–1970: A Case-Study of Limited Local War.* New York: Columbia University Press.

Bates, Robert H., Avner Greif, Margaret Levi, Jean-Laurent Rosenthal, and Barry R. Weingast. 1998. *Analytic Narratives.* Princeton, NJ: Princeton University Press.

Beardsley, Kyle C., David M. Quinn, Bidisha Biswas, and Jonathan Wilkenfeld. 2006. "Mediation Style and Crisis Outcome." *Journal of Conflict Resolution* 50(1): 58–86.

Bell, Coral. 1971. *The Conventions of Crisis: A Study in Diplomatic Management.* Oxford: Oxford University Press.

Bennett, D. Scott. 1996. "Security, Bargaining, and the End of Interstate Rivalry." *International Studies Quarterly* 40(2): 157–83.

———. 1997a. "Measuring Rivalry Termination." *Journal of Conflict Resolution* 41(2): 227–54.

———. 1997b. "Testing Alternative Models of Alliance Duration, 1816–1984." *American Journal of Political Science* 41(3): 846–78.

———. 1998. "Integrating and Testing Models of Rivalry." *American Journal of Political Science* 42(4): 1200–32.

———. 2000. *The Strategic Perspective in the Classroom.* Washington, DC: CQ Press.

Bennett, D. Scott, and Allan C. Stam III. 1996. "The Duration of Interstate Wars, 1816–1985." *American Political Science Review* 90(2): 239–57.

———. 1998. "The Declining Advantages of Democracy: A Combined Model of War Outcomes and Duration." *Journal of Conflict Resolution* 42(3): 344–66.

———. 2000a. "A Universal Test of an Expected Utility Theory of War." *International Studies Quarterly* 44(3): 451–80.

———. 2000b. "A Cross-Validation of Bueno de Mesquita and Lalman's International Interaction Game." *British Journal of Political Science* 30(4): 541–61.

———. 2004. *The Behavioral Origins of War.* Ann Arbor: University of Michigan Press.

———. 2006. "Predicting the Length of the 2003 US-Iraq War." *Foreign Policy Analysis* 2(2): 101–16.

Benson, Michelle. 2007. "Status Quo Preferences and Disputes Short of War." *International Interactions* 33(3): 271–88.

Bercovitch, Jacob; J. Theodore Anagnoson, and Donnette L. Wille. 1991. "Some Conceptual Issues and Empirical Trends in the Study of Successful Mediation in International Relations." *Journal of Peace Research* 28(1): 7–17.

Bernauer, Thomas, and Vally Koubi. 2009. "Effects of Political Institutions on Air Quality." *Ecological Economics* 68(5): 1355–65.

Betts, Richard K. 1987. *Nuclear Blackmail and Nuclear Balance.* Washington, DC: Brookings.

Biddle, Stephen. 2001. "Rebuilding the Foundations of Offense-Defense Theory." *Journal of Politics* 63(3): 741–74.

———. 2004. *Military Power: Explaining Victory and Defeat in Modern Battle.* Princeton, NJ: Princeton University Press.

———. 2007. "Strategy in War." *PS: Political Science and Politics* 40(3): 461–66.

Blainey, Geoffrey. 1988. *The Causes of War.* 3rd ed. London: Macmillan.

Blair, Bruce G. 1993. *The Logic of Accidental Nuclear War.* Washington, DC: Brookings.

Borzecki, Jerzy. 2008. *The Soviet-Polish Peace of 1921 and the Creation of Interwar Europe.* New Haven, CT: Yale University Press.

Boulding, Kenneth E. 1962. *Conflict and Defense: A General Theory.* New York: Harper and Brothers.

Box-Steffensmeier, Janet M., and Christopher J. W. Zorn. 2001. "Duration Models and Proportional Hazards in Political Science." *American Journal of Political Science* 45(4): 972–88.

Bracken, Paul J. 1983. *The Command and Control of Nuclear Forces.* New Haven, CT: Yale University Press.

Brams, Steven J., and D. Marc Kilgour. 1988. *Game Theory and National Security.* New York: Basil Blackwell.

Braumoeller, Bear F., and Gary Goertz. 2000. "The Methodology of Necessary Conditions." *American Journal of Political Science* 44(4): 844–58.

Braybrooke, David, and Charles E. Lindblom. 1963. *A Strategy of Decision.* New York: Free Press.

Brecher, Michael. 1993. *Crises in World Politics: Theory and Reality.* Oxford: Pergamon Press.

Brecher, Michael, Patrick James, and Jonathan Wilkenfeld. 2000. "Escalation and War in the Twentieth Century: Findings from the International Crisis Behavior Project." In *What Do We Know About War?,* ed. John A. Vasquez. Lanham, MD: Rowman and Littlefield.

Brecher, Michael, and Jonathan Wilkenfeld. 1997. *A Study of Crisis.* Ann Arbor: University of Michigan Press.

———. 2000. *A Study of Crisis.* Ann Arbor: University of Michigan Press.

Bremer, Stuart A. 1992. "Dangerous Dyads." *Journal of Conflict Resolution* 36(2): 309–41.

———. 1995. "Advancing the Scientific Study of War." In *The Process of War: Advancing the Scientific Study of War*, eds. Stuart A. Bremer and Thomas R. Cusack. Amsterdam: Gordon and Breach.

Bremer, Stuart A., and Thomas R. Cusack, eds. 1995. *The Process of War: Advancing the Scientific Study of War.* Amsterdam: Gordon and Breach.

Brodie, Bernard, ed. 1946. *The Absolute Weapon: Atomic Power and World Order.* New York: Harcourt Brace.

———. 1959. "The Anatomy of Deterrence." *World Politics* 11(2): 173–79.

Bueno de Mesquita, Bruce. 1975. "Measuring Systemic Polarity." *Journal of Conflict Resolution* 19(2): 187–216.

———. 1978. "Systemic Polarization and the Occurrence and Duration of War." *Journal of Conflict Resolution* 22(2): 241–67.

———. 1981. *The War Trap.* New Haven, CT: Yale University Press.

———. 1990a. "Big Wars, Little Wars: Avoiding Selection Bias." *International Interactions* 16(3): 159–69.

———. 1990b. "Pride of Place: The Origins of German Hegemony." *World Politics* 43(1): 28–52.

———. 2006. *Principles of International Politics.* 3rd ed. Washington, DC: CQ Press.

———. 2010. *Principles of International Politics.* 4th ed. Washington, DC: CQ Press.

Bueno de Mesquita, Bruce, Feryal Marie Cherif, George W. Downs, and Alastair Smith. 2005. "Thinking Inside the Box: A Closer Look at Democracy and Human Rights." *International Studies Quarterly* 49(3): 439–58.

Bueno de Mesquita, Bruce, and David Lalman. 1988. "Empirical Support for Systemic and Dyadic Explanations of International Conflict." *World Politics* 41(1): 1–20.

———. 1992. *War and Reason: Domestic and International Imperatives.* New Haven, CT: Yale University Press.

Bueno de Mesquita, Bruce, James D. Morrow, Randolph M. Siverson, and Alastair Smith. 1999. "An Institutional Explanation of the Democratic Peace." *American Political Science Review* 93(4): 791–807.

———. 2004. "Testing Novel Implications from the Selectorate Theory of War." *World Politics* 56(3): 363–88.

Bueno de Mesquita, Bruce, James D. Morrow, and Ethan R. Zorick. 1997. "Capabilities, Perception, and Escalation." *American Political Science Review* 91(1): 15–27.

Bueno de Mesquita, Bruce, and Randolph M. Siverson. 1995. "War and the Survival of Political Leaders: A Comparative Study of Regime Types and Political Accountability." *American Political Science Review* 89(4): 841–55.

Bueno de Mesquita, Bruce, Randolph M. Siverson, and Gary Woller. 1992. "War and the Fate of Regimes: A Comparative Analysis." *American Political Science Review* 86(3): 638–46.

Bueno de Mesquita, Bruce, and Alastair Smith. 2007. "Foreign Aid and Policy Concessions." *Journal of Conflict Resolution* 51(2): 251–84.

———. 2009. "A Political Economy of Aid." *International Organization* 63(2): 309–40.

———. 2010. "Leader Survival, Revolutions, and the Nature of Government Finance." *American Journal of Political Science* 54(4): 936–50.

———. 2011. *The Dictator's Handbook: Why Bad Behavior Is Almost Always Good Politics.* New York: PublicAffairs.

Bueno de Mesquita, Bruce, Alastair Smith, Randolph M. Siverson, and James D. Morrow. 2003. *The Logic of Political Survival.* Cambridge, MA: MIT Press.

Bull, Hedley. 1966. "International Theory: The Case for a Classical Approach." *World Politics* 18(3): 361–77.

———. 1977. *The Anarchical Society.* New York: Columbia University Press.

Bush, George W. 2004. "President and Prime Minister Blair Discussed Iraq, Middle East." http://georgewbush-whitehouse.archives.gov/news/releases/2004/11/20041112-5.html.

Butterworth, Robert L. 1978. "Do Conflict Managers Matter?: An Empirical Assessment of Interstate Security Disputes and Resolution Efforts, 1945–1974." *International Studies Quarterly* 22(2): 195–214.

Caprioli, Mary. 2004. "Feminist IR Theory and Quantitative Methodology: A Critical Analysis." *International Studies Review* 6(2): 253–69.

Caprioli, Mary, and Mark A. Boyer. 2001. "Gender, Violence, and International Crisis." *Journal of Conflict Resolution* 45(4): 503–18.

Carlson, Lisa J., and Raymond Dacey. 2006. "Sequential Analysis of Deterrence Games with a Declining Status Quo." *Conflict Management and Peace Science* 23(2): 181–98.

Carment, David. 1993. "The International Dimensions of Ethnic Conflict: Concepts, Indicators, and Theory." *Journal of Peace Research* 30(2): 137–50.

Carment, David, and Patrick James. 1995. "Internal Constraints and Interstate Ethnic Conflict: Toward a Crisis-Based Assessment of Irredentism." *Journal of Conflict Resolution* 39(1): 82–109.

Casper, Gretchen, and Claudiu Tufis. 2003. "Correlation Versus Interchangeability: The Limited Robustness of Empirical Findings on Democracy Using Highly Correlated Data Sets." *Political Analysis* 11(2): 196–203.

Chan, Steve. 1997. "In Search of Democratic Peace: Problems and Promise." *Mershon International Studies Review* 41(2): 59–91.

Chapman, Terrence L. and Scott Wolford. 2010. "International Organizations, Strategy, and Crisis Bargaining." *Journal of Politics* 72(1): 227–42.

Chase, John L. 1955. "Unconditional Surrender Reconsidered." *Political Science Quarterly* 70 (2): 258–79.

Chiozza, Giacomo, and H. E. Goemans. 2011. *Leaders and International Conflict*. Cambridge: Cambridge University Press.

Choi, Seung-Whan. 2011. "Re-Evaluating Capitalist and Democratic Peace Models." *International Studies Quarterly* 55(3): 759–69.

Christensen, Thomas J., and Jack Snyder. 1990. "Chain Gangs and Passed Bucks: Predicting Alliance Patterns in Multipolarity." *International Organization* 44(2): 137–68.

Claire, Rodger W. 2004. *Raid on the Sun: Inside Israel's Secret Campaign That Denied Saddam the Bomb*. New York: Broadway

Clary, David A. 2009. *Eagles and Empire: The United States, Mexico, and the Struggle for a Continent*. New York: Bantam Dell.

Clausewitz, Carl von. (1832)1976. *On War*. Edited and translated by Michael Howard and Peter Paret. Princeton, NJ: Princeton University Press.

Clinton, Bill. 1994. "State of the Union Address." http://www.washingtonpost.com/wp-srv/politics/special/states/docs/sou94.htm.

Clodfelter, Michael. 2008. *Warfare and Armed Conflicts: A Statistical Encyclopedia of Casualty and Other Figures, 1494–2007*. 3rd ed. Jefferson, NC: McFarland.

Colaresi, Michael P. 2001. "Shocks to the System: Great Power Rivalries and the Leadership Long Cycle." *Journal of Conflict Resolution* 45(5): 569–93.

Colaresi, Michael P., Karen Rasler, and William R. Thompson. 2007. *Strategic Rivalries in World Politics*. Cambridge: Cambridge University Press.

Colaresi, Michael P., and William R. Thompson. 2002. "Hot Spots or Hot Hands? Serial Crisis Behavior, Escalating Risks, and Rivalry." *Journal of Politics* 64(4): 1175–98.

Connaughton, Richard. 2003. *Rising Sun and Tumbling Bear: Russia's War with Japan*. London: Cassell.

Coox, Alvin D. 1977. *The Anatomy of a Small War: The Soviet-Japanese Struggle for Changkufeng-Khasan, 1938*. Westport, CT: Greenwood.

Copeland, Dale C. 2000. *The Origins of Major War*. Ithaca, NY: Cornell University Press.

Cornwell, Derekh, and Michael P. Colaresi. 2002. "Holy Trinities, Rivalry Termination, and Conflict." *International Interactions* 28(4): 325–53.

Corum, James S. 1992. *The Roots of Blitzkrieg: Hans von Seeckt and German Military Reform*. Lawrence: University Press of Kansas.

Cusack, Thomas R. 1995. "On the Theoretical Deficit in the Study of War." In *The Process of War: Advancing the Scientific Study of War*, eds. Stuart A. Bremer and Thomas R. Cusack. Amsterdam: Gordon and Breach.

Daalder, Ivo H., and Michael E. O'Hanlon. 2001. *Winning Ugly: NATO's War to Save Kosovo.* Washington, DC: Brookings.
Daddis, Gregory A. 2011. *No Sure Victory: Measuring U.S. Army Effectiveness and Progress in the Vietnam War.* Oxford: Oxford University Press.
———. 2014. *Westmoreland's War: Reassessing American Strategy in Vietnam.* New York: Oxford University Press.
Dafoe, Allan. 2011. "Statistical Critiques of the Democratic Peace: Caveat Emptor." *American Journal of Political Science* 55(2): 247–62.
Dahl, Robert. 1957. "The Concept of Power." *Behavioral Science* 2(3): 201–15.
———. 1971. *Polyarchy: Participation and Opposition.* New Haven, CT: Yale University Press.
Danilovic, Vesna. 2001. "Conceptual and Selection Bias Issues in Deterrence." *Journal of Conflict Resolution* 45(1): 97–125.
———. 2002. *When the Stakes Are High: Deterrence and Conflict among Major Powers.* Ann Arbor: University of Michigan Press.
Davies, Norman. 1972. *White Eagle, Red Star: The Polish-Soviet War, 1919–1920.* New York: St. Martin's Press.
Department of State. 1968. *Foreign Relations of the United States: The Conferences at Washington, 1941–1942, and Casablanca, 1943.* Washington, DC: US Government Printing Office.
Department of the Army. 2008. *FM 3-0 Operations.* Washington, DC: Department of the Army.
Deutsch, Karl W., and Dieter Senghaas. 1973. "The Steps to War: A Survey of System Levels, Decision Stages, and Research Results." In *International Yearbook of Foreign Policy Studies,* vol. 1, ed. Patrick J. McGowan. Beverly Hills, CA: Sage.
Deutsch, Karl W., and J. David Singer. 1964. "Multipolar Power Systems and International Stability." *World Politics* 16(3): 390–406.
Diehl, Paul F. 1983. "Arms Races and Escalation: A Closer Look." *Journal of Peace Research* 20(3): 205–12.
———. 1985. "Arms Races to War: Testing Some Empirical Linkages." *Sociological Quarterly* 26(3): 331–49.
———. 1992. "What are they Fighting for? The Importance of Issues in International Conflict Research." *Journal of Peace Research* 29(3): 333–44.
Diehl, Paul F., ed. 1998. *The Dynamics of Enduring Rivalries.* Urbana: University of Illinois Press.
Diehl, Paul F., and Gary Goertz. 2000. *War and Peace in International Rivalry.* Ann Arbor: University of Michigan Press.
———. 2012. "The Rivalry Process: How Rivalries Are Sustained and Terminated." In *What Do We Know About War?,* 2nd ed., ed. John A. Vasquez. Lanham, MD: Rowman and Littlefield.
Dixon, William J. 1994. "Democracy and the Peaceful Settlement of International Conflict." *American Political Science Review* 88(1): 1–17.
Dixon, William J., and Paul D. Senese. 2002. "Democracy, Disputes, and Negotiated Settlements." *Journal of Conflict Resolution* 46(4): 547–71.
Dobbs, Michael. 2008. *One Minute to Midnight: Kennedy, Khrushchev, and Castro on the Brink of Nuclear War.* New York: Knopf.
Donnelly, Thomas, Margaret Roth, and Caleb Baker. 1991. *Operation Just Cause: The Storming of Panama.* New York: Lexington.
Dorussen, Han, and Hugh Ward. 2010. "Trade Networks and the Kantian Peace." *Journal of Peace Research* 47(1): 29–42.
Downs, George W., and David M. Rocke. 1994. "Conflict, Agency, and Gambling for Resurrection: The Principal-Agent Problem Goes to War." *American Journal of Political Science* 38(2): 362–80.
Doyle, Michael W. 1986. "Liberalism and World Politics." *American Political Science Review* 80(4): 1151–69.

Dupuy, R. Ernest, and Trevor N. Dupuy. 1993. *The Harper Encyclopedia of Military History: From 3500 BC to the Present*. 4th ed. New York: HarperCollins.

Dupuy, Trevor N. 1979. *Numbers, Predictions, and War: Using History to Evaluate Combat Factors and Predict the Outcome of Battles*. Indianapolis, IN: Bobbs-Merrill.

———. 1987. *Understanding War: History and Theory of Combat*. New York: Paragon House.

———. 1989. "Combat Data and the 3:1 Rule." *International Security* 14(1): 195–201.

Echevarria, Antulio J. 2003. "Clausewitz's Center of Gravity: It's Not What We Thought." *Naval War College Review* 56(1): 108–23.

Embree, Michael. 2006. *Bismarck's First War: The Campaign of Schleswig and Jutland 1864*. Solihull, UK: Helion.

Epstein, Joshua M. 1989. "The 3:1 Rule, the Adaptive Dynamic Model, and the Future of Security Studies." *International Security* 13(4): 90–127.

Faber, David. 2008. *Munich, 1938: Appeasement and World War II*. New York: Simon and Schuster.

Farber, Henry S., and Joanne Gowa. 1995. "Polities and Peace." *International Security* 20(2): 123–46.

———. 1997. "Common Interests or Common Polities? Reinterpreting the Democratic Peace." *Journal of Politics* 59(2): 393–417.

Farcau, Bruce W. 1996. *The Chaco War: Bolivia and Paraguay, 1932–1935*. Westport, CT: Praeger.

Farrar, L. L., Jr. 1977. "Cycles of War: Historical Speculations on Future International Violence." *International Interactions* 3(2): 161–79.

Fearon, James D. 1994. "Domestic Political Audience Costs and the Escalation of International Disputes." *American Political Science Review* 88(3): 577–92.

———. 1995. "Rationalist Explanations for War." *International Organization* 49(3): 379–414.

———. 2002. "Selection Effects and Deterrence." *International Interactions* 28(1): 5–30.

Figes, Orlando. 2010. *The Crimean War: A History*. New York: Metropolitan.

Filson, Darren, and Suzanne Werner. 2002. "A Bargaining Model of War and Peace: Anticipating the Onset, Duration, and Outcome of War." *American Journal of Political Science* 46(4): 819–38.

———. 2004. "Bargaining and Fighting: The Impact of Regime Type on War Onset, Duration, and Outcomes." *American Journal of Political Science* 48(2): 296–313.

Foot, Rosemary. 1990. *A Substitute for Victory: The Politics of Peacemaking at the Korean Armistice Talks*. Ithaca, NY: Cornell University Press.

Fortna, Virginia Page. 2003. "Scraps of Paper? Agreements and the Durability of Peace." *International Organization* 57(2): 337–72.

———. 2004a. *Peace Time: Cease-Fire Agreements and the Durability of Peace*. Princeton, NJ: Princeton University Press.

———. 2004b. "Does Peacekeeping Keep Peace? International Intervention and the Duration of Peace After Civil War." *International Studies Quarterly* 48(2): 269–92.

———. 2004c. "Interstate Peacekeeping: Causal Mechanisms and Empirical Effects." *World Politics* 56(4): 481–519.

Frazier, Derrick V., and William J. Dixon. 2006. "Third-Party Intermediaries and Negotiated Settlements, 1946–2000." *International Interactions* 32(4): 385–408.

Freedman, Lawrence, ed. 1983. *The Troubled Alliance: Atlantic Relations in the 1980s*. London: Heinemann.

Freedman, Lawrence. 2002. "Conclusion: The Future of Strategic Studies." In *Strategy in the Contemporary World*, eds. John Baylis, James Wirtz, Eliot Cohen, and Colin S. Gray. Oxford: Oxford University Press.

Fromkin, David. 1970. "Entangling Alliances." *Foreign Affairs* 48(4): 688–700.

Fuller, J. F. C. 1949. *The Second World War, 1939–1945: A Strategical and Tactical History*. New York: Duell, Sloan, and Pearce.

Garnham, David. 1985. "The Causes of War: Systemic Findings." In *Polarity and War*, ed. Alan Ned Sabrosky. Boulder, CO: Westview Press.

Gartner, Scott Sigmund. 1998. "Opening Up the Black Box of War." *Journal of Conflict Resolution* 42(3): 252–8.
Gartner, Scott Sigmund, and Jacob Bercovitch. 2006. "Overcoming Obstacles to Peace: The Contribution of Short-Lived Conflict Settlements." *International Studies Quarterly* 50(4): 819–40.
Gartner, Scott Sigmund, and Gary M. Segura. 1998. "War, Casualties, and Public Opinion." *Journal of Conflict Resolution* 42(3): 278–300.
Gartner, Scott Sigmund, and Randolph M. Siverson. 1996. "War Expansion and War Outcome." *Journal of Conflict Resolution* 40(1): 4–15.
Gartzke, Erik. 1998. "Kant We All Get Along?: Opportunity, Willingness, and the Origins of the Democratic Peace." *American Journal of Political Science* 42(1): 1–27.
———. 2000. "Preferences and the Democratic Peace." *International Studies Quarterly* 44(2): 191–210.
———. 2007. "The Capitalist Peace." *American Journal of Political Science* 51(1): 166–91.
Gartzke, Erik, Quan Li, and Charles Boehmer. 2001. "Investing in the Peace: Economic Interdependence and International Conflict." *International Organization* 55(2): 391–438.
Gartzke, Erik, and Michael W. Simon. 1999. "'Hot Hand': A Critical Analysis of Enduring Rivalries." *Journal of Politics* 61(3): 777–98.
Gates, Scott, and Brian D. Humes. 1997. *Games, Information, and Politics: Applying Game Theoretic Models to Political Science*. Ann Arbor: University of Michigan Press.
Gati, Charles. 2006. *Failed Illusions: Moscow, Washington, Budapest, and the 1956 Hungarian Revolt*. Palo Alto, CA: Stanford University Press.
Geller, Daniel S. 1990. "Nuclear Weapons, Deterrence, and Crisis Escalation." *Journal of Conflict Resolution* 34(2): 291–310.
Geller, Daniel S., and J. David Singer. 1998. *Nations at War: A Scientific Study of International Conflict*. Cambridge: Cambridge University Press.
George, Alexander L. 1969. "The 'Operational Code': A Neglected Approach to the Study of Political Leaders and Decision-Making." *International Studies Quarterly* 13(2): 190–222.
George, Alexander L., and Juliette L. George. 1956. *Woodrow Wilson and Colonel House: A Personality Study*. New York: John Day.
George, Alexander L., and Richard Smoke. 1974. *Deterrence in American Foreign Policy: Theory and Practice*. New York: Columbia University Press.
———. 1989. "Deterrence and Foreign Policy." *World Politics* 41(2): 170–82.
Ghosn, Faten, Glenn Palmer, and Stuart A. Bremer. 2004. "The MID3 Data Set, 1993–2001." *Conflict Management and Peace Science* 21(2): 133–54.
Gibbons, Robert. 1992. *Game Theory for Applied Economists*. Princeton, NJ: Princeton University Press.
Gibler, Douglas M. 1996. "Alliances that Never Balance: The Territorial Settlement Treaty." *Conflict Management and Peace Science* 15(1): 75–97.
———. 1997. "Control the Issues, Control the Conflict: The Effects of Alliances That Settle Territorial Issues on Interstate Rivalries." *International Interactions* 22(4): 341–68.
———. 2000. "Alliances: Why Some Cause War and Why Others Cause Peace." In *What Do We Know About War?*, ed. John A. Vasquez. Lanham, MD: Rowman and Littlefield.
———. 2009. *International Military Alliances, 1648–2008*. 2 vols. Washington, DC: CQ Press.
———. 2012. *The Territorial Peace: Borders, State Development, and International Conflict*. Cambridge: Cambridge University Press.
Gibler, Douglas M., and Meredith Reid Sarkees. 2004. "Measuring Alliances: The Correlates of War Formal Interstate Alliance Dataset, 1816–2000." *Journal of Peace Research* 41(2): 211–22.
Gibler, Douglas M., and John A. Vasquez. 1998. "Uncovering the Dangerous Alliances, 1495–1980." *International Studies Quarterly* 42(4): 785–807.
Gibran, Daniel K. 1998. *The Falklands War: Britain Versus the Past in the South Atlantic*. Jefferson, NC: McFarland.
Gilbert, Martin. 1994. *The First World War: A Complete History*. New York: Henry Holt.

Glaser, Charles, and Chaim Kaufmann. 1998. "What Is the Offense-Defense Balance and Can We Measure It?" *International Security* 22(4): 44–82.

Gleditsch, Nils Petter, Peter Wallensteen, Mikael Eriksson, Margareta Sollenberg, and Håvard Strand. 2002. "Armed Conflict 1946–2001: A New Dataset." *Journal of Peace Research* 39(5): 615–37.

Gochman, Charles S. 1991. "Interstate Metrics: Conceptualizing, Operationalizing, and Measuring the Geographic Proximity of States since the Congress of Vienna." *International Interactions* 17(1): 93–112.

Gochman, Charles S., and Russell Leng. 1983. "Realpolitik and the Road to War: An Analysis of Attributes and Behavior." *International Studies Quarterly* 27(1): 97–120.

Gochman, Charles S., and Zeev Maoz. 1984. "Militarized Interstate Disputes, 1816–1976: Procedures, Patterns, and Insights." *Journal of Conflict Resolution* 28(4): 586–615.

Goda, Norman J. W. 1998. *Tomorrow the World: Hitler, Northwest Africa, and the Path toward America*. College Station: Texas A&M University Press.

Goemans, H. E. 2000. *War and Punishment: The Causes of War Termination and the First World War*. Princeton, NJ: Princeton University Press.

Goertz, Gary, and Paul F. Diehl. 1992. *Territorial Changes and International Conflict*. London: Routledge.

———. 1995. "The Initiation and Termination of Enduring Rivalries: The Impact of Political Shocks." *American Journal of Political Science* 39(1): 30–52.

Goertz, Gary, Bradford Jones, and Paul F. Diehl. 2005. "Maintenance Processes in International Rivalries." *Journal of Conflict Resolution* 49(5): 742–69.

Goldman, Stuart D. 2012. *Nomonhan, 1939: The Red Army's Victory that Shaped World War II*. Annapolis, MD: Naval Institute Press.

Goldstein, Erik. 1992. *Wars and Peace Treaties, 1816–1991*. London: Routledge.

Gordon, Michael R., and Bernard E. Trainor. 2006. *Cobra II: The Inside Story of the Invasion and Occupation of Iraq*. New York: Pantheon.

Gowa, Joanne. 1999. *Ballots and Bullets: The Elusive Democratic Peace*. Princeton, NJ: Princeton University Press.

———. 2011. "The Democratic Peace after the Cold War." *Economics and Politics* 23(2): 153–71.

Grauer, Ryan, and Michael C. Horowitz. 2012. "What Determines Military Victory? Testing the Modern System." *Security Studies* 21(1): 83–112.

Gray, Colin S. 1999. *Modern Strategy*. Oxford: Oxford University Press.

———. 2012. *War, Peace and International Relations: An Introduction to Strategic History*. 2nd ed. London: Routledge.

Grenfell, Russell. 1953. *Unconditional Hatred: German War Guilt and the Future of Europe*. New York: Devin-Adair.

Gurr, Ted Robert. 1988. "War, Revolution, and the Growth of the Coercive State." *Comparative Political Studies* 21(1): 45–65.

Halberstam, David. 2007. *The Coldest Winter: America and the Korean War*. New York: Hyperion.

Hall, Richard C. 2000. *The Balkan Wars 1912–1913: Prelude to the First World War*. London: Routledge.

Hamilton, Earl J. 1977. "The Role of War in Modern Inflation." *Journal of Economic History* 37(1): 13–19.

Hammel, Eric. 1992. *Six Days in June: How Israel Won the 1967 Arab-Israeli War*. New York: Scribner's.

Harrington, Daniel F. 2012. *Berlin on the Brink: The Blockade, the Airlift, and the Early Cold War*. Lexington: University Press of Kentucky.

Harsanyi, John C. 1977. "Advances in Understanding Rational Behavior." In *Foundational Problems in the Social Sciences*, eds. Robert E. Butts and Jaakko Hintikka. Dordrecht, Netherlands: D. Reidel.

Hart, Peter. 2013. *The Great War: A Combat History of the First World War*. Oxford: Oxford University Press.

Hartcup, Guy. 1988. *The War of Invention: Scientific Developments, 1914–1918*. London: Brassey's.
———. 2000. *The Effect of Science on the Second World War*. New York: St. Martin's Press.
Harvey, Frank P. 1998. "Rigor Mortis or Rigor, More Tests: Necessity, Sufficiency, and Deterrence Logic." *International Studies Quarterly* 42(4): 675–707.
———. 2012. *Explaining the Iraq War: Counterfactual Theory, Logic, and Evidence*. Cambridge: Cambridge University Press.
Hastings, Max. 1987. *The Korean War*. New York: Simon and Schuster.
———. 2013. *Catastrophe 1914: Europe Goes to War*. New York: Knopf.
Hastings, Max, and Simon Jenkins. 1984. *The Battle for the Falklands*. New York: Norton.
Hegre, Håvard. 2000. "Development and the Liberal Peace: What Does It Take to Be a Trading State?" *Journal of Peace Research* 37(1): 5–30.
Hensel, Paul R. 1994. "One Thing Leads to Another: Recurrent Militarized Disputes in Latin America, 1816–1986." *Journal of Peace Research* 31(3): 281–97.
———. 1996. "The Evolution of Interstate Rivalry." PhD diss., University of Illinois.
———. 1999. "An Evolutionary Approach to the Study of Interstate Rivalry." *Conflict Management and Peace Science* 17(2): 175–206.
———. 2012. "Territory: Geography, Contentious Issues, and World Politics." In *What Do We Know About War?*, 2nd ed., ed. John A. Vasquez. Lanham, MD: Rowman and Littlefield.
Hensel, Paul R., Sara McLaughlin Mitchell, Thomas E. Sowers II, and Clayton L. Thyne. 2008. "Bones of Contention: Comparing Territorial, Maritime, and River Issues." *Journal of Conflict Resolution* 52(1): 117–43.
Hermann, Margaret G., and Charles W. Kegley Jr. 1995. "Rethinking Democracy and International Peace: Perspectives from Political Psychology." *International Studies Quarterly* 39(4): 511–33.
Herzog, Chaim, and Shlomo Gazit. 2005. *The Arab-Israeli Wars: War and Peace in the Middle East*. 2nd ed. New York: Viking.
Hindmoor, Andrew. 2006. *Rational Choice*. New York: Palgrave Macmillan.
Hiro, Dilip. 1991. *The Longest War: The Iran-Iraq Military Conflict*. New York: Routledge.
Hoover, Kenneth, and Todd Donovan. 2004. *The Elements of Social Scientific Thinking*. 8th ed. Belmont, CA: Thomson Wadsworth.
Horowitz, Michael C., Erin M. Simpson, and Allan C. Stam III. 2011. "Domestic Institutions and Wartime Casualties." *International Studies Quarterly* 55(4): 909–36.
Horvath, William J. 1968. "A Statistical Model for the Duration of Wars and Strikes." *Behavioral Science* 13(1): 18–28.
Howard, Michael. 2001. *The Franco-Prussian War: The German Invasion of France, 1870–1871*. 2nd ed. London: Routledge.
———. 2002. *Clausewitz: A Very Short Introduction*. Oxford: Oxford University Press.
Huth, Paul K. 1988. *Extended Deterrence and the Prevention of War*. New Haven, CT: Yale University Press.
———. 1996. *Standing Your Ground: Territorial Disputes and International Conflict*. Ann Arbor: University of Michigan Press.
———. 2000. "Territory: Why Are Territorial Disputes between States a Central Cause of International Conflict?" In *What Do We Know About War?*, ed. John A. Vasquez. Lanham, MD: Rowman and Littlefield.
Huth, Paul K., and Todd L. Allee. 2002. *The Democratic Peace and Territorial Conflict in the Twentieth Century*. Cambridge: Cambridge University Press.
Huth, Paul K., D. Scott Bennett, and Christopher Gelpi. 1992. "System Uncertainty, Risk Propensity, and International Conflict among the Great Powers." *Journal of Conflict Resolution* 36(3): 478–517.
Huth, Paul K., and Bruce Russett. 1984. "What Makes Deterrence Work? Cases from 1900 to 1980." *World Politics* 36(4): 496–526.
———. 1990. "Testing Deterrence Theory: Rigor Makes a Difference." *World Politics* 42(4): 466–501.

———. 1993. "General Deterrence between Enduring Rivals: Testing Three Competing Models." *American Political Science Review* 87(1): 61–73.

Ikle, Fred Charles. 1971. *Every War Must End*. New York: Columbia University Press.

Intriligator, Michael D., and Dagobert L. Brito. 1984. "Can Arms Races Lead to the Outbreak of War?" *Journal of Conflict Resolution* 28(1): 63–84.

———. 1987. "The Stability of Mutual Deterrence." In *Exploring the Stability of Deterrence*, eds. Jacek Kugler and Frank C. Zagare. Boulder, CO: Lynne Rienner.

Jaggers, Keith, and Ted Robert Gurr. 1995. "Tracking Democracy's Third Wave with the Polity III Data." *Journal of Peace Research* 32(4): 469–82.

James, Patrick. 1988. *Crisis and War*. Montreal, QC: McGill-Queen's University Press.

Jefferson, Thomas. 1801. "First Inaugural Address." March 4. http://www.bartleby.com/124/pres16.html.

Jervis, Robert. 1976. *Perception and Misperception in International Politics*. Princeton, NJ: Princeton University Press.

———. 1978. "Cooperation Under the Security Dilemma." *World Politics* 30(2): 167–214.

———. 1982/83. "Deterrence and Perception." *International Security* 7(3): 3–30.

———. 1985. "Introduction." In *Psychology and Deterrence*, eds. Robert Jervis, Richard Ned Lebow, and Janice Gross Stein. Baltimore, MD: Johns Hopkins University Press.

———. 1988. "The Political Effects of Nuclear Weapons." *International Security* 13(2): 80–90.

———. 1989. "Rational Deterrence: Theory and Evidence." *World Politics* 41(2): 183–207.

Jervis, Robert, Richard Ned Lebow, and Janice Gross Stein, eds. 1985. *Psychology and Deterrence*. Baltimore, MD: Johns Hopkins University Press.

Jones, Daniel M., Stuart A. Bermer, and J. David Singer. 1996. "Militarized Interstate Disputes, 1816–1992." *Conflict Management and Peace Science* 15(2): 163–213.

Jones, Sean Lynn. 1995. "Offense-Defense Theory and Its Critics." *Security Studies* 4(4): 660–91.

Jones, Seth G. 2009. *In the Graveyard of Empires: America's War in Afghanistan*. New York: Norton.

Jordan, David, James D. Kiras, David J. Lonsdale, Ian Speller, Christopher Tuck, and C. Dale Walton. 2008. *Understanding Modern Warfare*. Cambridge: Cambridge University Press.

Kahn, Herman. 1962. *Thinking About the Unthinkable*. New York: Horizon Press.

Kalijarvi, Thorsten V. 1948. "Settlements of World Wars I and II Compared." *Annals of the American Academy of Political and Social Science* 257: 194–202.

Kane, Thomas M., and David J. Lonsdale. 2012. *Understanding Contemporary Strategy*. London: Routledge.

Kaufmann, William. 1956. "The Requirements of Deterrence." In *Military Policy and National Security*, ed. William Kaufmann. Princeton, NJ: Princeton University Press.

Kecskemeti, Paul. 1958. *Strategic Surrender: The Politics of Victory and Defeat*. Stanford, CA: Stanford University Press.

Keegan, John. 1990. *The Second World War*. New York: Penguin.

———. 1999. *The First World War*. New York: Knopf.

———. 2009. *The American Civil War: A Military History*. New York: Knopf.

Keeley, Lawrence H. 1996. *War before Civilization*. New York: Oxford University Press.

Keohane, Robert O. 1984. *After Hegemony: Cooperation and Discord in the World Political Economy*. Princeton, NJ: Princeton University Press.

Keohane, Robert O., ed. 1986. *Neorealism and Its Critics*. New York: Columbia University Press.

Kershaw, Ian. 2011. *The End: The Defiance and Destruction of Hitler's Germany, 1944–1945*. New York: Penguin Press.

Keynes, John M. 1920. *The Economic Consequences of the Peace*. New York: Harcourt, Brace, and Howe.

Kilcullen, David. 2010. *Counterinsurgency*. Oxford: Oxford University Press.

Kim, Woosang. 1989. "Power, Alliance, and Major Wars, 1816–1975." *Journal of Conflict Resolution* 33(2): 255–73.

Kimball, Anessa L. 2006. "Alliance Formation and Conflict Initiation: The Missing Link." *Journal of Peace Research* 43(4): 371–89.

King, Gary, Robert O. Keohane, and Sidney Verba. 1994. *Designing Social Inquiry: Scientific Inference in Qualitative Research*. Princeton, NJ: Princeton University Press.

Klein, James P., Gary Goertz, and Paul F. Diehl. 2006. "The New Rivalry Dataset: Procedures and Patterns." *Journal of Peace Research* 43(3): 331–48.

Korn, David A. 1992. *Stalemate: The War of Attrition and Great Power Diplomacy in the Middle East, 1967–1970*. Boulder, CO: Westview Press.

Koubi, Vally. 2005. "War and Economic Performance." *Journal of Peace Research* 42(1): 67–82.

Krauthammer, Charles. 2002. "The Obsolescence of Deterrence." *The Weekly Standard* (December 9): 13.

Kristensen, Hans M., and Robert S. Norris. 2013. "Global Nuclear Weapons Inventories, 1945–2013." *Bulletin of the Atomic Scientists* 69(5) 75–81.

Kugler, Jacek, and Marina Arbetman. 1989. "Exploring the 'Phoenix Factor' with the Collective Goods Perspective. *Journal of Conflict Resolution* 33(1): 84–112.

Kugler, Jacek, and Douglas Lemke, eds. 1996. *Parity and War: Evaluations and Extensions of The War Ledger.* Ann Arbor: University of Michigan Press.

Lai, Brian, and Dan Reiter. 2000. "Democracy, Political Similarity, and International Alliances, 1816–1992." *Journal of Conflict Resolution* 44(2): 203–27.

Lake, David A. 1992. "Powerful Pacifists: Democratic States and War." *American Political Science Review* 86(1): 24–37.

Lambeth, Benjamin S. 2001. *NATO's Air War for Kosovo: A Strategic and Operational Assessment*. Santa Monica, CA: RAND.

Lampton, David M. 1973. "The U.S. Image of Peking in Three International Crises." *Western Political Quarterly* 26(1): 28–50.

Lavoy, Peter R., ed. 2009. *Asymmetric Warfare in South Asia: The Causes and Consequences of the Kargil Conflict*. Cambridge: Cambridge University Press.

Lebovic, James H. 2007. *Deterring International Terrorism and Rogue States: US National Security Policy after 9/11*. New York: Routledge.

Lebow, Richard Ned. 1981. *Between Peace and War: The Nature of International Crises*. Baltimore, MD: Johns Hopkins University Press.

———. 1984. "Windows of Opportunity: Do States Jump Through Them?" *International Security* 9(1): 147–86.

Lebow, Richard Ned, and Janice Gross Stein. 1989. "Rational Deterrence Theory: I Think, Therefore I Deter." *World Politics* 41(2): 208–24.

———. 1990. "Deterrence: The Elusive Dependent Variable." *World Politics* 42(3): 336–69.

Leeds, Brett Ashley. 2003. "Do Alliances Deter Aggression? The Influence of Military Alliances on the Initiation of Militarized Interstate Disputes." *American Journal of Political Science* 47(3): 427–39.

Leeds, Brett Ashley, Andrew G. Long, and Sara McLaughlin Mitchell. 2000. "Reevaluating Alliance Reliability: Specific Threats, Specific Promises." *Journal of Conflict Resolution* 44(5): 686–99.

Leeds, Brett Ashley, Jeffrey M. Ritter, Sara McLaughlin Mitchell, and Andrew G. Long. 2002. "Alliance Treaty Obligations and Provisions, 1815–1944." *International Interactions* 28(3): 237–60.

Lemke, Douglas. 1995. "The Tyranny of Distance: Redefining Relevant Dyads." *International Interactions* 21(1): 23–38.

———. 2002. *Regions of War and Peace*. Cambridge: Cambridge University Press.

Lemke, Douglas, and William Reed. 1996. "Regime Types and Status Quo Evaluations." *International Interactions* 22(2): 143–64.

———. 2001. "War and Rivalry among Great Powers." *American Journal of Political Science* 45(2): 457–69.

Lemke, Douglas, and Suzanne Werner. 1996. "Power Parity, Commitment to Change, and War." *International Studies Quarterly* 40(2): 235–60.

Leng, Russell J. 1983. "When Will They Ever Learn?" *Journal of Conflict Resolution* 27(3): 379–419.

———. 1993. *Interstate Crisis Behavior, 1816–1980: Realism versus Reciprocity.* Cambridge: Cambridge University Press.

———. 2000a. *Bargaining and Learning in Recurring Crises: The Soviet-American, Egyptian-Israeli, and Indo-Pakistani Rivalries.* Ann Arbor: University of Michigan Press.

———. 2000b. "Escalation: Crisis Behavior and War." In *What Do We Know About War?*, ed. John Vasquez. Lanham, MD: Rowman and Littlefield.

Leuchars, Chris. 2002. *To the Bitter End: Paraguay and the War of the Triple Alliance.* Westport, CT: Greenwood.

Levy, Jack S. 1984. "The Offensive/Defensive Balance of Military Technology: A Theoretical and Historical Analysis." *International Studies Quarterly* 28(2): 219–38.

———. 1988. "Domestic Politics and War." *Journal of Interdisciplinary History* 18(4): 653–73.

———. 1994. "Learning and Foreign Policy: Sweeping a Conceptual Minefield." *International Organization* 48(2): 279–312.

———. 1997. "Prospect Theory, Rational Choice, and International Relations." *International Studies Quarterly* 41(1): 87–112.

Levy, Jack S., and T. Clifton Morgan. 1986. "The War-Weariness Hypothesis: An Empirical Test." *American Journal of Political Science* 30(1): 26–49.

Lo, Nigel, Barry Hashimoto, and Dan Reiter. 2008. "Ensuring Peace: Foreign-Imposed Regime Change and Postwar Peace Duration, 1914–2001." *International Organization* 62(4): 717–36.

Luce, Duncan R., and Howard Raiffa. 1957. *Games and Decisions: Introduction and Critical Survey.* New York: Wiley.

Lupfer, Timothy. 1981. *The Dynamics of Doctrine: The Changes in German Tactical Doctrine during the First World War.* Ft. Leavenworth, KS: US Army Combat Studies Institute.

Luttwak, Edward N. 1980. "The Operational Level of War." *International Security* 5(3): 61–79.

MacMillan, John. 1998. *On Liberal Peace.* London: Tauris.

———. 2003. "Beyond the Separate Democratic Peace." *Journal of Peace Research* 40(2): 233–43.

MacMillan, Margaret. 2013. *The War That Ended Peace: The Road to 1914.* New York: Knopf.

Mahnken, Thomas G. 2007. "Strategic Theory." In *Strategy in the Contemporary World*, 2nd. ed., eds. John Baylis, James Wirtz, Colin S. Gray, and Eliot Cohen. Oxford: Oxford University Press.

Mansfield, Edward D., and Jack Snyder. 1995. "Democratization and the Danger of War." *International Security* 20(1): 5–38.

Maoz, Zeev. 1984. "Peace by Empire? Conflict Outcomes and International Stability, 1816–1976." *Journal of Peace Research* 21(3): 227–41.

Maoz, Zeev. 1990. *Paradoxes of War: On the Art of National Self-Entrapment.* Boston: Unwin Hyman.

———. 1996. *Domestic Sources of Global Change.* Ann Arbor: University of Michigan Press.

———. 2000. "Alliances: The Street Gangs of World Politics—Their Origins, Management, and Consequences, 1816–1986." In *What Do We Know About War?*, ed. John Vasquez. Lanham, MD: Rowman and Littlefield.

———. 2005. "Dyadic MID Dataset, version 2.0." http://psfaculty.ucdavis.edu/zmaoz/dyadmid.html.

———. 2009. "The Effects of Strategic and Economic Interdependence on International Conflict Across Levels of Analysis." *American Journal of Political Science* 53(1): 223–40.

Maoz, Zeev, and Nasrin Abdolali. 1989. "Regime Types and International Conflict, 1816–1976." *Journal of Conflict Resolution* 33(1): 3–35.

Maoz, Zeev, and Ben D. Mor. 2002. *Bound by Struggle: The Strategic Evolution of Enduring International Rivalries.* Ann Arbor: University of Michigan Press.

Maoz, Zeev, and Bruce Russett. 1993. "Normative and Structural Causes of Democratic Peace, 1946–1986." *American Political Science Review* 87(3): 624–38.

Marley, David F. 2008. *Wars of the Americas: A Chronology of Armed Conflict in the Western Hemisphere, 1492 to the Present.* 2nd ed. 2 vol. Santa Barbara, CA: ABC-CLIO.

Marshall, Monty G., and Keith Jaggers. 2002. "Polity IV Project: Political Regime Characteristics and Transitions, 1800–2002." http://www.cidcm.umd.edu/inscr/polity.

Marshall, Monty G., Keith Jaggers, and Ted Robert Gurr. 2010. "Polity IV Project: Political Regime Characteristics and Transitions, 1800–2010." http://www.systemicpeace.org/inscr/p4manualv2010.pdf.

Marwick, Arthur. 1974. *War and Social Change in the Twentieth Century*. Basingstoke, UK: Palgrave Macmillan.

Maxwell, Stephen. 1968. *Rationality in Deterrence* (Adelphi Paper no. 50). London: International Institute for Strategic Studies.

McCarty, Nolan, and Adam Meirowitz. 2007. *Political Game Theory: An Introduction*. Cambridge: Cambridge University Press.

McConnell, Malcolm. 1991. *Just Cause: The Real Story of America's High-Tech Invasion of Panama*. New York: St. Martin's Press.

McPherson, James M. 1988. *Battle Cry of Freedom: The Civil War Era*. Oxford: Oxford University Press.

Mearsheimer, John J. 1983. *Conventional Deterrence*. Ithaca, NY: Cornell University Press.

———. 1989. "Assessing the Conventional Balance: The 3:1 Rule and Its Critics." *International Security* 13(4): 54–89.

———. 1990. "Back to the Future: Instability in Europe After the Cold War." *International Security* 15(1): 5–56.

———. 2001. *The Tragedy of Great Power Politics*. New York: Norton.

Middlebrook, Martin. 1987. *Task Force: The Falklands War, 1982*. Rev. ed. New York: Penguin.

———. 1989. *The Fight for the Malvinas: The Argentine Forces in the Falklands War*. New York: Viking.

Midlarsky, Manus I. 1988. *The Onset of World War*. Boston: Allen and Unwin.

———. 1990a. "Big Wars, Little Wars—A Single Theory?" *International Interactions* 16(3): 157–58.

———. 1990b. "Systemic Wars and Dyadic Wars: No Single Theory." *International Interactions* 16(3): 171–81.

Miksche, Ferdinand Otto. 1952. *Unconditional Surrender: The Roots of a World War III*. London: Faber and Faber.

Millett, Allan R., and Williamson Murray. 1988. *Military Effectiveness*, vol. 3. Boston: Allen and Unwin.

Mintz, Alex. 2004. "How Do Leaders Make Decisions? A Poliheuristic Perspective." *Journal of Conflict Resolution* 48(1): 3–13.

Mitchell, Sara McLaughlin. 2002. "A Kantian System? Democracy and Third Party Conflict Resolution." *American Journal of Political Science* 46(4): 749–59.

———. 2012. "Norms and the Democratic Peace." In *What Do We Know About War?*, 2nd ed., ed. John A. Vasquez. Lanham, MD: Rowman and Littlefield.

Molinari, M. Christina. 2000. "Military Capabilities and Escalation: A Correction to Bueno de Mesquita, Morro, and Zorick." *American Political Science Review* 94(2)L 425-27.

Morgan, Patrick M. 1983. *Deterrence: A Conceptual Analysis*. 2nd ed. Beverly Hills, CA: Sage.

Morgan, T. Clifton, and Sally Howard Campbell. 1991. "Domestic Structure, Decisional Constraints, and War—So Why Kant Democracies Fight?" *Journal of Conflict Resolution* 35(2): 187–211.

Morgenthau, Hans J. 1967. *Politics Among Nations: The Struggle for Power and Peace*. 4th ed. New York: Knopf.

Morillo, Stephen. 2006. *What Is Military History?* Cambridge: Polity Press.

Morris, Benny. 2008. *1948: A History of the First Arab-Israeli War*. New Haven, CT: Yale University Press.

Morrison, Donald G., and David C. Schmittlein. 1980. "Jobs, Strikes, and Wars: Probability Models for Duration." *Organizational Behavior and Human Performance* 25(2): 224–51.

Morrow, James D. 1989. "Capabilities, Uncertainty, and Resolve: A Limited Information Model of Crisis Bargaining." *American Journal of Political Science* 33(4): 941–72.

———. 1991. "Alliances and Asymmetry: An Alternative to the Capability Aggregation Model of Alliances." *American Journal of Political Science* 35(4): 904–33.
———. 1994. *Game Theory for Political Scientists*. Princeton, NJ: Princeton University Press.
———. 2000. "Alliances: Why Write Them Down?" *Annual Review of Political Science* 3: 63–83.
Morton, Rebecca B. 1999. *Methods and Models: A Guide to the Empirical Analysis of Formal Models in Political Science*. Cambridge: Cambridge University Press.
Most, Benjamin A., and Harvey Starr. 1989. *Inquiry, Logic and International Politics*. Columbia: University of South Carolina Press.
Moul, William B. 1988. "Balances of Power and the Escalation to War of Serious Disputes among the European Great Powers, 1815–1939: Some Evidence." *American Journal of Political Science* 32(2): 241–75.
Mousseau, Michael. 2000. "Market Prosperity, Democratic Consolidation, and Democratic Peace." *Journal of Conflict Resolution* 44(4): 472–507.
———. 2009. "The Social Market Roots of Democratic Peace." *International Security* 33(4): 52–86.
———. 2012. "A Market-Capitalist or a Democratic Peace?" In *What Do We Know About War?*, 2nd ed., ed. John A. Vasquez. Lanham, MD: Rowman and Littlefield.
Mousseau, Michael, Håvard Hegre, and John R. Oneal. 2003. "How the Wealth of Nations Conditions the Liberal Peace." *European Journal of International Relations* 9(2): 277–314.
Mueller, John E. 1973. *War, Presidents, and Public Opinion*. New York: Wiley.
———. 1989. *Retreat from Doomsday: The Obsolescence of Major War*. New York: Basic Books.
Mulligan, William. 2010. *The Origins of the First World War*. Cambridge: Cambridge University Press.
Munton, Don, and David A. Welch. 2011. *The Cuban Missile Crisis: A Concise History*. 2nd ed. Oxford: Oxford University Press.
Murray, Williamson. 1998. "Innovation: Past and Future." In *Military Innovation in the Interwar Period*, eds. Williamson Murray and Allan R. Millett. Cambridge: Cambridge University Press.
Murray, Williamson, Alvin Berstein, and MacGregor Knox, eds. 1994. *The Making of Strategy: Rulers, States, and War*. Cambridge: Cambridge University Press.
Murray, Williamson, and Allan R. Millett, eds. 1998a. *Military Innovation in the Interwar Period*. Cambridge: Cambridge University Press.
———. 1998b. "Introduction." In *Military Innovation in the Interwar Period*, eds. Williamson Murray and Allan R. Millett. Cambridge: Cambridge University Press.
———. 2000. *A War to Be Won: Fighting the Second World War*. Cambridge, MA: Belknap.
Nagl, John A. 2005. *Learning to Eat Soup with a Knife: Counterinsurgency Lessons from Malaya and Vietnam*. Chicago: University of Chicago Press.
Nakdimon, Shlomo. 1988. *First Strike: The Exclusive Story of How Israel Foiled Iraq's Attempt to Get the Bomb*. New York: HarperCollins.
Nef, John U. 1950. *War and Human Progress*. Cambridge, MA: Harvard University Press.
O'Connor, Raymond G. 1971. *Diplomacy for Victory: FDR and Unconditional Surrender*. New York: Norton.
Oneal, John R., and Bruce M. Russett. 1997. "The Classical Liberals Were Right: Democracy, Interdependence, and Conflict, 1950–1985." *International Studies Quarterly* 41(2): 267–93.
Oren, Michael B. 2002. *Six Days of War: June 1967 and the Making of the Modern Middle East*. Oxford: Oxford University Press.
Organski, A. F. K. 1968. *World Politics*. 2nd ed. New York: Knopf.
Organski, A. F. K., and Jacek Kugler. 1977. "The Costs of Major Wars: The Phoenix Factor." *American Political Science Review* 71(4): 1347–66.
———. 1980. *The War Ledger*. Chicago: University of Chicago Press.
O'Toole, G. J. A. 1984. *The Spanish War: An American Epic 1898*. New York: Norton.
Otterbein, Keith F. 2004. *How War Began*. College Station: Texas A&M University Press.

Oye, Kenneth A. 1986. *Cooperation under Anarchy*. Princeton, NJ: Princeton University Press.
Paine, S. C. M. 2003. *The Sino-Japanese War of 1894–1895: Perceptions, Power, and Primacy*. New York: Cambridge University Press.
Palmer, Glenn, and T. Clifton Morgan. 2006. *A Theory of Foreign Policy*. Princeton, NJ: Princeton University Press.
Pape, Robert A. 1996. *Bombing to Win: Air Power and Coercion in War*. Ithaca, NY: Cornell University Press.
Paret, Peter, ed. 1986. *Makers of Modern Strategy: From Machiavelli to the Nuclear Age*. Princeton, NJ: Princeton University Press.
Parrish, Thomas. 1998. *Berlin in the Balance, 1945–1949: The Blockade, the Airlift, the First Major Battle of the Cold War*. Reading, MA: Addison-Wesley.
Paul, T. V. 1994. *Asymmetric Conflicts: War Initiation by Weaker Powers*. Cambridge: Cambridge University Press.
Peacock, Alan T., and Jack Wiseman. 1961. *The Growth of Public Expenditure in the United Kingdom*. Princeton, NJ: Princeton University Press.
Peterson, Tim, and Stephen L. Quackenbush. 2010. "Not All Peace Years Are Created Equal: Trade, Imposed Settlements, and Recurrent Conflict." *International Interactions* 36(4): 363–83.
Pillar, Paul R. 1983. *Negotiating Peace: War Termination as a Bargaining Process*. Princeton, NJ: Princeton University Press.
Posen, Barry R. 1984. *The Sources of Military Doctrine: France, Britain, and Germany between the World Wars*. Ithaca, NY: Cornell University Press.
Powell, Robert. 1985. "The Theoretical Foundations of Strategic Nuclear Deterrence." *Political Science Quarterly* 100(1): 75–96.
———. 1990. *Nuclear Deterrence Theory: The Search for Credibility*. Cambridge: Cambridge University Press.
———. 1999. *In the Shadow of Power*. Princeton, NJ: Princeton University Press.
———. 2002. "Bargaining Theory and International Conflict." *Annual Review of Political Science* 5: 1–30.
———. 2003. "Nuclear Deterrence Theory, Nuclear Proliferation, and National Missile Defense." *International Security* 27(4): 86–118.
———. 2004. "Bargaining and Learning While Fighting." *American Journal of Political Science* 48(2): 344–61.
Prados, John. 2009. *Vietnam: The History of an Unwinnable War, 1945–1975*. Lawrence: University Press of Kansas.
Preston, Diana. 1999. *The Boxer Rebellion: The Dramatic Story of China's War on Foreigners That Shook the World in the Summer of 1900*. New York: Berkley.
Quackenbush, Stephen L. 2004. "The Rationality of Rational Choice Theory." *International Interactions* 30(2): 87–107.
———. 2006a. "Identifying Opportunity for Conflict: Politically Active Dyads." *Conflict Management and Peace Science* 23(1): 37–51.
———. 2006b. "Not only Whether but Whom: Three-party Extended Deterrence." *Journal of Conflict Resolution* 50(4): 562–83.
———. 2006c. "National Missile Defense and Deterrence." *Political Research Quarterly* 59(4): 533–41.
———. 2010a. "General Deterrence and International Conflict: Testing Perfect Deterrence Theory." *International Interactions* 36(1): 60–85.
———. 2010b. "Territorial Issues and Recurrent Conflict." *Conflict Management and Peace Science* 27(3): 239–52.
———. 2011a. "Deterrence Theory: Where Do We Stand?" *Review of International Studies* 37(2): 741–62.
———. 2011b. *Understanding General Deterrence: Theory and Application*. New York: Palgrave Macmillan.
———. 2014. *Peace Through Victory: Imposed Settlements and Recurrent Conflict*. Book Manuscript.

Quackenbush, Stephen L., and A. Cooper Drury. 2011. "National Missile Defense and Satisfaction." *Journal of Peace Research* 48(4): 469–80.

Quackenbush, Stephen L., and Michael Rudy. 2009. "Evaluating the Monadic Democratic Peace." *Conflict Management and Peace Science* 26(3): 268–85.

Quackenbush, Stephen L., and Jerome F. Venteicher. 2008. "Settlements, Outcomes, and the Recurrence of Conflict." *Journal of Peace Research* 45(6): 723–42.

Quackenbush, Stephen L., and Frank C. Zagare. 2006. "Game Theory: Modeling Interstate Conflict." In *Making Sense of International Relations Theory*, ed. Jennifer Sterling-Folker. Boulder, CO: Lynne Rienner.

Quester, George. 1977. *Offense and Defense in the International System*. New York: Wiley.

Rabinovich, Abraham. 2004. *The Yom Kippur War: The Epic Encounter that Transformed the Middle East*. New York: Schocken.

Raghavan, Srinath. 2013. *1971: A Global History of the Creation of Bangladesh*. Cambridge, MA: Harvard University Press.

Ramsay, Kristopher. 2008. "Settling It on the Field: Battlefield Events and War Termination." *Journal of Conflict Resolution* 52(6): 850–79.

Rapoza, Kenneth. 2011. "By 2020, China No. 1, US No. 2." *Forbes*. 26 May. http://www.forbes.com/sites/kenrapoza/2011/05/26/by-2020-china-no-1-us-no-2/.

Rasler, Karen A. 1986. "War, Accommodation, and Violence in the United States, 1870–1970." *American Political Science Review* 80(3): 921–45.

Rasler, Karen, A., and William R. Thompson. 1985. "War and the Economic Growth of Major Powers." *American Journal of Political Science* 29(3): 513–38.

———. 1989. *War and Statemaking: The Shaping of Global Powers*. Boston: Unwin Hyman.

Rasmusen, Eric. 1989. *Games and Information*. Oxford, UK: Basil Blackwell.

Rauchhaus, Robert W. 2009. "Evaluating the Nuclear Peace Hypothesis: A Quantitative Approach." *Journal of Conflict Resolution* 53(2): 258–77.

Ray, James Lee. 2000. "On the Level(s), Does Democracy Correlate with Peace?" In *What Do We Know About War?*, ed. John Vasquez. Lanham, MD: Rowman and Littlefield.

Reed, William. 2000. "A Unified Statistical Model of Conflict Onset and Escalation." *American Journal of Political Science* 44(1): 84–93.

Regan, Patrick M., and Allan C. Stam III. 2000. "In the Nick of Time: Conflict Management, Mediation Timing, and the Duration of Interstate Disputes." *International Studies Quarterly* 44(2): 239–60.

Reiter, Dan. 2003. "Exploring the Bargaining Model of War." *Perspectives on Politics* 1(1): 27–43.

———. 2009. *How Wars End*. Princeton, NJ: Princeton University Press.

Reiter, Dan, and Curtis Meek. 1999. "Determinants of Military Strategy, 1903–1994: A Quantitative Empirical Test." *International Studies Quarterly* 43(2): 363–87.

Reiter, Dan, and Allan C. Stam III. 1998. "Democracy, War Initiation, and Victory." *American Political Science Review* 92(2): 377–89.

———. 2002. *Democracies at War*. Princeton, NJ: Princeton University Press.

Resnik, Michael D. 1987. *Choices: An Introduction to Decision Theory*. Minneapolis: University of Minnesota Press.

Rhodes, Richard. 1986. *The Making of the Atomic Bomb*. New York: Simon and Schuster.

Rioux, Jean-Sebastien. 1998. "A Crisis Based Evaluation of the Democratic Peace Proposition." *Canadian Journal of Political Science* 31(2): 263–83.

Rothgeb, John M., Jr. 1993. *Defining Power: Influence and Force in the Contemporary International System*. New York: St. Martin's Press.

Rousseau, David L., Christopher Gelpi, Dan Reiter, and Paul K. Huth. 1996. "Assessing the Dyadic Nature of the Democratic Peace, 1918–88." *American Political Science Review* 90(3): 512–33.

Royle, Trevor. 2000. *Crimea: The Great Crimean War, 1854–1856*. New York: Palgrave Macmillan.

Rummell, R. J. 1979. *Understanding Conflict and War: Volume 4, War, Power, Peace*. Beverly Hills, CA: Sage.

———. 1983. "Libertarianism and International Violence." *Journal of Conflict Resolution* 27(1) 27–71.

———. 1985. "Libertarian Propositions on Violence within and between Nations: A Test against Published Research Results." *Journal of Conflict Resolution* 29(3): 419–55.

———. 1995. "Democracies ARE Less Warlike Than Other Regimes." *European Journal of International Relations* 1(4): 457–79.

Russett, Bruce M. 2010. "Capitalism *or* Democracy? Not So Fast." *International Interactions* 36(2): 198–205.

Russett, Bruce M., and John R. Oneal. 2001. *Triangulating Peace*. New York: Norton.

Sabrosky, Alan Ned. 1980. "Interstate Alliances: Their Reliability and the Expansion of War." In *The Correlates of War II: Testing Some Realpolitik Models,* ed. J. David Singer. New York: Free Press.

Sagan, Scott D. 1993. *The Limits of Safety.* Princeton, NJ: Princeton University Press.

Sample, Susan G. 1997. "Arms Races and Dispute Escalation: Resolving the Debate." *Journal of Peace Research* 34(1): 7–22.

Sander, Gordon F. 2013. *The Hundred Day Winter War: Finland's Gallant Stand against the Soviet Army*. Lawrence: University Press of Kansas.

Sarkees, Meredith Reid. 2000. "The Correlates of War Data on War: An Update to 1997." *Conflict Management and Peace Science* 18(1): 123–44.

Sarkees, Meredith Reid, and Frank Whelon Wayman. 2010. *Resort to War: A Data Guide to Inter-state, Extra-state, Intra-state, and Non-state Wars, 1816–2007*. Washington, DC: CQ Press.

Schelling, Thomas C. 1960. *The Strategy of Conflict.* Cambridge, MA: Harvard University Press.

———. 1966. *Arms and Influence*. New Haven, CT: Yale University Press.

Schneider, Gerald, and Nils Petter Gleditsch. 2010. "The Capitalist Peace: The Origins and Prospects of a Liberal Idea." *International Interactions* 36(2): 107–14.

Schultz, Kenneth A. 1998. "Domestic Opposition and Signaling in International Crises." *American Political Science Review* 92(4): 829–44.

———. 1999. "Do Democratic Institutions Constrain or Inform? Contrasting Two Institutional Perspectives on Democracy and War." *International Organization* 53(2): 233–66.

Schultz, Kenneth A. 2001. *Democracy and Coercive Diplomacy.* Cambridge: Cambridge University Press.

Schulzinger, Robert D. 1997. *A Time for War: The United States and Vietnam, 1941–1975*. New York: Oxford University Press.

Selten, Reinhard. 1975. "A Re-examination of the Perfectness Concept for Equilibrium Points in Extensive Games." *International Journal of Game Theory* 4(1): 25–55.

Senese, Paul D. 1996. "Geographical Proximity and Issue Salience: Their Effects on the Escalation of Militarized Interstate Conflict." *Conflict Management and Peace Science* 15(2): 133–61.

———. 1997. "Between Dispute and War: The Effect of Joint Democracy on Interstate Conflict Escalation." *Journal of Politics* 59(1): 1–27.

———. 1999. "Democracy and Maturity: Deciphering Conditional Effects on Levels of Dispute Intensity." *International Studies Quarterly* 43(3): 483–502.

———. 2005. "Territory, Contiguity, and International Conflict: Assessing a New Joint Explanation." *American Journal of Political Science* 49(4): 769–79.

Senese, Paul D., and Stephen L. Quackenbush. 2003. "Sowing the Seeds of Conflict: The Effect of Dispute Settlements on Durations of Peace." *Journal of Politics* 65(3): 696–717.

Senese, Paul D., and John A. Vasquez. 2003. "A Unified Explanation of Territorial Conflict: Testing the Impact of Sampling Bias, 1919–1992." *International Studies Quarterly* 47(2): 275–98.

———. 2008. *The Steps to War: An Empirical Study.* Princeton, NJ: Princeton University Press.

Shimshoni, Johnathan. 1988. *Israel and Conventional Deterrence.* Ithaca, NY: Cornell University Press.

Shy, John. 1986. "Jomini." In *Makers of Modern Strategy: From Machiavelli to the Nuclear Age,* ed. Peter Paret. Princeton, NJ: Princeton University Press.

Signorino, Curtis S. 1999. "Strategic Interaction and the Statistical Analysis of International Conflict." *American Political Science Review* 93(2): 279–97.

Signorino, Curtis S., and Jeffrey Ritter. 1999. "Tau-b or Not Tau-b: Measuring the Similarity of Foreign Policy Positions." *International Studies Quarterly* 43(1): 115–44.

Signorino, Curtis S., and Ahmer Tarar. 2006. "A Unified Theory and Test of Extended Immediate Deterrence." *American Journal of Political Science* 50(3): 586–605.

Silbey, David J. 2013. *The Boxer Rebellion and the Great Game in China.* New York: Hill and Wang.

Simon, Herbert A. 1957. *Models of Man.* New York: Wiley.

———. 1976. "From Substantive to Procedural Rationality." In *Method and Appraisal in Economics,* ed. Spiro J. Latsis. Cambridge: Cambridge University Press.

———. 1997. *Administrative Behavior.* 4th ed. New York: Free Press.

Simowitz, Roslyn L., and Barry L. Price. 1986. "Progress in the Study of International Conflict: A Methodological Critique." *Journal of Peace Research* 23(1): 29–40.

Singer, J. David. 1969. "The Incompleat Theorist: Insight Without Evidence." In *Contending Approaches to International Politics,* eds. James N. Rosenau and Klaus Knorr. Princeton, NJ: Princeton University Press.

———. 1988. "Reconstructing the Correlates of War Dataset on Material Capabilities of States, 1816–1985." *International Interactions* 14(2): 115–32.

Singer, J. David, Stuart A. Bremer, and John Stuckey. 1972. "Capability Distribution, Uncertainty, and Major Power War, 1820–1965." In *Peace, War and Numbers,* ed. Bruce M. Russett. New York: Free Press.

Singer, J. David, and Melvin Small. 1966. "Formal Alliances, 1815–1939." *Journal of Peace Research* 3(1): 1–31.

———. 1972. *The Wages of War, 1816–1965: A Statistical Handbook.* New York: Wiley.

Siverson, Randolph M. 1980. "War and Change in the International System." In *Change in the International System,* eds. Ole R. Holsti, Randolph M. Siverson, and Alexander L. George. Boulder, CO: Westview Press.

Siverson, Randolph M., and Ross A. Miller. 1995. "The Escalation of Disputes to War." In *The Process of War: Advancing the Scientific Study of War,* eds. Stuart A. Bremer and Thomas R. Cusack. Amsterdam: Gordon and Breach.

Siverson, Randolph M., and Harvey Starr. 1991. *The Diffusion of War: A Study of Opportunity and Willingness.* Ann Arbor: University of Michigan Press.

Siverson, Randolph M., and Michael R. Tennefoss. 1984. "Power, Alliance, and the Escalation of International Conflict, 1815–1965." *American Political Science Review* 78(4): 1057–69.

Slantchev, Branislav L. 2003. "The Principle of Convergence in Wartime Negotiations." *American Political Science Review* 97(4): 621–32.

———. 2005."Military Coercion in Interstate Crises." *American Political Science Review* 99(4): 533–47.

Sloan, Elinor C. 2012. *Modern Military Strategy: An Introduction.* London: Routledge.

Small, Melvin, and J. David Singer. 1976. "The War Proneness of Democratic Regimes." *Jerusalem Journal of International Relations* 1(1): 50–69.

———. 1982. *Resort to Arms.* Beverly Hills, CA: Sage.

Smith, Alastair. 1995. "Alliance Formation and War." *International Studies Quarterly* 39(4): 405–25.

———. 1996. "To Intervene or Not to Intervene." *Journal of Conflict Resolution* 40(1): 16–40.

———. 1998. "Fighting Battles, Winning Wars." *Journal of Conflict Resolution* 42(3): 301–20.

Smith, Alastair, and Allan C. Stam III. 2004. "Bargaining and the Nature of War." *Journal of Conflict Resolution* 48(6): 783–813.

Smith, Hugh. 2005. *On Clausewitz: A Study of Military and Political Ideas.* New York: Palgrave Macmillan.

Smoke, Richard. 1987. *National Security and the Nuclear Dilemma*. Reading, MA: Addison-Wesley.
Snyder, Glenn H. 1961. *Deterrence and Defense: Toward a Theory of National Security.* Princeton, NJ: Princeton University Press.
———. 1997. *Alliance Politics*. Ithaca, NY: Cornell University Press.
Snyder, Glenn H., and Paul Diesing. 1977. *Conflict Among Nations: Bargaining, Decision Making, and System Structure in International Crises*. Princeton, NJ: Princeton University Press.
Snyder, Jack. 1984. "Civil-Military Relations and the Cult of the Offensive, 1914 and 1984." *International Security* 9(1): 108–46.
Snyder, Richard C., Henry W. Bruck, and Burton M. Sapin. 1954. *Decision-Making as an Approach to the Study of International Politics*. Princeton, NJ: Princeton University Press.
Snyder, Richard C., Henry W. Bruck, and Burton M. Sapin, eds. 1962. *Foreign Policy Decision-Making: An Approach to the Study of International Politics*. New York: Free Press.
Sorokin, Gerald L. 1994. "Alliance Formation and General Deterrence: A Game-Theoretic Model and the Case of Israel." *Journal of Conflict Resolution* 38(2): 298–325.
Stam, Allan C. III. 1996. *Win, Lose, or Draw: Domestic Politics and the Crucible of War.* Ann Arbor: University of Michigan Press.
Starr, Harvey, and Benjamin A. Most. 1976. "The Substance and Study of Borders in International Relations Research." *International Studies Quarterly* 20(4): 581–620.
Stein, Arthur A. 1980. *The Nation at War*. Baltimore, MD: Johns Hopkins University Press.
Stinnett, Douglas, and Paul F. Diehl. 2001. "The Path(s) to Rivalry: Behavioral and Structural Explanations of Rivalry Development." *Journal of Politics* 63(3): 717–40.
Stinnett, Douglas M., Jaroslav Tir, Philip Schafer, Paul F. Diehl, and Charles Gochman. 2002. "The Correlates of War Project Direct Contiguity Data, Version 3." *Conflict Management and Peace Science* 19(2): 58–66.
Stoll, Richard J. 1984. "Bloc Concentration and Dispute Escalation among the Major Powers, 1830–1965." *Social Science Quarterly* 65(1): 48–59.
———. 1995. "The Evolution of War." In *The Process of War: Advancing the Scientific Study of War*, eds. Stuart A. Bremer and Thomas R. Cusack. Amsterdam: Gordon and Breach.
Sullivan, Michael P., and Randolph M. Siverson. 1981. "Theories of War: Problems and Prospects." In *Cumulation in International Relations Research*, eds. P. Terrance Hopmann, J. David Singer, and Dina A. Zinnes. Denver, CO: University of Denver Press.
Sullivan, Patricia L. 2007. "War Aims and War Outcomes: Why Powerful States Lose Limited Wars." *Journal of Conflict Resolution* 51(3): 496–524.
———. 2012. *Who Wins? Predicting Success and Failure in Armed Conflict*. Oxford: Oxford University Press.
Sun Tzu. 1963. *The Art of War,* trans. Samuel B. Griffith. Oxford: Oxford University Press.
Tammen, Ronald L., Jacek Kugler, Douglas Lemke, Allan C. Stam III, Carole Alsharabati, Mark Andrew Abdollahian, Brian Efird, and A. F. K. Organski. 2000. *Power Transitions: Strategies for the 21st Century.* New York: Chatham House.
Tarar, Ahmer, and Bahar Leventoglu. 2009. "Public Commitment in Crisis Bargaining." *International Studies Quarterly* 53(3): 817–39.
Tellis, Ashley J., C. Christine Fair, and Jamison Jo Medby. 2001. *Limited Conflicts Under the Nuclear Umbrella: Indian and Pakistani Lessons from the Kargil Crisis*. Santa Monica, CA: RAND.
Thakur, Ramesh C. 1982. "Tacit Deception Reexamined: The Geneva Conference of 1954." *International Studies Quarterly* 26(1): 127–39.
Thompson, William R. 1995. "The Consequences of War." In *The Process of War: Advancing the Scientific Study of War*, eds. Stuart A. Bremer and Thomas R. Cusack. Amsterdam: Gordon and Breach.
———. 2001. "Identifying Rivals and Rivalries in World Politics." *International Studies Quarterly* 45(4): 557–86.
Thompson, William R., and David R. Dreyer. 2012. *Handbook of International Rivalries, 1494–2010*. Washington, DC: CQ Press.

Thompson, William R., and Richard Tucker. 1997. "A Tale of Two Democratic Peace Critiques." *Journal of Conflict Resolution* 41(3): 428–54.

Thompson, William R., and Gary Zuk. 1982. "War, Inflation, and the Kondratieff Long Wave." *Journal of Conflict Resolution* 26(4): 621–44.

Thomson, David, E. Meyer, and A. Briggs. 1945. *Patterns of Peacemaking*. New York: Oxford University Press.

Tickner, J. Ann. 1997. "You Just Don't Understand: Troubled Engagements Between Feminists and IR Theorists." *International Studies Quarterly* 41(4): 611–32.

Tilly, Charles. 1975. "Reflections on the History of European State Making." In *The Formation of National States in Western Europe*, ed. Charles Tilly. Princeton, NJ: Princeton University Press.

———. 1978. *From Mobilization to Revolution*. Boston: Addison-Wesley.

Tir, Jaroslav, and Paul F. Diehl. 2002. "Geographic Dimensions of Enduring Rivalries." *Political Geography* 21(2): 263–86.

Toland, John. 1991. *In Mortal Combat: Korea, 1950–1953*. New York: William Morrow.

Trachtenberg, Marc. 1991. *History and Strategy*. Princeton, NJ: Princeton University Press.

Trager, Robert F., and Dessislava P. Zagorcheva. 2005/06. "Deterring Terrorism: It Can Be Done." *International Security* 30(3): 87–123.

Trask, David F. 1981. *The War with Spain in 1898*. New York: Free Press.

Trotter, William R. 1991. *A Frozen Hell: The Russo-Finnish Winter War of 1939–1940*. Chapel Hill, NC: Algonquin.

Turner, Barry. 2006. *Suez 1956: The Inside Story of the First Oil War*. London: Hodder and Stoughton.

Valeriano, Brandon. 2012. *Becoming Rivals: The Process of Interstate Rivalry Development*. New York: Routledge.

Van Evera, Stephen. 1984. "The Cult of the Offensive and the Origins of the First World War." *International Security* 9(1): 58–107.

———. 1998. "Offense, Defense, and the Causes of War." *International Security* 22(4): 5–43.

Vasquez, John A. 1993. *The War Puzzle*. Cambridge: Cambridge University Press.

———. 1995. "Why Do Neighbors Fight? Proximity, Interaction, or Territoriality." *Journal of Peace Research* 32(3): 277–93.

———. 1996. "Distinguishing Rivals That Go to War from Those That Do Not: A Quantitative Comparative Case Study of the Two Paths to War." *International Studies Quarterly* 40(4): 531–58.

Vasquez, John A., and Brandon Valeriano. 2010. "Classification of Interstate Wars." *Journal of Politics* 72(2): 292–309.

Verba, Sidney. 1961. "Assumptions of Rationality and Non-Rationality in Models of the International System." In *The International System: Theoretical Essays*, eds. Klaus Knorr and Sidney Verba. Princeton, NJ: Princeton University Press.

Voeten, Erik. 2000. "Clashes in the Assembly." *International Organization* 54(2): 185–215.

———. 2004. "Resisting the Lonely Superpower: Responses of States in the United Nations to U.S. Dominance." *Journal of Politics* 66(3): 729–54.

Vuchinich, Samuel, and Jay Teachman. 1993. "The Duration of Wars, Strikes, Riots, and Family Arguments." *Journal of Conflict Resolution* 37(3): 544–68.

Wagner, R. Harrison. 2000. "Bargaining and War." *American Journal of Political Science* 44(3): 469–84.

Walker, Stephen G., Mark Schafer, and Michael D. Young. 1998. "Systematic Procedures for Operational Code Analysis: Measuring and Modeling Jimmy Carter's Operational Code." *International Studies Quarterly* 42(1): 175–90.

Wallace, Michael D. 1979. "Arms Races and Escalation: Some New Evidence." *Journal of Conflict Resolution* 23(1): 3–16.

———. 1981. "Old Nails in New Coffins: The Para Bellum Hypothesis Revisited." *Journal of Peace Research* 18(1): 91–95.

———. 1982. "Armaments and Escalation: Two Competing Hypotheses." *International Studies Quarterly* 26(1): 37–56.

Walt, Stephen M. 1987. *The Origins of Alliances*. Ithaca, NY: Cornell University Press.

Walter, Barbara. 1997. "The Critical Barrier to Civil War Settlement." *International Organization* 51(3): 335–64.

———. 2002. *Committing to Peace: The Successful Settlement of Civil Wars*. Princeton, NJ: Princeton University Press.

Waltz, Kenneth N. 1959. *Man, the State, and War: A Theoretical Analysis*. New York: Columbia University Press.

———. 1964. "The Stability of a Bipolar World." *Daedalus* 93(3): 881–909.

———. 1979. *Theory of International Politics*. New York: Random House.

———. 1988. "The Origins of War in Neorealist Theory." *Journal of Interdisciplinary History* 18(4): 615–28.

Warner, Denis, and Peggy Warner. 2002. *The Tide at Sunrise: A History of the Russo-Japanese War, 1904–1905*. London: Routledge.

Washington, George. 1796. "Farewell Address." September 17. http://avalon.law.yale.edu/18th_century/washing.asp.

Wawro, Geoffrey. 1996. *The Austro-Prussian War: Austria's War with Prussia and Italy in 1866*. Cambridge: Cambridge University Press.

———. 2003. *The Franco-Prussian War: The German Conquest of France in 1870–1871*. Cambridge: Cambridge University Press.

Wayman, Frank Whelon. 2000. "Rivalries: Recurrent Disputes and Explaining War." In *What Do We Know about War?*, ed. John A. Vasquez. Lanham, MD: Rowman and Littlefield.

Weede, Erich. 1976. "Overwhelming Preponderance as a Pacifying Condition Among Contiguous Asian Dyads, 1950–1969." *Journal of Conflict Resolution* 20(3): 395–411.

———. 1983. "Extended Deterrence by Superpower Alliance." *Journal of Conflict Resolution* 27(2): 231–54.

———. 2003. "Globalization: Creative Destruction and the Prospect of a Capitalist Peace." In *Globalization and Armed Conflict*, eds. Gerald Schneider, Katherine Barbieri, and Nils Petter Gleditsch. Lanham, MD: Rowman and Littlefield.

———. 2010. "The Capitalist Peace and the Rise of China: Establishing Global Harmony by Economic Interdependence." *International Interactions* 36(2): 206–13.

Weinberg, Gerhard. 1994. *A World at Arms: A Global History of World War II*. Cambridge: Cambridge University Press.

———. 1998. *Germany, Hitler, and World War II: Essays in Modern German and World History*. Cambridge: Cambridge University Press.

Werner, Suzanne. 1998. "Negotiating the Terms of Settlement: War Aims and Bargaining Leverage." *Journal of Conflict Resolution* 42(3): 321–43.

———. 1999. "The Precarious Nature of Peace." *American Journal of Political Science* 43(3): 912–34.

Werner, Suzanne, and Amy Yuen. 2005. "Making and Keeping Peace." *International Organization* 59(2): 262–93.

Wheelan, Joseph. 2007. *Invading Mexico: America's Continental Dream and the Mexican War, 1846–1848*. New York: Carroll and Graf.

Wheeler, Hugh. 1975. "Effects of War on Industrial Growth." *Society* 12(4): 48–52.

———. 1980. "Postwar Industrial Growth." In *The Correlates of War II: Testing Some Realpolitik Models*, ed. J. David Singer. New York: Free Press.

Wheeler-Bennett, John W. 1956. *Brest-Litovsk: The Forgotten Peace, March 1918*. London: Macmillan.

Willmott, H. P. 1984. *June 1944*. Poole, UK: Blandford.

Williamson, Samuel R. Jr., and Russel Van Wyk. 2003. *July 1914: Soldiers, Statesmen, and the Coming of the Great War*. Boston: Bedford/St. Martin's Press.

Wilmot, Chester. 1952. *The Struggle for Europe*. New York: Harper.

Wittman, Donald. 1979. "How a War Ends: A Rational Choice Approach." *Journal of Conflict Resolution* 23(4): 743–63.

Wohlstetter, Albert. 1959. "The Delicate Balance of Terror." *Foreign Affairs* 37(2): 211–34.

Woodward, Llewellyn. 1970. *British Foreign Policy in the Second World War.* 3 vols. London: Her Majesty's Stationary Office.

Wright, Jonathan. 2007. *Germany and the Origins of the Second World War.* New York: Palgrave.

Yamamoto, Yoshinobu. 1990. "Rationality or Chance: The Expansion and Control of War." In *Prisoners of War?*, eds. Charles S. Gochman and Alan Ned Sabrosky. Lexington, MA: Lexington Books.

Zagare, Frank C. 1979. "The Geneva Conference of 1954: A Case of Tacit Deception." *International Studies Quarterly* 23(3): 390–411.

———. 1982. "Competing Game-Theoretic Explanations: The Geneva Conference of 1954." *International Studies Quarterly* 26(1): 141–6.

———. 1983. "A Game-Theoretic Evaluation of the Cease-Fire Alert Decision of 1973." *Journal of Peace Research* 20(1): 73–86.

———. 1990. "Rationality and Deterrence." *World Politics* 42(2): 238–60.

———. 1996. "Classical Deterrence Theory: A Critical Assessment." *International Interactions* 21(4): 365–87.

———. 2004. "Reconciling Rationality with Deterrence: A Re-examination of the Logical Foundations of Deterrence Theory." *Journal of Theoretical Politics* 16(2): 107–41.

———. 2007. "Toward a Unified Theory of Interstate Conflict." *International Interactions* 33(3): 305–27.

———. 2009. "Explaining the 1914 War in Europe: An Analytic Narrative." *Journal of Theoretical Politics* 21(1): 63–95.

———. 2011. *The Games of July: Explaining the Great War.* Ann Arbor: University of Michigan Press.

Zagare, Frank C., and D. Marc Kilgour. 2000. *Perfect Deterrence.* Cambridge: Cambridge University Press.

Zinnes, Dina A. 1976. "The Problem of Cumulation." In *In Search of Global Patterns*, ed. James N. Rosenau. New York: Free Press.

Index

CQ Press, an imprint of SAGE, is the leading publisher of books, periodicals, and electronic products on American government and international affairs. CQ Press consistently ranks among the top commercial publishers in terms of quality, as evidenced by the numerous awards its products have won over the years. CQ Press owes its existence to Nelson Poynter, former publisher of the *St. Petersburg Times,* and his wife Henrietta, with whom he founded Congressional Quarterly in 1945. Poynter established CQ with the mission of promoting democracy through education and in 1975 founded the Modern Media Institute, renamed The Poynter Institute for Media Studies after his death. The Poynter Institute (*www.poynter.org*) is a nonprofit organization dedicated to training journalists and media leaders.

In 2008, CQ Press was acquired by SAGE, a leading international publisher of journals, books, and electronic media for academic, educational, and professional markets. Since 1965, SAGE has helped inform and educate a global community of scholars, practitioners, researchers, and students spanning a wide range of subject areas, including business, humanities, social sciences, and science, technology, and medicine. A privately owned corporation, SAGE has offices in Los Angeles, London, New Delhi, and Singapore, in addition to the Washington DC office of CQ Press.